D1520739

INORGANIC SYNTHESES

Volume XX

Board of Directors

FRED BASOLO *Northwestern University*
WILLIAM L. JOLLY *University of California*
HENRY F. HOLTZCLAW *University of Nebraska*
JAY H. WORRELL *University of South Florida*
BODIE E. DOUGLAS *University of Pittsburgh*
DUWARD F. SHRIVER *Northwestern University*

Future Volumes

XXI **JOHN P. FACKLER, JR.** *Case Western Reserve University*
XXII **SMITH L. HOLT** *University of Georgia*
XXIII **STANLEY KIRSCHNER** *Wayne State University*

International Associates

E. O. FISCHER *Technische Universität München*
JACK LEWIS *Cambridge University*
LAMBERTO MALATESTA *University of Milan*
F. G. A. STONE *University of Bristol*
GEOFFREY WILKINSON *Imperial College of Science and Technology London*
AKIO YAMAMOTO *Tokyo Institute of Technology Yokohama*
MARTIN A. BENNETT *Australian National University*

Editor-in-Chief
DARYLE H. BUSCH
Department of Chemistry
The Ohio State University

INORGANIC SYNTHESES

Volume XX

A Wiley-Interscience Publication
JOHN WILEY & SONS
New York   Chichester   Brisbane   Toronto

Published by John Wiley & Sons, Inc.

Copyright © 1980 Inorganic Syntheses, Inc.

All rights reserved. Published simultaneously in Canada.

Reproduction or translation of any part of this work beyond that permitted by Sections 107 or 108 of the 1976 United States Copyright Act without the permission of the copyright owner is unlawful. Requests for permission or further information should be addressed to the Permissions Department, John Wiley & Sons, Inc.

Library of Congress Catalog Number: 39-23015
ISBN 0-471-07715-1

Printed in the United States of America

10 9 8 7 6 5 4 3 2 1

To
Paolo Chini
1928–1980

PREFACE

Continuing the function of *Inorganic Syntheses*, this volume contains 57 numbered sections including almost 150 individual preparations, among which are some important intermediates, covering a number of broad, currently active areas of inorganic chemistry. These areas are divided into seven chapters. First, a series of offerings, featuring the double infusion technique and crystal growth in gels, treat solid materials and techniques of crystal growth. A variety of interesting and useful stoichiometrically simple compounds are presented in Chapter 2. They include such main group element derivatives as TeF_4 and SeO_2F_2, and a number of transition element derivatives (e.g., trirhenium nonahalides) the important metal chromium(III) oxides (chromites), a neat preparation for copper(I) acetate, and the simplest method yet for anhydrous $PtCl_2$ (the third for this compound in *Inorganic Syntheses*). Chapter 3 contains an array of classic coordination compounds (ethylenediamine complexes of rhodium, β-diketonates of platinum(II)), and also some materials that have been of interest only in the more recent past (phosphine complexes of iron, a CO_2 adduct, phosphite complexes of several metal ions). Syntheses are given for a number of rather exotic ligand complexes in Chapter 4. These include the clathrochelates of Goedken, the sepulchrates of Sargeson, the superphthalocyanine of Marks, and examples of compartmental ligands. A lengthy section is devoted to compounds of biological interest. Emphasis ranges from metal intercalation reagents, over synthetic macrocycles, corrin complexes and analogues, and porphyrins (including the O_2 adduct of Collman's picket-fence porphyrin) to complexes of dinitrogen. David Dolphin contributed a large group of syntheses on the complexes of the natural macrocycles and their close relatives. Organometallic compounds are soundly represented in Chapters 6 and 7. Chapter 6 contains a number of representative element organometallic compounds (hexamethyldisilaselenane, digermatellurane, digermoxane) and varieties of transition metal derivatives (e.g., halocyclopentadienyl complexes of manganese and rhodium, examples of the novel metalla-β-diketone compounds). The final chapter is devoted to the timely subject of metal clusters and it includes preparations for several of Chini's rhodium clusters, some iron clusters by Volume XIX Editor D. Shriver, and μ_3-alkylidyne (tricarbonylcobalt) compounds by Seyfreth.

I want to thank the many authors and checkers who have produced the substance of this volume. The Editorial Board has provided constant aid in evaluation and editing of original manuscripts. Special thanks go to John Bailar and

Therald Moeller in this regard. Invaluable guidance came from past Editors Durward Shriver, Bodie Douglas, George Parshall, and Aaron Wold. Warren H. Powell and Thomas E. Sloan of Chemical Abstracts Service have continued their key role in assuring the use of appropriate current nomenclature. Tom Sloan also prepared the indices. Jay Worrell, Therald Moeller, and Conard Fernelius helped with proofreading. I must add a word of sincere appreciation for my late secretary, Mrs. Denise Hughes, who provided the order and attention necessary to monitor the progress of the many manuscripts from which this volume was drawn. I have enjoyed serving as your Editor partly because I first encountered *Inorganic Syntheses* while I was in graduate school, when I was an editorial assistant, checker, and author, as well as user, and have remained an appreciative subscriber over the years.

Daryle H. Busch

Columbus, Ohio
January 1980

NOTICE TO CONTRIBUTORS

The *Inorganic Syntheses* series is published to provide all users of inorganic substances with detailed and foolproof procedures for the preparation of important and timely compounds. Thus the series is the concern of the entire scientific community. The Editorial Board hopes that all chemists will share in the responsibility of producing *Inorganic Syntheses* by offering their advice and assistance in both the formulation of and the laboratory evaluation of outstanding syntheses. Help of this kind will be invaluable in achieving excellence and pertinence to current scientific interests.

There is no rigid definition of what constitutes a suitable synthesis. The major criterion by which syntheses are judged is the potential value to the scientific community. An ideal synthesis is one that presents a new or revised experimental procedure applicable to a variety of related compounds, at least one of which is critically important in current research. However, syntheses of individual compounds that are of interest or importance are also acceptable. Syntheses of compounds that are available commercially at reasonable prices are not acceptable.

The Editoral Board lists the following criteria of content for submitted manuscripts. Style should conform with that of previous volumes of *Inorganic Syntheses*. The introductory section should include a concise and critical summary of the available procedures for synthesis of the product in question. It should also include an estimate of the time required for the synthesis, an indication of the importance and utility of the product, and an admonition if any potential hazards are associated with the procedure. The Procedure should present detailed and unambiguous laboratory directions and be written so that it anticipates possible mistakes and misunderstandings on the part of the person who attempts to duplicate the procedure. Any unusual equipment or procedure should be clearly described. Line drawings should be included when they can be helpful. All safety measures should be stated clearly. Sources of unusual starting materials must be given, and, if possible, minimal standards of purity of reagents and solvents should be stated. The scale should be reasonable for normal laboratory operation, and any problems involved in scaling the procedure either up or down should be discussed. The criteria for judging the purity of the final product should be delineated clearly. The section on Properties should supply and discuss those physical and chemical characteristics that are relevant to judging the purity of the product and to permitting its handling and use in an intelligent manner. Under References, all pertinent literature citations should be listed in order. A style sheet is available from the Secretary of the Editorial Board.

The Editorial Board determines whether submitted syntheses meet the general specifications outlined above. Every synthesis must be satisfactorily reproduced in a laboratory other than the one from which it was submitted.

Each manuscript should be submitted in duplicate to the Secretary of the Editorial Board, Professor Jay H. Worrell, Department of Chemistry, University of South Florida, Tampa, FL 33620. The manuscript should be typewritten in English. Nomenclature should be consistent and should follow the recommendations presented in *Nomenclature of Inorganic Chemistry*, 2nd Ed., Butterworths & Co, London, 1970 and in *Pure Appl. Chem.*, **28**, No. 1 (1971). Abbreviations should conform to those used in publications of the American Chemical Society, particularly *Inorganic Chemistry*.

TOXIC SUBSTANCES AND LABORATORY HAZARDS

The attention of the user is directed to the notices under this heading on page ix of Volume XIX and pages xv–xvii in Volume XVIII. It cannot be redundant to stress the ever-present need for the experimental chemist to evaulate procedures and anticipate and prepare for hazards. Obvious hazards associated with the preparations in this volume have been delineated, but it is impossible to forsee and discuss all possible sources of danger. Therefore the synthetic chemist should be familiar with general hazards associated with toxic, flammable, and explosive materials. In light of the primitive state of knowledge of the biological effects of chemicals, it is prudent that all the syntheses reported in this and other volumes of *Inorganic Syntheses* be conducted with rigorous care to avoid contact with all reactants, solvents, and products.

CONTENTS

Chapter One SOLID STATE

1. Crystal Growth in Gels ... 1
   - A. General Experimental Considerations ... 2
   - B. Crystals Grown by the Standard Technique ... 3
   - C. Tartrates ... 4
   - D. Carbonates ... 4
   - E. Sulfides ... 4
   - F. Other Experiments Using the Standard Technique ... 5
   - G. The Complex-Dilution Method ... 6
   - H. Liesegang Rings ... 7
2. Crystal Growth in Gels ... 9
   - A. Calcium Tartrate Crystals Grown by the Standard Technique ... 9
   - B. Copper(II) Chloride Crystals Grown by the Complex-Dilution Method ... 10
3. Single-Crystal Growth by a Double-Infusion Technique ... 11
   - A. The Double-Infusion Technique ... 11
   - B. Single Crystals of Rare Earth Transition Element Cyanides $Ln[M(CN)_6] \cdot nH_2O$ ... 12
   - C. Silver(I) Periodate $Ag_2H_3IO_6$ and Ammonium Periodate $(NH_4)_2H_3IO_6$ ... 15
4. Crystalline Silver Chloride ... 18
5. The Partially Oxidized Platinum Complex Rubidium Tetracyanoplatinate [Hydrogen Bis(sulfate)] (3:1:0.46): $Rb_3[Pt(CN)_4](SO_4-H-O_4S)_{0.46} \cdot H_2O$ ... 20
6. The Rubidium–Cesium Partially Oxidized Tetracyanoplatinate (Hydrogen Difluoride)–Fluoride Systems ... 23
   - A. Rubidium Tetracyanoplatinate (Hydrogen Difluoride) (2:1:0.29) Hydrate: $Rb_2[Pt(CN)_4](FHF)_{0.29} \cdot 1.67H_2O$ ... 24
   - B. Rubidium Tetracyanoplatinate (Hydrogen Difluoride) (2:1:0.38): $Rb_2[Pt(CN)_4](FHF)_{0.38}$ ... 25
   - C. Cesium Tetracyanoplatinate (Hydrogen Difluoride) (2:1:0.23): $Cs_2[Pt(CN)_4](FHF)_{0.23}$ ... 26
   - D. Cesium Tetracyanoplatinate (Hydrogen Difluoride) (2:1:0.38): $Cs_2[Pt(CN)_4](FHF)_{0.38}$ ... 28

E. Cesium Tetracyanoplatinate Fluoride (2:1:0.19):
   $Cs_2[Pt(CN)_4]F_{0.19}$ 29

Chapter Two STOICHIOMETRICALLY SIMPLE COMPOUNDS

7. Tellurium Tetrafluoride 33
8. Sulfur Tetrafluoride Oxide 34
9. Selenonyl Difluoride 36
10. Hydrogen Pentafluorooxoselenate(VI) 38
11. Rhenium Pentachloride and Volatile Metal Chlorides by Direct Chlorination Using a Vertical-Tube Reactor 41
12. Trirhenium Nonachloride 44
13. Trirhenium Nonahalides 46
    A. Tetraacetatodichlorodirhenium 46
    B. Trirhenium Nonachloride 47
    C. Trirhenium Nonabromide and Trirhenium Nonaiodide 47
14. β-Platinum(II) Chloride 48
15. Divalent Metal Chromium(III) Oxides (Chromites) 50
    A. Lithium Chromium(III) Oxide, $LiCrO_2$ 50
    B. Potassium Metal(II) Chlorides 51
    C. Divalent Metal Chromium(III) Oxides 52
16. Copper(I) Acetate 53

Chapter Three COORDINATION COMPOUNDS

17. Compounds Containing the *cis*-Bis(ethylenediamine)rhodium(III) Species 57
    A. Bis(ethylenediamine) (oxalato) rhodium(III) Perchlorate 58
    B. *cis*-Bis(ethylenediamine) Dinitrorhodium(III) Nitrate 59
    C. *cis*-Dichlorobis(ethylenediamine)rhodium(III) Chloride Hydrate . 60
18. Potassium Tetrahydroxodioxoosmate(VI) and *trans*-Bis(ethylenediamine)dioxoosmium(VI) Chloride 61
    A. Potassium Tetrahydroxodroxoosmate(VI) 61
    B. *trans*-Bis(ethylenediamine)dioxoosmium(VI) Chloride 62
19. Sodium Bis[2-ethyl-2-hydroxybutyrato(2-)] oxochromate(V) 63
20. The Bis(β-diketonato)platinum(II) Complexes 65
    A. Preparation of Stock Solution 66
    B. Bis[2,4-Pentanedionato(1-)]platinum(II) 66
    C. Bis[1,1,1-trifluoro-2,4-pentanedionato(1-)]platinum(II) 67
    D. Bis[1,1,1,5,5,5-Hexafluoro-2,4-pentanedionato(1-)]-platinum(II) 67

21. Trimethylphosphine Iron Complexes 69
    A. Dichlorobis(trimethylphosphine)iron(II) 70
    B. Tetrakis(trimethylphosphine)iron 71
    C. (Carbon dioxide–C,O)tetrakis(trimethylphosphine)iron 73
22. Pentakis(trimethyl phosphite)complexes of the d^8 transition metals . 76
    A. Pentakis(trimethyl phosphite)nickel(II) Bis(tetraphenylborate) . 76
    B. Pentakis(trimethyl phosphite)palladium(II)
       Bis(tetraphenylborate) 77
    C. Pentakis(trimethyl phosphite)rhodium(I) Tetraphenylborate ... 78
    D. Pentakis(trimethyl phosphite)iron(O) 79
    E. Pentakis(trimethyl phosphite)ruthenium(O) 80
    F. Pentakis(trimethyl phosphite)cobalt(I) Tetraphenylborate 81
23. Complexes of Aluminum Iodide with Pyridine and Related Bases ... 82
    A. Preparation of Triiodotris (pyridine) aluminum 83

Chapter Four COMPLEXES WITH COMPLICATED CHELATE
              LIGANDS

24. Caged Metal Ions: Cobalt Sepulchrates 85
    A. [Co(bicyclo[6.6.6]ane-1,3,6,8,10,13,16,19-N_8)]Cl_3 86
25. Clathrochelates Formed by the Reaction of Tris(2,3-butanedione
    Dihydrazone)metal Complexes with Formaldehyde: [5,6,14,15,20,21-
    Hexamethyl-1,3,4,7,8,10,12,13,16,17,19,22-dodecaazatetracyclo-
    [8.8.4$1^{3,17} \cdot 1^{8,12}$]tetracosa-4,6,13,15,19,21-
    hexaene-N^4,N^7,N^{13},N^{16},N^{19},N^{22}]metal(n+) Tetrafluoroborate . . 87
26. Compartmental Ligands 90
    A. 5,9,14,18-Tetramethyl-1,4,10,13-Tetraazacyclooctadeca-
       5,8,14,17-tetraene-7,16-dione, H_4 daen(I) 91
    B. [5,9,14,18-Tetramethyl-1,4,10,13-tetraazacyclootadeca-
       5,8,14,17-tetraene-7,16-donato(2-)N^1,N^4,O^7,O^{16}] copper(II),
       Cu(H_2 daen)(II) 92
    C. [[6,6'-(Ethylenedinitrilo)bis(2,4-heptanedionato)]-
       (2-)-N,N',O^4,$O^{4'}$] copper(II), Cu(H_2 daaen-N_2O_2)(III) 93
    D. μ\{[6,6'-(Ethylenedinitrilo)bis(2,4-heptanedionato)]-
       (4-)-N,N',O^4,$O^{4'}$:O^2,$O^{2'}$,O^4,$O^{4'}$] dicopper(II), Cu_2(daaen)(IV) . 94
    E. Aqua-μ-[[6,6'-(ethylenedinitrilo)bis(2,4-heptanedionato)]-
       (4-)-N,N',O^4,$O^{4'}$:O^2,$O^{2'}$,O^4,$O^{4'}$] copper(II)oxovanadium(IV),
       CuVO(daaen)H_2O, (V) 95
27. [7,12:21,26-Diimino-19,14;28,33:35,5-trinitrilo-5H-
    pentabenzo[c,h,m,r,w] [1,6,11,16,21] pentaazacyclopenta-
    cosinato(2-)]-dioxouranium(VI) (Uranyl Superphthalocyanine) 97

Chapter Five COMPOUNDS OF BIOLOGICAL INTEREST

28. Metallointercalation Reagents: Thiolato Complexes of (2,2′:6′,2″-terpyridine)platinum(II) 101
    A. Chloro(2,2′:6′,2″-terpyridine)platinum(II) Chloride Dihydrate, [Pt(terpy)Cl]Cl·2H$_2$O 102
    B. (2-Mercaptoethanolato-S)(2,2′:6′,2″-terpyridine)platinum(II) Nitrate, [Pt(terpy)(SCH$_2$CH$_2$OH)]NO$_3$ 103
    C. (2-Ammonioethanethiolato-S)(2,2′:6′,2″-terpyridine)-platinum(II) Nitrate [Pt(terpy)(SCH$_2$CH$_2$NH$_3$)](NO$_3$)$_2$ and other (2,2′:6′,2″-terpyridine)thiolatoplatinum(II) Complexes ... 104
29. Saturated, Unsubstituted Tetraazamacrocyclic Ligands and Their Cobalt(III) Complexes 105
    A. 1,4,7,10-Tetraazacyclotridecane ([13]aneN$_4$) 106
    B. 1,4,8,12-Tetraazacyclopentadecane ([15]aneN$_4$) 108
    C. 1,5,9,13-Tetraazacyclohexadecane ([16]aneN$_4$) 109
    D. *trans*-Dichloro([13]aneN$_4$) cobalt(III) Chloride, (I)-and (II)-*trans*-Dichloro([15]aneN$_4$) cobalt(III) Chloride, (I)-and (II)-*trans*-Dichloro([16]aneN$_4$) cobalt(III) Perchlorate .. 111
30. [7,16-Dihydro-6,8,15,17-tetramethyldibenzo[*b,i*] [1,4,8,11]-tetraazacyclotetradecinato(2-)]nickel(II), ([5,7,12,14-Me$_4$-2,3:9,10-Bzo$_2$-[14]hexaenato(2-)N$_4$]Ni(II) and the Neutral Macrocyclic Ligand 115
    A. [5,7,12,14-Me$_4$-2,3:9,10-Bzo$_2$[14]hexaenato(2-)N$_4$]Ni(II) ... 116
    B. The 5,7,12,14-Me$_4$-2,3:9,10-Bzo[14]hexaene N$_4$ Ligand 117
31. The Synthesis of Molybdenum and Tungsten Dinitrogen Complexes . 119
    A. Bis(acetonitrile)tetrachloromolybdenum(IV), [MoCl$_4$(CH$_3$CN)$_2$] 120
    B. Tetrachlorobis(tetrahydrofuran)molybdenum(IV), [MoCl$_4$(thf)$_2$] 121
    C. Trichlorotris(tetrahydrofuran)molybdenum(III), [MoCl$_3$(thf)$_3$] 121
    D. *trans*-Bis(dinitrogen)bis[ethylenebis(diphenylphosphine)]-molybdenum(0), *trans*-[Mo(N$_2$)$_2$(Ph$_2$PCH$_2$CH$_2$PPh$_2$)$_2$] 122
    E. Tetrachlorobis(triphenylphosphine)tungsten(IV), [WCl$_4$(PPh$_3$)$_2$] 124
    F. Tetrachloro[ethylenebis(diphenylphosphine)]-tungsten(IV), [WCl$_4$(Ph$_2$PCH$_2$CH$_2$PPh$_2$)] 125
    G. *trans*-Bis(dinitrogen)bis[ethylenebis(diphenylphosphine)]-tungsten(O), *trans*-[W(N$_2$)$_2$(Ph$_2$PCH$_2$CH$_2$PPh$_2$)$_2$] 126
32. Bis[2,3-butanedione dioximato(1-)] cobalt Complexes 127
    A. Bromobis[2,3-butanedione dioximato(1-)](dimethyl sulfide) cobalt(III) 128

| | | Contents | xvii |

    B. Bromobis[2,3-butanedione dioximato(1-)](4-*tert*-butylpyridine)
       cobalt(III) 130
    C. Bis[2,3-butanedione dioximato(1-)](4-*tert*-butylpyridine)-
       (2-ethoxyethyl) cobalt(III) 131
    D. Bis[2,3-butanedione dioximato(1-)](4-*tert*-butylpyridine)-
       (ethoxymethyl) cobalt(III) 132
33. Cobalamins and Cobinamides 134
    A. Methylcobalamin 136
    B. Hydroxocobalamin 138
    C. (2,2-Diethoxyethyl)cobalamin [I, R = $-CH_2CH(OEt)_2$] 138
    D. Dicyanocobyrinic Acid Heptamethyl Ester(II, R = R' = -CN) ... 139
    E. Aquacyanocobrinic Acid Heptamethyl Ester Perchlorate 141
    F. Aquamethylcobrinic Acid Heptamethyl Ester Perchlorate 141
34. Metalloporphines 143
    A. [5,10,15,20-Tetraphenyl-21H,23H-porphinato(2-)] nickel(II) ... 143
    B. Oxo[5,10,15,20-tetraphenyl-21H,23H-porphinato(2-)]-
       vanadium(IV) 144
    C. [2,3,7,8,12,13,17,18-Octaethyl-21H,23H-porphinato(2-)]-
       magnesium(II) 145
35. Iron Porphines 147
    A. Chloro[dimethyl protoporphyrinato(2-)]iron(III),
       chloro(dimethyl-7,12-diethenyl-3,8,13,17-tetramethyl-21H,23H-
       porphine-2,18-dipropionato(2-)]iron(III) 148
    B. Chloro[octaethyl-21H,23H-porphinato(2-)] iron(III) 151
    C. Chloro[mesoporphyrinato(2-)] iron(III) (mesohemin IX),
       chloro[7,12-diethyl-3,8,13,17-tetramethyl-21H,23H-porphine-
       2,18-dipropionato(2-)] iron(III) 152
    D. The Insertion of ^{57}Fe into Porphyrins 153
36. Metallophthalocyanins and Benzoporphines 155
    A. [1,4,8,11,15,18,22,25-Octamethyl-29H,31H-tetrabenzo[b,g,l,q]-
       porphinato(2-)] cobalt(II) 156
    B. 1,4,8,11,15,18,22,25-Octamethyl-29H,31H-tetrabenzo[b,g,l,q]-
       porphine 158
    C. Dilithium Phthalocyanine 159
    D. Phthalocyaninato(2-)iron(II) 160
37. (Dioxygen)(N-methylimidazole)[(*all-cis*)-5,10,15,20-
    tetrakis[2-(2,2-dimethylpropionamido)phenyl]porphyrinato(2-)-
    iron(II) 161
    A. 5,10,15,20-tetrakis(2-nitrophenyl)porphyrin, H_2TNPP 162
    B. 5,10,15,20-Tetrakis(2-aminylphenyl)porphyrin, H_2TAPP 163
    C. Separation of *all-cis*-H_2TAPP 164
    D. (*all-cis*)-5,10,15,20-Tetrakis[2,2-dimethylpropionamido)-
       phenyl]porphyrin, H_2TpivPP 165

xviii   Contents

E. Bromo{(*all-cis*)-5,10,15,20-tetrakis[2-(2,2-dimethylpropionamido)phenyl] porphyrinato(2-]}iron(III), Fe[(*all-cis*)TpivPP]Br 166

F. Bis(N-methylimidazole)[(*all-cis*)-5,10,15,20-tetrakis-[2-(2,2-dimethylpropionamido)phenyl] porphyrinato(2-)]-iron(II), Fe[(*all-cis*)TpivPP](N-MeC$_3$H$_3$N$_2$)$_2$ 167

G. (Dioxygen)(N-methylimidazole)[(*all-cis*)-5,10,15,20-tetrakis[2-(2,2-dimethylpropionamido)phenyl] porphyrinato-(2-)]iron(II), Fe[(*all-cis*)TpivPP](N-MeC$_3$H$_3$N$_2$)(O$_2$) 168

Chapter Six ORGANOMETALLIC COMPOUNDS

38. Silyl and Germyl Selenides and Tellurides 171
    A. Hexamethyldisilaselenane and Hexamethyldisilatellurane-[Bis(trimethylsilyl) selenide and Bis(trimethylsilyl) telluride] ... 173
    B. Digermaselenane and Digermatellurane(Digermyl selenide and Digermyl telluride) 175
39. Digermoxane and 1,3-Dimethyl-, 1,1,3,3-Tetramethyl-, and Hexamethyldigermoxane 176
    A. Digermoxane 178
    B. 1,3-Dimethyl-, 1,1,3,3-Tetramethyl-, and Hexamethyl Digermoxane 179
40. Di-μ-chloro-dichlorobis(ethylene)diplatinum(II), Di-μ-chloro-dichlorobis(styrene)diplatinum(II), and Di-μ-chloro-dichlorobis-(1-dodecene)diplatinum(II) 181
    A. *trans*-Dichloro(ethylene)(pyridine)platinum(II) 181
    B. Di-μ-chloro-dichlorobis(ethylene)diplatinum(II) 182
    C. Di-μ-chloro-dichlorobis(styrene)diplatinum(II) 182
    D. Di-μ-chloro-dichlorobis(1-dodecene)diplatinum(II) 183
41. Dibromomethylplatinum(IV) and Dihydroxodimethylplatinum(IV) Sesquihydrate 185
    A. Dibromodimethylplatinum(IV) 185
    B. Dihydroxodimethylplatinum(IV) Sesquihydrate 186
42. Halocyclopentadienyl Complexes of Manganese and Rhodium 188
    A. Tetrachloro-5-diazo-1,3-cyclopentadiene 189
        1. Procedures for Tetrachloro-2,4-cyclopentadiene-1-one Hydrazone, C$_5$Cl$_4$NNH$_2$ 190
        2. Procedures for Tetrachloro-5-diazo-1,3-cyclopentadiene, C$_5$Cl$_4$N$_2$ 190
    B. 5-Diazo-1,3-Cyclopentadiene, C$_5$H$_4$N$_2$ 191
    C. (Chlorotetraphenyl-η^5-cyclopentadienyl)(η^4-1,5-cyclooctadiene)-rhodium, Rh(η^5-C$_5$Ph$_4$Cl)(cod) 191
    D. (Bromotetraphenyl-η^5-cyclopentadienyl)(η^4-1,5-cyclooctadiene)

rhodium, $Rh(\eta\text{-}C_5Ph_4Br)(cod)$ and dicarbonyl
(chlorotetraphenyl-η^5-cyclopentadienyl) rhodium,
$Rh(\eta^5\text{-}C_5Ph_4Cl)(CO)_2$ 192

E. (Chloro-η^5-cyclopentadienyl)(η^4-1,5-cyclooctadiene)rhodium,
$Rh(\eta^5\text{-}C_5H_4Cl)(cod)$ 192

F. (Tricarbonyl(chloro-η^5-cyclopentadienyl)-manganese,
$Mn(\eta^5\text{-}C_5H_4Cl(CO)_3$ 192

G. (Bromo-η^5-cyclopentadienyl)tricarbonyl-manganese,
$Mn(\eta^5\text{-}C_5H_4Br)(CO)_3$ 193

H. Tricarbonyl(iodo-η^5-cyclopentadienyl)manganese,
$Mn(\eta^5\text{-}C_5H_4I)(CO)_3$ 193

I. Pentacarbonyl(pentachloro-η^1-cyclopentadienyl)manganese,
$Mn(\eta^1\text{-}C_5Cl_5)(CO)_5$ 193

J. (Bromotetrachloro-η^1-cyclopentadienyl)pentacarbonylmanganese,
$Mn(\eta^1\text{-}C_5Cl_4Br)(CO)_5$ 194

K. (η^4-1,5-cyclooctadiene)(pentachloro-η^5-cyclopentadienyl)-
rhodium, $Rh(\eta^5\text{-}C_5Cl_5)(cod)$ 194

L. Tricarbonyl(pentachloro-η^5-cyclopentadienyl)manganese,
$Mn(\eta^5\text{-}C_5Cl_5)(CO)_3$ 194

M. (Bromotetrachloro-η^5-cyclopentadienyl)tricarbonylmanganese,
$Mn(\eta^5\text{-}C_5Cl_4Br)(CO)_3$ 195

43. (η^6-Benzene)(η^5-cyclopentadienyl)molybdenum and (η^6-Benzene)-
(η^5-cyclopentadienyl)molybdenum Derivatives 196

A. (η^6-Benzene)(η^5-cyclopentadienyl)molybdenum 197
B. (η^6-Benzene)chloro(η^5-cyclopentadienyl)molybdenum 198
C. (η^6-Benzene)bromo(η^5-cyclopentadienyl)molybdenum 199
D. (η^6-Benzene)(η^5-cyclopentadienyl)iodomolybdenum 199

44. cis-[Acetyltetracarbonyl(1-hydroxyethylidene)rhenium]
(cis-Diacetyltetracarbonylrhenium hydrogen) 200
A. Acetylpentacarbonylrhenium 201
B. cis-[Acetyltetracarbonyl-(1-hydroxyethylidene)rhenium
(cis-Diacetyltetracarbonylrhenium hydrogen) 202

45. cis-[Acetyl(1-aminoethylidene)tetracarbonylrhenium]-
[cis-(Acetimidoyl)(acetyl)tetracarbonylrhenium hydrogen)] 204

46. Hydrido Complexes of Cobalt with Bis(phosphines) 206
A. Hydridobis[cis-vinylenebis(diphenylphosphine)]cobalt(I) 207
B. Hydridobis[ethylenebis(diphenylphosphine)]cobalt(I) 208

Chapter Seven METAL CLUSTER COMPLEXES

47. Tri-μ-carbonyl-nonacarbonyltetrarhodium, $Rh_4(CO)_9(\mu\text{-}CO)_3$ 209
48. Dipotassium μ_6-Carbido-nona-μ-carbonyl-hexacarbonylhexarhodate-
(2-), $K_2[Rh_6(CO)_6(\mu\text{-}CO)_9(\mu_6\text{-}C)]$ 212

xx Contents

49. Disodium Di-μ-carbonyl-octa-μ_3-carbonyl-icosacarbonyldodecarhodate(2-), $Na_2[Rh_{12}(CO)_{20}(\mu\text{-}CO)_2(\mu_3\text{-}CO)_8]$ 215
50. μ-Nitrido bis(triphenylphosphorus)(1+) Undecacarbonylhydridotriferrate(1-), $[(Ph_3P)_2N][Fe_3H(CO)_{11}]$ 218
51. Bis[μ-nitrido-bis(triphenylphosphorus)(1+)] Undecacarbonyltriferrate, $[(Ph_3P)_2N]_2[Fe_3(CO)_{11}]$ 222
52. μ_3-Alkylidyne-tris(tricarbonylcobalt) Compounds: Organocobalt Cluster Complexes 224
53. μ_3-Methylidyne- and μ_3-Benzylidyne-tris(tricarbonylcobalt) 226
    A. μ_3-Methylidyne-tris(tricarbonylcobalt) 227
    B. μ_3-Benzylidyne-tris(tricarbonylcobalt) 228
54. μ_3[(Ethoxycarbonyl)methylidyne]- and μ_3-{[(Methylamino)carbonyl]methylidyne}-tris(tricarbonylcobalt) 230
    A. μ_3-[(Ethoxycarbonyl)methylidyne]-tris(tricarbonylcobalt) 230
    B. μ_3-{[(Methylamino)carbonyl]methylidyne}-tris(tricarbonylcobalt) 232
55. μ_3-(Chloromethylidyne)-and μ_3-[$tert$-butoxycarbonyl)methylidyne]-tris(tricarbonylcobalt) 234
    A. μ_3-(Chloromethylidyne)-tris(tricarbonylcobalt) 234
    B. μ_3-[($tert$-Butoxycarbonyl)methylidyne]-tris(tricarbonylcobalt) . 235
56. Tetracarbonyl bis-μ-(2-methyl-2-propane thiolato)diiridium(I) 237
57. Decacarbonyl-μ-[diborane(6)]-μ-hydrido-trimanganese 240

Correction 243
Index of Contributors 245
Subject Index 249
Formula Index 281

INORGANIC SYNTHESES

Volume XX

Chapter One
SOLID STATE

1. CRYSTAL GROWTH IN GELS

Submitted by A. F. ARMINGTON* and J. J. O'CONNOR*

The growth of crystals in gel media is a relatively old method of crystal growth that gained some prominence about 60 years ago. Much of the earlier work on the subject has been reviewed by Holmes.[1] After this initial interest the subject remained nearly dormant, except as a curiosity, until a paper by Henisch et al.[2] revived interest in this type of crystal growth. Since then, approximately 100 papers have been published on the subject.

Gel growth is the easiest and least expensive of all crystal-growth techniques in that it consists merely of allowing two materials to react in a gel medium, forming a product that crystallizes in the gel. The simplest arrangement consists of mixing one reactant with the gel and then allowing the gel to set in a test tube that it partially fills. An aqueous solution of the second reactant is added above the set gel in the tube. The reactants diffuse through the gel until they come into contact; they then form an insoluble product that precipitates in crystalline form. In the second common method, a U-tube is partially filled with neutral gel and allowed to set. In this context, neutral means no reactants are added to the gel. The two reactants are placed above the set gel on the opposite sides of the U-tube, and the reaction again proceeds by diffusion. The quality and size of the crystals are similar for the two methods.

*HQ Rome Air Development Center (AFSC), Deputy for Electronic Technology, Solid State Sciences Division, Hanscom, AFB, MA 01731.

The most commonly used medium is silica gel. Numerous other gels have been used but with limited success, apparently because of the structural differences in the gels. A more detailed description of the various gels is given by Henisch.[3]

There are several chemical requirements that must be fulfilled to grow crystals by this method. First, the reactants employed must be soluble in the solvent (usually water), and the product must be relatively insoluble in the same solvent. Some solubility of the product crystal is required to grow crystals of appreciable size. While it is believed that this is related to supersaturation of the product, the exact mechanism is not yet understood. A second requirement is that the gel must remain stable in the presence of the reacting solutions and must not react with these solutions or the product formed. Silica gels are most stable in acid solutions and tend to dissolve at pH values much greater than 8. This can create problems with chelating agents, since these agents are most useful in highly basic solutions where the gel is unstable. If the pH is only slightly higher than 8 (up to 10) some crystals may still form, but the gel near the alkaline solution-gel interface will start to decompose and sometimes one must race with time to get the crystals grown and out of the gel before the gel itself dissolves.

The influence of the gel in crystal growth is not fully understood. It appears that the main effect of the gel is to slow down the rate of diffusion, which allows the crystal to grow at a slower rate. It is generally believed that the gel does not actually react with the solutions involved, but there is a possibility that it may have some effect on the degree of saturation that is attainable in the solution. The role the gel plays on the nucleation process is also unknown, but some observations on nucleation have been made by Henisch[3] and Armington et al.[4]

The sizes of the crystals grown in a simple experiment are rather small. With the exception of tartrates, which can be grown at least 2 cm long, most experiments yield crystals less than 5 mm along their major axes. Somewhat larger crystals have been grown by using more exotic equipment,[5] or in some cases by using larger tubes and allowing more time for growth to occur.[6] For most purposes, 50-mL test tubes are adequate.

In spite of some difficulties, the gel method has two distinct advantages. First, since the crystals are grown at or near room temperature, thermal strains are absent. Thus gel-grown crystals should be better for some purposes than are crystals grown at higher temperatures. Second, the method is simple and inexpensive.

A. GENERAL EXPERIMENTAL CONSIDERATIONS

Silica gels are formed in two ways. The most used method is that of mixing commercial waterglass (sp gr 1.06) with an equal amount of 1 N acid, usually hydrochloric or acetic acid, and allowing the gel to set for 2 days. A better reagent than waterglass[3] is a laboratory stock solution of 500 mL of water mixed with 244 g of $Na_2SiO_3 \cdot 9H_2O$, since the composition and purity of the commercial solution does vary. The authors prefer a variation of this method[7] in which the acid solution is titrated with sodium metasilicate solution to produce a specific pH. The pH has two effects on the gel.[8] First, the gelling rate is strongly affected by pH. Gels with a pH above 5 tend to set in a few minutes, while gels with a pH of 3 can require several days of setting time. Second, a gel with a pH of 3 will never set as firmly as one with a slightly higher pH. As a general rule, the firmness (or pH) does not significantly alter growth results. There are exceptions to this, however, notable in the growth of tartrate crystals.[3] It should be noted that gels that are formed too rapidly are often brittle. Gel aging, allowing the gel to set for a time before adding the reactant solution, has been investigated,[9] but not with conclusive results.

The purity of most crystals produced by this method is quite high; crystals generally contain only a few parts per million (atomic) of impurity except when doped. The sodium ion, which is present in the solutions, does not tend to enter the growing crystal. Some workers remove the sodium prior to the reaction by allowing it to diffuse into distilled water above the solution, but this is not necessary. The authors filter all solutions[7] through 0.45-μ "millipore" filters to eliminate larger particles that might serve as nucleation sites, but there is no evidence that this greatly affects the results.

One final caution concerns adding the solution above the gel when initiating the reaction. The gel is very fragile, particularly if it is formed at low pH. It is best to add the top solution with a pipette. In this way one can first add a few drops to cover the gel, which tends to protect its surface, and then slowly add the rest of the solution.

Three types of reactions that have been performed in gels are covered in the following sections: (1) standard growth, (2) the complex dilution method, and (3) reduction to form metallic crystals or dendrites.

B. CRYSTALS GROWN BY THE STANDARD TECHNIQUE

The preceding comments apply to this technique and it is by far the most used. The experiments can be done in a test tube, where one of the reactants is added to the gel before setting, or in a U-tube using a neutral gel to separate the reactants. The technique giving the better result varies from case to case. The tartrates, for example, produce better results in a test tube.

C. TARTRATES

While calcium tartrate has been chosen to illustrate the standard technique (see following synthesis), a host of other tartrates have been studied in the past few years with varying degrees of success.[3,8] The growth of iron, copper, and zinc tartrates has been reported by Kachi,[8] who essentially followed the procedure for calcium tartrate outlined here. These experiments have been performed at room temperature using the same molarities as the calcium tartrate. Crystals of about centimeter size are produced, but the time of reaction is unknown. This work is a follow-up of work previously reported by Henisch,[3] who also mentions cobalt, strontium, and the ammonium ion as possible reagents for tartrate crystal growth. Small crystals of lead tartrate have been obtained[10] using the same concentrations at room temperature.

D. CARBONATES

Several workers have reported on the growth of carbonates. Most attention has been given to calcium carbonate[11-13]; other carbonates have been reported[13] to produce only very small crystals. The growth of the calcite modification of calcium carbonate is well documented. The most satisfactory method involves the use of a U-tube at room temperature. A neutral gel is formed by titrating 0.5 M sodium metasilicate with 1 M acetic acid. The pH must be fairly high (6-8) to prevent the formation of calcium silicate. Crystals form during the first week and reach their maximum size in about 6 weeks. The crystals grow and remain clear until they reach a size of about 2 mm. Thereafter they tend to become cloudy (but still essentially single crystal) because of inclusion of the silica gel. During the growth, a few spherulites of aragonite and vaterite (calcium carbonate modifications) appear dispersed in the gel. The formation of aragonite is favored at high temperatures (70°).

E. SULFATES

The growth of both barium and strontium sulfate has been reported by Barta, et al.[12] and substantiated in other work.[14-18] The reaction is:

$$BaCl_2 + (NH_4)_2SO_4 \longrightarrow BaSO_4 + 2NH_4Cl$$

The experiment is generally conducted in a U-tube, and several variations have been reported. Strong acid may be used to titrate the metasilicate, or a strongly acid cation exchange resin may be used. Crystals of potassium–rare earth sulfates have also been reported.[18]

F. OTHER EXPERIMENTS USING THE STANDARD TECHNIQUE

Numerous other materials have been grown by the standard gel technique. The procedures are generally quite similar to those outlined and illustrated in later sections. Experiments with iodates often lead to interesting and unexpected results. Armington and O'Connor[19] attempted to grow potassium, rubidium, and cesium iodates in gel. No crystals resulted in the potassium iodate experiments and only microcrystalline platelets were produced in the cesium iodate experiments. One-centimeter crystals have been produced in the rubidium experiments along with several smaller ones. These have been identified as $RbIO_3 \cdot 2HIO_3$. Arend and Perison[20] grew 2-mm $Ag_2H_3IO_6$ crystals in test tubes at room temperature and a pH of 4.1. The pH varies as a function of depth in the gel as a result of the release of acid by the reaction: $2AgNO_3 + H_5IO_6 \rightarrow Ag_2H_3IO_6 + 2HNO_3$. This causes three iodates to form in the gel: Ag_3IO_5 which forms as a brown layer near the surface, rings of red-brown Ag_2HIO_5, and the 2-mm $Ag_2H_3IO_6$ yellow crystals at the lower end of the gel. The red-brown rings are Liesegang rings and are discussed at the end of this chapter.

Nassau et al.,[21] using an elaborate U-tube arrangement, experimented with copper iodate and found six different forms of copper iodate in different locations in the gel. The largest of these were $Cu(IO_3)_2 \cdot 2H_2O$, which grew to 2 mm.

Brushite, $CaHPO_4 \cdot 2H_2O$, has been grown in test tubes[22] at room temperature and up to 60°, with best results with the pH between 4 and 6. The crystals grew to 2 mm in about 4 weeks. Crystals of other phosphates, that is, KH_2PO_4,[23] $PbHPO_4$ and $Pb_4(NO_3)_2(PO_4)_2 \cdot 2H_2O$,[24] have been reported to grow in organic gels, but apparently not in silica gels because of the high solubilities of the compounds in water.

Lead and zinc sulfides have been produced by the standard technique in gels. The lead compounds are toxic and should be handled with care. For lead sulfide,[25,26] the gel is prepared by mixing sodium metasilicate (sp gr 1.05) with 2.26 N HCl to reach a pH of 6. A few drops of lead acetate or chloride solution are added to the gel before it is allowed to set. Setting requires about 1 week. The solution is then covered with a 1 M solution of thioacetamide. The thioacetamide slowly releases hydrogen sulfide into the solution according to the reaction:

$$H_2O + CH_3-\underset{\underset{S}{\|}}{C}-NH_2 \xrightarrow{H^+} CH_3-\underset{\underset{O}{\|}}{C}-NH_2 + H_2S$$

The sulfide ion then reacts with the lead ion in the gel to precipitate lead sulfide. Crystals up to a millimeter in length can be produced in 2 weeks. Temperatures range from room temperature to 35°, with best results reported at the higher temperature. The same sulfide source is used for the growth of zinc sulfide crystals.[25,27] The crystals grow to a size of 3-4 mm in 2 months. It should be mentioned that a modified U-tube system has been used to produce this size crystal and it is expected that smaller crystals will be produced if a simple U-tube is employed.

Lead iodide and lead hydroxide iodide[3,28,29] are popular experimental systems. The lead iodide is best prepared at 45°, but room temperature can also be used. Thin hexagonal platelets of PbI_2 up to 8 mm in diameter have been grown in 3 weeks at 45°. Lower temperatures produce smaller, but thicker, crystals. Lead hydroxide iodide (PbIOH) is formed in a basic gel in a test tube.[30] If the gel is not alkaline, some PbI_2 crystals are also formed.

Other crystals grown by this technique include lead molybdate,[31] barium and strontium tungstate,[32] silver acetate,[33] potassium perchlorate,[34] and copper(I) iodide.[35]

G. THE COMPLEX DILUTION METHOD

This technique differs from the standard technique in that the crystals are formed by dilution of a solution containing a complex that is stable only at higher concentrations of a common ion. The example used here (following synthesis) is copper(I) chloride, which is soluble only with high concentrations of chloride ion.[36] The role of the gel in this case is dilution of the complexed solution as it diffuses through the gel. This causes slow decomposition of the complex in the gel resulting in precipitation of the desired material. While copper(I) chloride has been the most successful system for application of this technique, some success has been noted with other systems, including copper(I) bromide and copper(I) iodide.[37] Silver iodide crystals have been produced by the diffusion of an AgI-HI complex into an acid (HI) gel.[2,3,38] Other silver halides have been grown in the same manner.[39,40] These experiments should be run in the dark. The growth of HgS (cinnabar) also has been reported[39] using this technique. The experiments are carried out at room temperature in a test tube using a complex of Na_2S-HgS. Crystals of black metacinnabar appear after 1 day, but on standing several days, these crystals convert to the red cinnabar.

Silver phosphate (Ag_3PO_4) has also been grown in silica gel[41] (0.5 M sodium metasilicate plus 1 M acetic acid). The experiments are performed at room temperature, in the dark, using test tubes. The solution above the gel contains Ag_3PO_4 and nitric acid which is added to keep the phosphate in solution. As the growth proceeds, the nitric acid diffuses more rapidly into the gel and the silver phosphate precipitates in the regions of lower acidity. Well-formed dodecahedra are produced (1.5 mm).

Small crystals or dendrites of metals can be formed in a similar way by reducing a metal salt contained in a gel. Lead dendrites are formed rapidly when metallic zinc is used as the reducing agent.[42,43] Similarly, small copper tetrahedra[1] and gold platelets[3,44] can be formed by reduction in gels.

H. LIESEGANG RINGS

No discussion of gel growth could be complete without a passing reference to the phenomenon of Liesegang rings, which were observed even before crystal growth was found in gels. The rings are named after the German chemist[3] who first observed them in 1896. Liesegang observed that concentric rings of silver dichromate are formed (in the gel) after silver nitrate solution is dropped on a gel containing potassium dichromate. The same result occurs when a gel containing potassium dichromate in a test tube is covered with a solution of silver nitrate. The rings appear periodically, usually with no evidence of crystallization between them. In the older work, agar or gelatin was used most often. The same phenomena can be observed in silica gels, however.[42] The true conditions causing these rings are not known, but they have led to much speculation. Liesegang ring formation is generally attributed to the diffusion process in a gel and probably has some relation to the solubility of the product that forms microcrystals in the rings. Using gelatin gels, Hatschek[45] found these rings when producing calcium sulfate, calcium carbonate, barium carbonate, barium oxalate, lead sulfate, lead chloride, and silver chromate. Holmes[42] gives details of several cases using silica gel. Perhaps the most impressive banding was observed by Holmes,[1] with gold as the product. In these experiments, silica gel was formed by mixing sodium metasilicate (sp gr 1.16) with an equal amount of 3 N H_2SO_4. One milliliter of a 1% solution of $AuCl_3$ was added to 25 mL of the mixture before the gel was allowed to set. An 8% oxalic acid solution was added above the gel. The upper bands in the gel consisted of a red layer on top, a blue layer in the middle, and a green layer on the bottom. A dozen or more bands were produced within the test tube. Small golden crystals were also present throughout the gel. A recent description of Liesegang rings has been presented by Strong.[46]

References

1. H. N. Holmes, *J. Phys. Chem.*, 21, 709 (1917).
2. H. K. Henisch, J. I. Hanoka, and J. Dennis, *J. Phys. Chem. Solids*, 26, 493 (1965).

3. H. K. Henisch, *Crystal Growth in Gels*, The Pennsylvania State University Press, University Park, PA, (1976).
4. A. F. Armington, J. J. O'Connor, and M. A. DiPietro, paper presented at the 155th National Meeting of the American Chemical Society, San Francisco (1968).
5. A. F. Armington and J. J. O'Connor, *J. Cryst. Growth*, 3,4, 367 (1968).
6. J. Nikl and H. K. Henisch, *J. Electrochem. Soc.*, 116, 1258 (1969).
7. J. J. O'Connor, M. A. DiPietro, A. F. Armington, and B. Rubin, *Nature*, 212, 68 (1966).
8. Sukeji Kachi, *J. Crystallogr. Soc. Jap.*, 16, 115 (1974).
9. E. S. Halberstat, H. K. Henisch, J. Nikl, and E. W. White, *J. Colloid Interface Sci.*, 29, 469 (1969).
10. M. Abdulkadhar and M. A. Ittyachen, *J. Cryst. Growth*, 39, 365 (1977).
11. A. Schwatz, D. Eckart, J. O'Connell, and K. Francis, *Mater. Res. Bull.*, 6, 1341 (1971).
12. C. Barta, J. Zemlicka, and V. Reni, *J. Cryst. Growth*, 10, 158 (1971).
13. J. W. McCauley and R. Roy, *Am. Mineral*, 59, 947 (1974).
14. G. M. Van Rosmalen and W. G. J. Marchee, *J. Cryst. Growth*, 35, 169 (1976).
15. R. D. Cody and H. R. Shanks, *J. Cryst. Growth*, 23, 275 (1974).
16. A. R. Patel and H. L. Bhat, *J. Cryst. Growth*, 12, 288 (1972).
17. G. Brouwer, G. M. Van Rosmalen, and P. Bennema, *J. Cryst. Growth*, 23, 228 (1974).
18. D. Bloor, *J. Cryst. Growth*, 7, 1 (1970).
19. A. F. Armington and J. J. O'Connor, *Mater. Res. Bull.*, 6, 653 (1971).
20. H. Arend and J. Perison, *Mater. Res. Bull.*, 6, 1205 (1971).
21. K. Nassau, A. S. Cooper, J. W. Shiever, and B. E. Prescott, *J. Solid State Chem.*, 8, 260 (1973).
22. R. Z. Legeros and J. P. Legeros, *J. Cryst. Growth*, 13/14, 476 (1972).
23. B. Brezina and M. Havrankova, *Mater. Res. Bull.*, 6, 537 (1971).
24. B. Brezina and M. Havrankova, *J. Cryst. Growth*, 34, 248 (1976).
25. W. Brenner, Z. Blank, and Y. Okamoto, *Nature*, 212, 392 (1966).
26. K. Sangwal and A. R. Patel, *J. Cryst. Growth*, 23, 282 (1974).
27. S. M. Patel and V. George, *J. Elect. Mater.*, 6, 499 (1977).
28. M. Chand and G. C. Trigunayat, *J. Cryst. Growth*, 30, 61 (1975).
29. M. Chand and G. C. Trigunayat, *J. Cryst. Growth*, 35, 307 (1976).
30. J. Dennis, H. F. Henisch, and P. Cherin, *J. Electrochem. Soc.*, 112, 1240 (1965).
31. K. S. Pillai and M. A. Ittyachen, *J. Cryst. Growth*, 39, 287 (1977).
32. A. R. Patel and S. K. Arora, *J. Cryst. Growth*, 18, 199 (1973).
33. D. Boulin and W. C. Ellis, *J. Cryst. Growth*, 6, 290 (1970).
34. A. R. Patel and A. V. Rao, *J. Cryst. Growth*, 38, 288 (1977).
35. I. Nakada, H. Ishizuku, and N. Ishihara, *Jap. J. Appl. Phys.*, 15, 919 (1976).
36. A. F. Armington and J. J. O'Connor, *Mater. Res. Bull.*, 2, 907 (1967).
37. J. J. O'Connor and A. F. Armington, *Mater. Res. Bull.*, 6, 765 (1971).
38. S. K. Suri, H. K. Henisch, and J. W. Faust, Jr., *J. Cryst. Growth*, 7, 277 (1970).
39. J. C. Murphy, H. A. Kues, and J. Bohandy, *Nature*, 218, 165 (1968).
40. Z. Blank, D. M. Speyer, W. Brenner, and Y. Okamoto, *Nature*, 216, 1103 (1967).
41. S. Mennicke and W. Dittmar, *J. Cryst. Growth*, 26, 197 (1974).
42. H. Holmes, *Colloid Chem.*, 1, 196 (1926).
43. H. M. Liaw and J. W. Faust, Jr., *J. Cryst. Growth*, 13/14, 471 (1972).
44. P. Kratochvil, B. Sprusil, and M. Heyrovsky, *J. Cryst. Growth*, 3/4, 360 (1968).
45. E. Hatschek, *Z. Kolloidchem.*, 8, 193 (1911).
46. C. L. Strong, *Sci. Am.*, 206, 155 (1962).

2. CRYSTAL GROWTH IN GELS

Submitted by A. F. ARMINGTON* and J. J. O'CONNOR*
Checked by AARON WOLD†

A. CALCIUM TARTRATE CRYSTALS GROWN BY THE STANDARD TECHNIQUE

$CaCl_2 \cdot 2H_2O + 2H_2O + HO_2C(CHOH)_2CO_2H \longrightarrow$

$Ca[O_2C(CHOH)_2CO_2] \cdot 4H_2O + 2HCl$

Calcium tartrate is the easiest of all gel-growth experiments and yields the best results of all materials.[1,2] Thus it has been the most used system for gel-growth studies, including nucleation, aging, and pH studies.

Procedure

The gel in this case is formed containing sodium tartrate and sodium hydrogen tartrate. About 50 mL of 0.5 M tartaric acid is titrated with 0.5 M sodium metasilicate to a pH of 4. After titration the solution is placed into 50-mL test tubes and allowed to set. About 30 mL of solution is added to each tube. The tubes should set for 1 or 2 days. Fifteen milliliters of a 0.5 M $CaCl_2$ solution is then added above the gel in each tube and the tubes are lightly stoppered. They are then placed in a controlled-temperature bath at 40°C in the dark. Close temperature control is not needed. The reaction time is about 2 weeks.

Properties

Clear crystals up to 1 cm in length have been produced in test tubes, but 5-mm crystals are more common. Crystals twice this size can be produced using 500-mL graduates in place of small test tubes. In this case 2 to 3 months of reaction time is required.

*HQ Rome Air Development Center (AFSC), Deputy for Electronic Technology, Solid State Sciences Division, Hanscom, AFB, MA 01731.
†Chemistry Department, Brown University, Providence, RI 02912.

B. COPPER(I) CHLORIDE CRYSTALS GROWN BY THE COMPLEX DILUTION METHOD

$$\text{CuCl}_{x+1}{}^{x-} \xrightarrow{\text{dilution}} \text{CuCl}_{(s)} + x\text{Cl}^-$$

Copper(I) chloride is precipitated from a concentrated solution of HCl by dilution during the diffusion process. This process is illustrative of the complex dilution method of growing crystals.

Procedure

A 50-mL test tube is used for the experiment.[3,4] A stock solution of CuCl in 5 N HCl is first prepared by mixing an excess of CuCl (18 g, 0.18 mole) with 200 mL of 5 N acid. Copper turnings are kept under the solution to prevent the formation of CuCl_2. If this is present the solution has a blue-green or brown color that disappears with continuous shaking; the solution becomes clear after the excess CuCl settles to the bottom. Mixing and then allowing the solution to stand overnight is generally sufficient to obtain a clear solution. The gel is then formed by titrating 50 mL of 1 N HCl with a 0.5 M sodium metasilicate solution to a pH of 4.5. The gel solution is then poured into 50-mL test tubes (about 30 mL in each tube) and allowed to set overnight. Fifteen milliliters of the CuCl stock solution is added to each tube. This must be done carefully to prevent damage to the gel at the interface. Slowly adding the first few milliliters with an eyedropper gives the best results. A few milliliters of mineral oil are added above the solution to prevent oxidation of the CuCl. The experiment is best run in the dark. In about 2 weeks, clear, 3-mm tetrahedral crystals are found in the gel.

Properties

Crystals of copper(I) chloride tend to oxidize in air when removed from the gel, but a coating of oil delays the process. The crystals are also stable in vacuum.

Larger CuCl crystals up to 8 mm have been produced using more complicated equipment.[3] The growth of CuBr has also been reported using approximately the same conditions except that HBr is used in the stock solution in place of HCl.[3] Copper(I) iodide can also be grown by this method,[5] although not as successfully as by the conventional gel method.[6] The complex appears to be less stable in that large amounts of CuI precipitate form at the gel-solution interface blocking the reaction. Some success is obtained when the stock solution is 7.5 M in HI, but the crystals are still poorly formed.

References

1. A. K. Henisch, *Crystal Growth in Gels*, The Pensylvania State University Press, University Park, PA, 1976.
2. A. F. Armington, J. J. O'Connor and M. A. Di Pietro, paper presented at the 155th National Meeting of the American Chemical Society, San Francisco, (1968).
3. A. F. Armington and J. J. O'Connor, *J. Cryst. Growth*, 3,4, 367 (1968).
4. A. F. Armington and J. J. O'Connor, *Mater. Res. Bull.*, 2, 907 (1967).
5. J. J. O'Connor and A. F. Armington, *Mater. Res. Bull.*, 6, 765 (1971).
6. I. Nakada, H. Ishizuku, and N. Ishihara, *Jap. J. Appl. Phys.*, 15, 919 (1976).

3. SINGLE-CRYSTAL GROWTH BY A DOUBLE-INFUSION TECHNIQUE

Submitted by W. HUBER,* F. HULLIGER,* and H. VETSCH*
Checked by C. A. LIEN,[†] J. C. PREBLE,[†] S. R. LEE[†] and A. WOLD[†]

A. THE DOUBLE-INFUSION TECHNIQUE

This method is appropriate for growing single crystals of compounds that have a medium or low solubility and cannot be recrystallized because of a disproportionation in the mother liquor. The reaction is not carried out in a gel medium (sodium metasilicate or an organic gel), as in the well-known gel technique;[1] rather a saturated solution of the compound itself is used as reaction medium.[2] Thus crystals of a compound AX can be grown by slowly and simultaneously adding a solution of a compound AY and a solution of a compound BX to a saturated solution of AX according to the reaction

$$AY + BX \longrightarrow AX + BY$$

Of course, the solubilities of AY, BX, and BY have to be considerably larger than that of AX. As is shown in Fig. 1, the double-infusion crystallizer consists of a thermostated (filled with oil) Pyrex vessel containing the saturated solution and eventually some seed crystals mounted on a stirrer. The solutions AY and BX are introduced through glass capillary tubes by means of a micropump[‡]

---

*Laboratorium für Festkörperphysik ETH, CH-8093 Zürich, Switzerland.
[†]Department of Chemistry, Brown University, Providence, RI 02912.
[‡]A product of INFORS AG, CH-4149 Hofstetten (Switzerland), was used in our experiments. The speed of the pump was reduced by a factor of 100 from that of the commercial model.

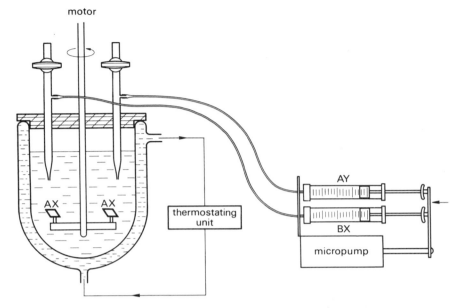

Fig. 1. *Schematic representation of the double-infusion device.*

of the type used in medical practice. The infusion rate thus can be varied in a ratio of $1:10^4$, with a maximum input of 4 mL/hr with 50-mL syringes.* The capillary tubes dip into the growth solution. A diameter of about 10 μ permits a flow rate sufficiently high to avoid back diffusion and precipitation within the tubes. If necessary, however, the capillaries may be placed above the solution. For example, in a variant of the method one reactant may be infused into the other reactant, which has been dissolved in a gel. Nicolau et al.[3] have proposed a simple mathematical treatment for a dropwise input of feed solution in a single-infusion technique.

B. SINGLE CRYSTALS OF RARE EARTH TRANSITION ELEMENT CYANIDES Ln[M(CN)$_6$]·nH$_2$O

Prandtl and Mohr[15] probably were the first to prepare rare earth hexacyanoferrates Ln[Fe(CN)$_6$]·nH$_2$O in microcrystalline form. Some of these cyanides

*Fifty-milliliter polypropylene syringes from Becton, Dickinson & Co. Ltd., Dun Laoghaire, Dublin, Ireland.

form semipermeable membranes that are of interest in connection with the desalination of seawater. $La[Fe(CN)_6] \cdot 5H_2O$ crystallizes in a hexagonal structure[4,5] in which each Fe^{3+} is octahedrally surrounded by six carbon atoms and each Ln^{3+} is surrounded by a trigonal prism formed by six nitrogen atoms. The five H_2O molecules are located outside the five prism faces. The structure becomes less stable with decreasing size of the rare earth ion. The compounds lose one H_2O per formula and their structure becomes orthorhombic[6,7]. In these phases Fe is in a low-spin state ($S = \frac{1}{2}$), which gives rise to the reddish color of all hexacyanoferrate(III) salts. Similar phases exist with cobalt and chromium.[7,8] Cobalt in the cobalt(III) cyano complexes is nonmagnetic, and the colors of these salts are similar to those of the corresponding rare-earth oxides Ln_2O_3. The same holds for the chromium(III) compounds, since the d-electron transitions in $[Cr(CN)_6]^{3-}$ lie beyond the visible range. Magnetic ordering at low temperatures was observed in the chromium(III) and iron(III) cyano complexes[7] but not in the cobalt(III) derivatives.[8] The highest ordering temperature is found in ferrimagnetic $Tb[Cr(CN)_6] \cdot 4H_2O$, namely, 11.7 K. Antiferromagnetic $Lu[Cr(CN_6] \cdot 4H_2O$ has a Néel temperature of 2.8 K, which is quite remarkable for a superexchange by way of five nonmagnetic ions:

$$Cr-C-N-Lu-N-C-Cr$$

Whereas the double-infusion technique is adequate for the growth of hexacyanoferrate(III) and hexacyanocobaltate(III) crystals the solubilities of the chromium(III) salts are so high that crystals can be obtained by crystallization from a saturated solution.[7]

Procedure for the Preparation of Samarium(III) Hexacyanoferrate(III) Tetrahydrate Crystals

Since these hexacyano complexes are hydrated salts, aqueous solutions are used. To prepare the saturated $Sm[Fe(CN)_6]$ solution 10 mL of 1 M neutral $SmCl_3$ solution is first added, followed by the same amount of 1 M $K_3[Fe(CN)_6]$ solution, to 500 mL of distilled water kept at 65°.* In this way a supersaturated solution is obtained that (if everything is clean) can remain in this metastable state for hours before the first crystallites separate. To accelerate the procedure, microcrystals of $Sm[Fe(CN)_6] \cdot 4H_2O$ are prepared by mixing 1 M solutions of 1 mL of $SmCl_3$ and 1 mL of $K_3[Fe(CN)_6]$. The tiny seed crystals that are obtained are used to initiate precipitation in the growth solution. After 10 minutes, still at 65°, the solution is filtered and 50 mL of hot dis-

---

*The checkers used 5 mL of the $SmCl_3$ solution and 5 mL of the $K_3[Fe(CN)_6]$ solution. They used only 250 mL of distilled water and other amounts were changed proportionately.

tilled water is added; the solution is kept for 30 min at 65° and then filtered. This nearly saturated solution is transferred to the growth vessel, which has been carefully cleaned to avoid nucleation (Fig. 2), and is kept at a constant temperature of 65°. The two syringes are filled with 1 M solutions of $SmCl_3$ and $K_3[Fe(CN)_6]$, prepared at room temperature. The $SmCl_3$ solution may be prepared from $SmCl_3 \cdot 6H_2O$ or by reacting Sm_2O_3 with HCl. In the latter case, care should be taken to avoid the presence of free hydrochloric acid, as it would quickly disproportionate the potassium hexacyanoferrate(III) solution. The solution slowly darkens anyway, since at 65° the $K_3[Fe(CN)_6]$ solution is not stable. The glass capillaries dip into the growth solution so that the infused solutions automatically come to the required temperature. With an infusion speed of 0.14 mL/hr single crystals of $Sm[Fe(CN)_6] \cdot 4H_2O$ with edge lengths up to 3 mm grow on the walls of the growth vessel within 4 days. Introduction of seed crystals and stirring may reduce the number and increase the size of the crystals. Seed crystals may be mounted in plastic holders on the stirrer. One should not use a magnetically driven stirrer but a motor-driven glass stirrer with 30–60 rotations/min.

Fig. 2. Crystal-growth unit for rare earth hexacyanoferrates(III): (1) micropump, (2) syringes, (3) details of a syringe, (4) crystal-growth vessel surrounded by a copper-coil heat exchanger (connected with (6)), (5) heater for heating the oil in the outer vessel to slightly below the desired growth temperature, (6) thermostating unit for the heat exchanger.

Properties of $Sm[Fe(CN)_6] \cdot 4H_2O$

The resulting crystals are dark red and are partly of a pseudohexagonal habit as indicated by b/a being close to $\sqrt{3}$. The dimensions of the orthorhombic cell at room temperature are: $a = 7.433$, $b = 12.875$, $c = 13.730$ Å.[6] Below Curie temperature $T_C = 3.5$ K the compound transforms to a ferrimagnet exhibiting a distinct hysteresis loop and a marked anisotropy. The Sm salt is fairly stable in air. With acids poisonous HCN gas is evolved.

C. SILVER(I) PERIODATE $Ag_2H_3IO_6$ AND AMMONIUM PERIODATE $(NH_4)_2H_3IO_6$

These orthoperiodates are of interest because they undergo phase transitions connected with dielectric anomalies.[9-12] Silver(I) periodate $Ag_2H_3IO_6$ is a compound of very limited solubility and low thermal stability, and $(NH_4)_2H_3IO_6$ is stable only over a small temperature range. Since the Ag^+ and NH_4^+ ions are very similar in size, it is not surprising that at room temperature $(NH_4)_2H_3IO_6$ and $Ag_2H_3IO_6$ crystallize in the same structure (space group $R\bar{3}$ if we neglect the H atoms). This structure shows a close relationship to the well-known ferro- and antiferroelectric compounds KH_2PO_4 and $(NH_4)H_2PO_4$. Both periodates undergo two-step phase transitions to an antiferroelectric phase below room temperature. The rhombohedral room-temperature cell contains one formula unit. The structure contains nearly regular isolated $[IO_6]$ octahedra. Along the threefold axis, two Ag^+ or NH_4^+ ions are inserted between the iodate octahedra. The O—O distance between adjacent octahedra is 2.6 Å, which is indicative of strong hydrogen bridges.

At 254 and 252 K for the ammonium salt and between 250 and 210 K for the silver salt, the protons within these bridges become localized and the crystals reveal antiferroelectric behavior.[11] The space-group sequence on going from high to lower temperatures is $R\bar{3}$ with $Z = 1$ to $P\bar{1}$ with $Z = 2$ to $P\bar{1}$ with $Z = 4$ for the Landau transitions in $Ag_2H_3IO_6$ and $R\bar{3}$ with $Z = 1$ to an unknown intermediate phase and finally to $R3$ with $Z = 8$ in $(NH_4)_2H_3IO_6$.[10] At room temperature (20°) the following lattice constants of the hexagonal cell have been determined:

$(NH_4)_2H_3IO_6$: $a = 6.928$ (2) Å, $c = 11.159$ (5) Å[9] *

$Ag_2H_3IO_6$: $a = 5.9366$ (4) Å, $c = 12.7190$ (5) Å[12]

Preparation of $Ag_2H_3IO_6$ and $(NH_4)_2H_3IO_6$ Single Crystals

$$2AgNO_3 + H_5IO_6 \longrightarrow Ag_2H_3IO_6 + 2HNO_3$$

*The hydrogen positions were determined by Tichý et al.[13]

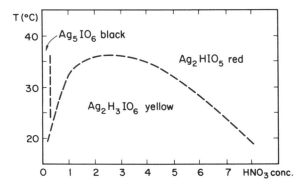

Fig. 3. Phase diagram[14] showing the different silver iodates that crystallize from HNO_3 solutions near room temperature. Acid concentration given in grams of HNO_3 per 100 mL of solution.

A saturated solution of $Ag_2H_3IO_6$ in roughly 2% nitric acid is prepared. The starting solutions must be sufficiently acidified. Otherwise, instead of the yellow $Ag_2H_3IO_6$, a different silver periodate, black Ag_5IO_6, precipitates. Moreover, the reaction temperature should not be too high, since at higher temperatures the red Ag_2HIO_5 is favored, as can be seen in Fig. 3. In detail, 6.79 g (0.04 mole) of $AgNO_3$ and 4.55 g (0.02 mole) of H_5IO_6 are each dissolved in 100 mL of 1.5% HNO_3 solution. The solutions are thoroughly mixed and the resulting solution is diluted to 250 mL. The silver solution must be protected from direct illumination. The solution is allowed to stand for 2 days and then the residual precipitate is filtered off. The saturated $Ag_2H_3IO_6$ solution is cooled to 5° and transferred to the carefully cleaned Pyrex growth vessel. The two 50-mL plastic syringes* are filled with an aqueous H_5IO_6 solution (7.6 mg/mL) and a silver nitrate–nitric acid solution (11.3 mg/mL in 1.8% HNO_3), respectively. The glass capillaries connected with these syringes are dipped into the growth solution and the reaction is started with an infusion speed of 0.3 mL/day. A first run is made to gain some seed crystals. Seed crystals of about 1 mm in edge length are then mounted on the stirrer and the procedure is repeated while the crystals are rotated at a speed of about 0.5 Hz. With this infusion speed the resulting crystal growth rate is near 0.05 mm/day. Thus perfect single crystals 10–15 mm in edge length require a growth time of about 1 year. Less perfect crystals may be grown at a slightly higher infusion speed. All other crystal-growth techniques that have been tried yielded only small crystals less than 2

---

*Fifty-milliliter polypropylene syringes were obtained from Becton, Dickinson & Co. Ltd., Dun Laoghaire, Dublin, Ireland.

mm in edge length. Characteristically, the yellow $Ag_2H_3IO_6$ crystals always adopt the shape of rhombohedra.

The ammonium salt $(NH_4)_2H_3IO_6$ is prepared similarly. One of the solutions in the 50-mL plastic syringes is 20% aqueous NH_4OH; that in the other syringe consists of aqueous H_5IO_6 (43.7 mg H_5IO_6/mL H_2O). These two solutions are infused into a 250-mL saturated solution of $(NH_4)_2H_3IO_6$ in 20% NH_4OH kept at 20°. The saturated growth solution is prepared in the following way: 4.56 g of H_5IO_6 in 30 mL of H_2O is mixed with 50 mL of 15% aqueous NH_4OH. About 4 g of $(NH_4)_2H_3IO_6$ precipitates immediately and about three-quarters of this is redissolved in the solution when its volume is diluted to 250 mL. Even after extensive stirring a 1-g fraction remains undissolved. The saturated $(NH_4)_2H_3IO_6$ solution is filtered and transferred to the carefully cleaned growth vessel. A considerable excess of ammonia is required because $(NH_4)_2H_3IO_6$ is incongruently soluble.[9] Infusion is started at a speed of 0.5 mL/day. The procedure with the seed crystals is the same as in the case of the silver salt. Within about 2 months colorless hexagonal plates approximately 1 mm thick and up to 2 cm in length and width grow.

To prevent growth of additional small crystals on the walls of the growth vessel, the bath temperature may be modulated within a range of 3-4° so that new tiny seed crystals are immediately redissolved.

References

1. H. K. Henish, *Crystal Growth in Gels*, Pennsylvania State University Press, University Park, PA, 1970.
2. W. Huber, H. Arend, and G. Kahr, *Mater. Res. Bull.*, 9, 1573 (1974).
3. I. F. Nicolau, M. Ittu, and R. Dabu, *J. Cryst. Growth*, 13/14, 462 (1972).
4. W. E. Bailey, R. J. Williams, and W. O. Milligan, *Acta Crystallogr.*, B29, 1365 (1973).
5. H. Kietaibl and W. Petter, *Helv. Phys. Acta*, 47, 425 (1974).
6. H. Kietaibl and W. Petter, to be published.
7. F. Hulliger, M. Landolt, and H. Vetsch, *J. Solid State Chem.*, 18, 283 (1976).
8. F. Hulliger, M. Landolt, and H. Vetsch, *J. Solid State Chem.*, 18, 307 (1976).
9. H. Gränicher, R. Kind, W. M. Meier, and W. Petter, *Helv. Phys. Acta*, 41, 843 (1968).
10. J. Petzelt, J. Roos, and H. Gränicher, *Ferroelectrics*, 13, 437 (1976).
11. J. Roos, R. Kind, and J. Petzelt, *Z. Phys.*, B24, 99 (1976).
12. F. Herlach, D. Aboav, H. Gränicher, and W. Petter, *Helv. Phys. Acta*, 30, 252 (1957).
13. K. Tichý, J. Beneš, H. Arend, and A. Rüegg, Report AF-SSP-93, Institut für Reaktortechnik ETH, Würenlingen, 1975, p. 48; *Acta Crystallogr.*, B36, 000 (1980) in press.
14. W. M. Meier, Diploma work ETH, Zurich 1954.
15. W. Prantle and S. Mohr, *Z. anorg. allg. Chem.*, 236, 243 (1938).

4. CRYSTALLINE SILVER CHLORIDE

$$HAgCl_2 \longrightarrow AgCl + HCl$$
$$2HCl + CaCO_3 \longrightarrow CaCl_2 + CO_2 + H_2O$$

Submitted by JOHN C. BAILAR, JR.*
Checked by AARON WOLD†

Because of the photochemical importance of silver chloride, its chemistry and physics have been investigated extensively and many methods have been devised for the formation of silver chloride crystals. Berry et al. describe some of these.[1] Microscopic crystals can be grown by mixing solutions of Ag^+ and Cl^- in gelatin or silica gel,[2] or by slow diffusion of solutions of these into each other.[3] It is also possible to get crystals by the slow decomposition of silver complexes in a solution containing chloride ion.[4] Masses of large crystals can be formed by the slow solidification of molten silver chloride[5] or by pressing the solid between chromium or quartz plates at high pressure. These latter methods have been used for making silver chloride sheets and lenses for optical devices.[6]

The method described here gives individual, single crystals of up to 2 mm in diameter. It depends on the slow loss of hydrogen chloride from a solution of the complex $HAgCl_2$.

Procedure

The apparatus consists of a vacuum desiccator, in which the space normally occupied by the drying agent has been filled with chips of limestone (roughly 1 cm in diameter) and enough water to cover them. Freshly precipitated silver chloride is dissolved in concentrated hydrochloric acid (the solubility is about 3 g/100 mL) and the solution is filtered or decanted into a beaker, which is then put on the plate of the desiccator. The desiccator is closed, but the outlet is left open. The desiccator is put in a dark place and allowed to stand quietly. No temperature regulation is necessary beyond that ordinarily maintained in a laboratory. No vacuum is applied—a vacuum desiccator is used only because it has an outlet for the carbon dioxide formed. There is little or no evaporation of the hydrochloric acid solution—only a decrease in HCl concentration.

Tiny crystals of silver chloride form on the surface of the solution, and larger ones form on the bottom and sides of the beaker. Those that form on the walls of the beaker seem to be the most perfect. They are clear, colorless cubes and

---
*School of Chemical Sciences, University of Illinois, Urbana, IL 61801.
†Chemistry Department, Brown University, Providence, RI 02912.

may reach 2 mm in diameter. The yield and size of the crystals depend on the number and surface area of the limestone chips, the amount of water covering the chips, and the duration of the experiment. The largest crystals have been obtained by allowing the crystallization to proceed for about 5 months.

Properties

The crystals formed by this procedure are white cubes, but they tend to aggregate into needlelike formations. They darken only slowly when exposed to diffuse daylight, and somewhat more rapidly in sunlight. They dissolve rapidly in solutions of strong complexing agents such as ammonia, sodium thiosulfate, and sodium cyanide. If made from pure silver nitrate and hydrochloric acid they are exceedingly pure, because silver chloride rejects H^+ and NO_3^{1-} during the crystallization process.[1]

References

1. C. Berry, W. West, and W. Moser, in *The Art and Science of Growing Crystals*, Wiley, New York, 1963, p. 214.
2. Z. Blank, D. M. Speyer, W. Bredner, and Y. Okamoto, *Nature*, **216**, 1103 (1967).
3. J. Sivadjian, *J. Pharm. Chim.* **17**, [8], 457 (1933).
4. S. Makishima, T. Tomotsu, and S. Hayakawa, *Photogr. Sensitivity Symp. Tokyo*, Bd. 2, 125 (1958); *Chem. Abstr.*, **52**, 16100 (1958); E. Hayek, M. Hohenlohe-Profanter, B. Marcic, and E. Beetz, *Congr. Int. Chim. Pure Appl. 16e⁻ Paris, 1957, Mém Sect. Chim. Minér.*, **1960**, 881.
5. B. M. Brainin, *Sb. Nauchn. Rabot Aspirantov Molodykh, Spets. Petrozavodsk. Univ. Gumaritarn. Fiz. Nauk*, 148 (1965); F. Zoergiebel, G. Haase, G. Henig, E. Schoepper, and J. U. Schott, *Z. Angew, Phys.*, **30**, 316 (1970); H. Kanzaki, *J. Phys. Soc. Jap.*, **11**, 120 (1956); M. T. Speckling, *Phil. Mag.*, **18**, 691 (1968); A Neuhaus, *Chem. Ingr. -Tech.*, **28**, 350 (1956).
6. W. J. Luhman and A. E. Gorum, *Acta Metall.*, **7**, 685 (1959); E. Klein and R. Matejec, *Z. Elektrochem.*, **61**, 1127 (1957); R. Matejec, *Z. Phys.*, **148**, 454 (1957); Englehard Industries, Inc., Br. Pat. 946,504 (1964); *Chem. Abstr.*, **60**, 11652 (1964).

5. THE PARTIALLY OXIDIZED PLATINUM COMPLEX RUBIDIUM TETRACYANOPLATINATE [HYDROGEN BIS(SULFATE)] (3:1:0.46): $Rb_3[Pt(CN)_4]$- $(SO_4$-H-$O_4S)_{0.46} \cdot H_2O$*

$$Rb_2[Pt(CN)_4] \cdot 1.5H_2O + 0.5\ Rb_2SO_4 + 0.42\ H_2SO_4 \xrightarrow[0.8\ V\ (DC)]{H_2O}$$

$$Rb_3[Pt(CN)_4](SO_4\text{-}H\text{-}O_4S)_{0.46} \cdot H_2O + 0.5H_2O + 0.84H^+$$

Submitted by RICHARD BESINGER,[†] DAVID A. VIDUSEK,[†] DANIEL P. GERRITY,[†] and JACK M. WILLIAMS[‡]
Checked by KRISTIN BOWMAN MERTES[§]

The partially oxidized tetracyanoplatinate salts continue to be of high current interest because of their one-dimensional properties. These salts,[1] containing Cl^-, Br^- (see Reference 2), and $(FHF)^-$, have previously been reported in the literature. We have now prepared the first such compound containing the hydrogen bis(sulfate) $(SO_4\text{-}H\text{-}O_4S)^{3-}$ anion. Levy[3] has reported the preparation of what appeared to be a sulfate complex. The electrolytic procedure described here is one that we have reported previously[1] for the preparation of related salts of $(FHF)^-$. This procedure was originally reported by Terry[4] for the preparation of $K_{1.75}[Pt(CN)_4] \cdot 1.5H_2O$ and was extended by Miller[5] for the preparation of $K_2[Pt(CN)_4]X_{0.3} \cdot 3H_2O$, where $X = Br^-$ or Cl^-. The electrolytic synthesis described here is suitable for the preparation of high-purity triclinic crystals of $Rb_3[Pt(CN)_4](SO_4\text{-}H\text{-}O_4S)_{0.46} \cdot H_2O$ with a Pt—Pt intrachain repeat separation of 2.826(1) Å.

Materials

The $Rb_2[Pt(CN)_4] \cdot 1.5H_2O$ used in this preparation is prepared as described by Koch et al.[6] All other chemicals used are ACS reagent grade. Distilled water is

---

*Work performed under the auspices of the Office of Basic Energy Sciences of the U.S. Department of Energy. *By acceptance of this article, the publisher and/or recipient acknowledges the U.S. Governments right to retain a nonexclusive, royalty-free license in and to any copyright covering this paper.*

[†]Associated Colleges of the Midwest research participants sponsored by the Argonne Center for Educational Affairs: Richard Besinger from Carleton College, Northfield, MN., David A. Vidusek from Illinois State University, Normal, IL, and Daniel P. Gerrity from Cornell College, Mount Vernon, IA.

[‡]Correspondence: Chemistry Division, Argonne National Laboratory, Argonne, IL 60439.
[§]Department of Chemistry, University of Kansas, Lawrence, KS 66045.

used throughout the entire preparation, and all of the reactions are carried out in polyethylene beakers.

Procedure

■ **Caution** *Because of the extremely poisonous nature of cyanide, these steps should be carried out in a well-ventilated area using protective gloves and clothing and a face shield.*

An 0.8 M solution of Rb_2SO_4 is prepared by adding 1.34 g (5.01 × 10^{-3} mole) of Rb_2SO_4 to 6 mL of water. The solution is then saturated with 1.08 g (2.15 × 10^{-3} mole) of $Rb_2[Pt(CN)_4] \cdot 1.5H_2O$ and acidified to a pH of less than 1 with 0.25 mL of 9 M H_2SO_4. The sulfuric acid is added in small increments while the solution is stirred to minimize possible generation of HCN gas. The clear solution is transferred to an electrolytic cell consisting of a 50-mL polyethylene beaker with two small opposing holes drilled 5 mm from the bottom; two platinum electrodes are cemented in place by epoxy cement through the holes with the electrode tips being separated by about 4 mm.

After 72 hr of electrolysis at 0.8 V the reddish-bronze crystals are suction filtered through a medium-porosity fritted filter, washed with two 3-mL portions of cold water, and allowed to dry in air. Yield: 1.5 g (nearly 100%) based on $Rb_2[Pt(CN)_4] \cdot 1.5H_2O$.

Anal. Calcd. for $Rb_3[Pt(CN)_4](SO_4\text{-}H\text{-}O_4S)_{0.46} \cdot H_2O$: Rb, 38.71; Pt, 29.45; C, 7.25; N, 8.46; H, 0.37; S, 4.45. Found[7]: Rb, 39.08, 39.29; Pt, 29.53, 29.90; C, 7.02; N, 8.36; H, 0.21; S, 4.46, 4.50. Found[8]: C, 7.27, 7.20; N, 8.18, 8.35; H, 0.47, 0.39; S, 4.34, 4.48.

A platinum oxidation state of +2.30 was determined by iodine-thiosulfate titrations[9] assuming a molecular weight of 662.44. The calculated Pt oxidation state, based on the chemical analyses and crystallographic analysis, is +2.38 ± 0.02.[10]

Thermogravimetric analysis (25-300°) shows a weight loss of 0.126% between 40° and 115°. This weight loss corresponds to 0.47 mole of H_2O per mole of compound.

Emission spectrographic analyses indicate the product to be of high purity and to contain the following impurities: 0.5% Ba, 0.05% K, 0.005% Li, 0.005% Ca, and a faint trace (<0.01%) of Na.[11] Attempts at preparing this compound by a chemical preparation using a tenfold or threefold excess of 30% H_2O_2 have failed to produce a similar product.

Physical Properties

The compound $Rb_3[Pt(CN)_4](SO_4\text{-}H\text{-}O_4S)_{0.46} \cdot H_2O$ forms triclinic crystals that have a reddish-metallic luster. The cell constants as determined from single-crystal X-ray diffractometer measurements are as follows: $a = 5.652(1)$ Å, $b = 9.372(2)$ Å, $c = 13.762(2)$ Å, $\alpha = 71.19(1)°$, $\beta = 81.94(2)°$ $\gamma = 73.44(1)°$, $V = 660.40$ Å^3. The observed crystal density of 3.24(1) g/cm^3 is in excellent agreement with the calculated density of 3.18 g/cm^3 based on two formula weights per unit cell.

The first 10 reflections of the powder pattern correspond to the following d-spacings: 12.80 (s), 8.42 (s), 6.27 (m), 4.45 (m), 4.29 (s), 4.16 (m), 4.01 (w), 3.85 (w), 3.73 (w), 3.63 (w) Å.[12] An infrared study[13] of the product suggests the presence of a sulfate anion, and preliminary crystallographic studies indicate that it is present as the complex $(SO_4\text{-}H\text{-}O_4S)^{3-}$ species.[14]

The hydrogen bis(sulfate) compound exhibits a room-temperature single-crystal conductivity (four probe) of ~2000-2100 ohms^{-1} cm^{-1}. Thus it can be categorized as a highly conductive species within the tetracyanoplatinate family of compounds.

References and Notes

1. A. J. Schultz, C. C. Coffey, G. C. Lee, and J. M. Williams, *Inorg. Chem.*, **16**, 2129 (1977).
2. G. D. Stucky, A. J. Schultz, and J. M. Williams, *Annu. Rev. Mater. Sci.*, **7**, 310 (1977).
3. L. A. Levy, *J. Chem. Soc.*, **1912**, 1081.
4. H. Terrey, *J. Chem. Soc.*, **1928**, 202.
5. J. S. Miller, *Science*, **194**, 189 (1976).
6. T. R. Koch, J. A. Abys, and J. M. Williams, *Inorg. Synth.*, **19**, 9 (1979).
7. Galbraith Laboratories, Knoxville, TN.
8. Midwest Microlabs, Indianapolis, IN. This laboratory does not perform Rb or Pt analyses.
9. Iodine-thiosulfate titrations were performed by Ms. E. Streets of Argonne National Laboratory.
10. It is difficult to assess the true oxidation state of the platinum atom because the structure likely contains (small amounts of) additional species such as H_3O^+ or HSO_4^-. This is not entirely unexpected since the compound was prepared in a highly acidic medium.
11. Emission spectrographic analyses were performed by J. P. Faris of Argonne National Laboratory.
12. We wish to thank Ms. E. Sherry for obtaining the X-ray powder patterns.
13. The infrared studies were performed by Dr. J. Ferraro of Argonne National Laboratory.
14. J. M. Williams, A. J. Schultz, and R. Besinger, work in progress.

6. THE RUBIDIUM-CESIUM PARTIALLY OXIDIZED TETRACYANOPLATINATE (HYDROGEN DIFLUORIDE)-FLUORIDE SYSTEMS*

The partially oxidized tetracyanoplatinate (POTCP) salts continue to be of high interest because of their anisotropic metallic properties. By using either electrolytic[1] or chemical[2] synthesis procedures, markedly different alkali metal tetracyanoplatinate (hydrogen difluoride) or fluoride materials may be prepared. Also, by utilizing the known equilibrium constants for the aqueous hydrogen fluoride system, it is possible to vary the $[FHF]^-$ and F^- concentration by adjusting the solution pH's. This results in the selective preparation of $[FHF]^-$ or F^- containing POTCP salts.

The compounds differ in their Pt—Pt intrachain separations, electrical conductivities, degree of partial oxidation of the platinum atom, and metallic color. These major differences occur even though two pairs of these four compounds contain the same anion, $(FHF)^-$, and cation, Rb^+ or Cs^+. This provides the unique opportunity to explore only the effects of structural, not elemental, changes on the anisotropic electrical behavior of POTCP salts.

■ **Caution.** *Because of the extremely poisonous nature of cyanide, and the very corrosive nature of HF, these steps should be carried out in a well-ventilated fume hood using protective gloves and clothing and face shield. Should any solution containing HF spill on gloves or clothing they should be removed immediately. At no time should any solutions containing HF be exposed to glass.*

Materials

The $Cs_2[Pt(CN)_4]\cdot H_2O$ and $Rb_2[Pt(CN)_4]\cdot 1.5H_2O$ starting materials for these preparations are synthesized as described by Maffly et al.[3] and Koch et al.,[4] respectively. The cesium fluoride and rubidium fluoride are 99.9% pure. All other chemicals are ACS reagent grade. Distilled water is used throughout the procedures.

---

*Work performed under the auspices of the Office of Basic Energy Sciences of the U.S. Department of Energy. *By acceptance of this article, the publisher and/or recipient acknowledges the U.S. Government's right to retain a nonexclusive, royalty-free license in and to any copyright covering this paper.*

A. RUBIDIUM TETRACYANOPLATINATE (HYDROGEN DIFLUORIDE) (2:1:0.29) HYDRATE: $Rb_2[Pt(CN)_4](FHF)_{0.29} \cdot 1.67H_2O$

$$60Rb_2[Pt(CN)_4] \cdot 1.5H_2O + 9H_2O_2 + 36HF \xrightarrow{RbF, H_2O}$$
$$60Rb_2[Pt(CN)_4](FHF)_{0.29} \cdot 1.67H_2O + 8H_2O$$

Submitted by KIM L. STEARLEY* and JACK M. WILLIAMS[†]
Checked by KRISTIN BOWMAN MERTES[‡]

Procedure

A solution of $Rb_2[Pt(CN)_4]$ is prepared by dissolving 1.1 g (2.2 × 10^{-3} mole) of $Rb_2[Pt(CN)_4] \cdot 1.5H_2O$ in 6 mL of distilled water. The solution is then transferred to a 50-mL polyethylene beaker. To this is added a 2.0 g (1.9 × 10^{-2} mole) of rubidium fluoride. A fine green-white precipitate appears immediately. To this solution and precipitate is added 3 mL of concentrated HF and upon stirring dissolution is complete. To the clear solution is added 2 mL of 15% H_2O_2 solution and the beaker is allowed to stand (uncovered) in a plastic desiccator. After 1-2 days the formation of greenish-bronze metallic crystals is observed. The crystals are harvested 1 week after initial formation. They are washed with two 5-mL portions of ice-cold water and are allowed to dry in the air. Yield: 0.38 g (37%) based on $Rb_2[Pt(CN)_4] \cdot 1.5H_2O$. The crystals should be stored in a polyethylene vial in a refrigerator to avoid possible loss of HF from the complex. Larger crystals may be grown by diluting the original peroxide solution by a factor of 10 and allowing the crystals to form through slow evaporation coupled with seeding over a period of several weeks or months. Chemical analyses[5] are given in Table I. Thermal gravimetric analysis showed the complex to contain 1.67 water molecules of hydration per molecule.[6]

Physical Properties[7]

The compound may be identified with X-ray diffraction powder pattern data.[8] The first 12 reflections occur at the following d-spacings: 8.95 (vs), 8.13 (vw), 6.47 (m), 6.13 (m), 4.94 (vw), 4.43 (s), 4.21 (vw), 4.06 (s), 3.88 (s), 3.79 (vw), 3.54 (vw), 3.42 (vw) Å. Preliminary X-ray diffraction studies indicate that for $Rb_2[Pt(CN)_4](FHF)_{0.29} \cdot 1.67H_2O$, $c = 5.78$ Å and Pt—Pt = 2.89 Å.

The compound $Rb_2[Pt(CN)_4](FHF)_{0.29} \cdot 1.67H_2O$ appears to be stable to-

---

*Research participant sponsored by the Argonne Center for Education Affairs, Illinois State University, Normal, IL.
[†]Chemistry Division, Argonne National Laboratory, Argonne, IL 60439.
[‡]Department of Chemistry, University of Kansas, Lawrence, KS 66045.

TABLE I Chemical Analyses of $Rb_2[Pt(CN)_4](FHF)_{0.29} \cdot 1.67H_2O^5$

Elements	Percentage Composition					Average of Five Samples	Theoretical
	Sample 1	Sample 2	Sample 3	Sample 4	Sample 5		
C	9.28	9.44	9.41	10.04	9.92	9.62	9.39
N	10.93	11.27	11.32	10.91	11.92	11.27	10.96
H	0.32	0.22	0.25	0.32	0.26	0.27	0.65
F	2.53, 2.35	2.23, 2.11, 2.09	1.99, 1.94, 1.92	2.14, 2.04, 2.00	2.28, 2.26, 2.10	2.14	2.15
total halogen, (Cl, Br, I)	1.28 as Cl	trace or none	trace or none	trace or none	trace or none	–	–

ward hydration and dehydration and may be stored over a saturated solution of NaCl that has a relative humidity of 75.1% at 25°.

B. RUBIDIUM TETRACYANOPLATINATE (HYDROGEN DIFLUORIDE) (2:1:0.38): $Rb_2[Pt(CN)_4](FHF)_{0.38}$

$$100Rb_2[Pt(CN)_4] \cdot 1.5H_2O + 76HF \xrightarrow[1.5\ V]{RbF, H_2O} 100Rb_2[Pt(CN)_4](FHF)_{0.38} + 150H_2O + 19H_2$$

Submitted by C. COFFEY* and JACK M. WILLIAMS†
Checked by KRISTIN BOWMAN MERTES‡

Procedure

A solution of $Rb_2[Pt(CN)_4]$ is prepared by dissolving 1.1 g (2.2×10^{-3} mole) of $Rb_2[Pt(CN)_4] \cdot 1.5H_2O$ in 6 mL of distilled water. The solution is then transferred to a 50-mL polyethylene beaker. To this is added 2.0 g (1.9×10^{-2} mole) of rubidium fluoride. A green-white precipitate appears immediately. To this solution and precipitate 3 mL of concentrated HF is added with stirring until dissolution is complete. The clear solution is transferred to a polyethylene cell consisting of a 50-mL polyethylene beaker with two opposing holes drilled near the bottom to contain two platinum electrodes (each approximately 25 mm long). The electrode tips are separated by about 2 mm and are attached to a 1.5 V dry cell. The solution is electrolyzed for 10 hours. By the end of this time lustrous metallic yellow-gold colored crystals fill the beaker. They are harvested

---

*Research participant sponsored by the Argonne Center for Educational Affairs, Mercyhurst College, Erie, PA.
†Chemistry Division, Argonne National Laboratory, Argonne, IL 60439.
‡Department of Chemistry, University of Kansas, Lawrence, KS 66045.

and washed with two 6-mL portions of distilled water and allowed to dry in air. Yield: 0.63 g (61%) based on the $Rb_2[Pt(CN)_4] \cdot 1.5H_2O$. The crystals should be stored in a polyethylene vial in a refrigerator to avoid possible loss of HF from the complex. Larger crystals can be grown using a variable-voltage source set at 1 V DC. Because of the apparent decomposition of $Rb_2[Pt(CN)_4](FHF)_{0.38}$ when exposed to air at room temperature for long periods of time, the electrolysis should be stopped while there is still some solution remaining. *Anal.* Calcd. for $Rb_2[Pt(CN)_4](FHF)_{0.38}$: C, 9.54; N, 11.13; F, 2.94; H, 0.47. Found[5]: C, 9.66; N, 10.53; F, 3.05–3.15; H, 0.15. A platinum oxidation state of +2.36 is determined by iodine–thiosulfate titration.[10] The Pt oxidation state, as determined in an X-ray crystallographic study, is approximately 2.38 and is reported as such in this synthesis.

Physical Properties[7]

The first 10 reflections in the X-ray diffraction pattern[8] occur at the following d-spacings: 9.28 (w), 6.57 (vw), 4.60 (m), 4.13 (s), 4.05 (m), 2.91 (s), 2.83 (m), 2.76 (m), 2.56 (vw), 2.41 (vw) Å.

Single-crystal X-ray studies indicate that $Rb_2[Pt(CN)_4](FHF)_{0.38}$ is body-centered tetragonal with lattice constants a = 12.66 Å and c = 5.58 Å, as compared to $Rb_{1.7}[Pt(CN)_4] \cdot 2H_2O$, which is monoclinic. Thermogravimetric analyses indicate that $Rb_2[Pt(CN)_4](FHF)_{0.38}$ is anhydrous.

C. CESIUM TETRACYANOPLATINATE (HYDROGEN DIFLUORIDE) (2:1:0.23): $Cs_2[Pt(CN)_4](FHF)_{0.23}$

$$200Cs_2[Pt(CN)_4] \cdot H_2O + 23H_2O_2 + 92HF \xrightarrow[H_2O]{CsF} 200Cs_2[Pt(CN)_4](FHF)_{0.23} + 246H_2O$$

Submitted by MICHAEL J. MICHALCZYK,* DAVID A. VIDUSEK,* and JACK M. WILLIAMS†
Checked by KRISTIN BOWMAN MERTES‡

Procedure

Initially, 1.25 g (2.13×10^{-3} mole) of $Cs_2[Pt(CN)_4] \cdot H_2O$ is dissolved in 10 mL of water to produce a saturated solution. The solution is transferred to a 50-mL

---

*Research participants sponsored by the Argonne Center for Educational Affairs: Michael J. Michalczyk from Knox College, Galesburg, IL and David A. Vidusek from Illinois State University, Normal, IL.
†Chemistry Division, Argonne National Laboratory, Argonne, IL 60439.
‡Department of Chemistry, University of Kansas, Lawrence, KS 66045.

TABLE I Chemical Analyses[5] of $Cs_2[Pt(CN)_4](FHF)_{0.23}$

	Percentage Composition				Average of Four Samples	Theoretical
Elements	Sample 1	Sample 2	Sample 3	Sample 4		
C	7.53	9.37	8.23	8.03	8.29	8.36
N	9.74	9.42	9.41	9.39	9.49	9.76
H	0.25	0.21	0.18	2.10	0.69	0.05
F	1.51, 1.37, 1.30	1.60, 1.50, 1.44	1.75, 1.70, 1.64	1.63, 1.49	1.54	1.52
Total halogen (Cl, Br and I)	trace or none	trace or none	trace or none	trace or none	trace or none	–

polyethylene beaker and 2.0 g (1.32×10^{-2} mole) of cesium fluoride is added. The white cesium fluoride appears to become yellow upon addition and then dissolves leaving a clear solution (slight warming may be necessary to dissolve the cesium fluoride). Two milliliters of concentrated HF (28.9 M) is added to acidify the solution, followed by 1 mL of 15% H_2O_2 solution. The beaker is placed in a polyethylene desiccator containing anhydrous $CaSO_4$ and the solution is allowed to evaporate at 25°.

After 24–36 hours, bronze crystals fill the bottom of the beaker. They are filtered (plastic funnel), washed with two 2-mL portions of ice cold water and allowed to dry in air. Yield: 0.80–0.90 g (78–88%), based on $Cs_2[Pt(CN)_4] \cdot H_2O$. For analyses see Table I.

Anal. Calcd. for $Cs_2[Pt(CN)_4](FHF)_{0.23}$: Cs, 46.34; Pt, 33.97; C, 8.36; N, 9.76; H, 0.05; F, 1.52. Thermogravimetric analysis[6] of $Cs_2[Pt(CN)_4](FHF)_{0.23}$ showed that the compound contained no water. A platinum oxidation state of +2.27 was determined by iodine-thiosulfate titrations.[10] The Pt oxidation state reported here was that determined from the fluorine analyses.

Physical Properties[7]

The compound $Cs_2[Pt(CN)_4](FHF)_{0.23}$ forms body-centered tetragonal crystals that have a distinctive bronzish-brown metallic luster. The cell constants as determined from a preliminary single-crystal X-ray diffraction analysis are as follows: $a = 13.051(2)$ Å, $b = 13.049(2)$ Å, $c = 5.744(1)$ Å. The Pt—Pt separation (c/2) is 2.87 Å. The first 12 reflections of the powder pattern correspond to the following d-spacings:[8] 12.35 (vs), 8.62 (s), 4.47 (m), 4.00 (m), 3.09 (m), 3.00 (m), 2.85 (m), 2.72 (m), 2.50 (vw), 2.32 (vw), 2.26 (vw), 2.20 (vw) Å. Although $Cs_2[Pt(CN)_4](FHF)_{0.23}$ is anhydrous, it appears to be relatively stable toward hydration and dehydration and may be stored safely over a saturated solution of NaCl that has a relative humidity of 75.1% at 25°. However, it should be stored in a plastic container to avoid reactions with glass.

D. CESIUM TETRACYANOPLATINATE (HYDROGEN DIFLUORIDE) (2:1:0.38): $Cs_2[Pt(CN)_4](FHF)_{0.38}$

$$100Cs_2[Pt(CN)_4] \cdot H_2O + 76HF \xrightarrow[1.5\ V]{CsF, H_2O}$$
$$100Cs_2[Pt(CN)_4](FHF)_{0.38} + 100H_2O + 19H_2$$

Submitted by D. P. GERRITY* and JACK M. WILLIAMS[†]
Checked by KRISTIN BOWMAN MERTES[‡]

Procedure

Initially 1.3 g (2.2×10^{-3} mole) of $Cs_2[Pt(CN)_4] \cdot H_2O$ is dissolved in 6 mL of water to produce a saturated solution. It may be necessary to warm the solution slightly to dissolve the $Cs_2[Pt(CN)_4] \cdot H_2O$. The solution is transferred to a 50-mL polyethylene beaker and 4.0 g (2.6×10^{-2} mole) of cesium fluoride is added. The white cesium fluoride appears to become yellow after it is added. It then dissolves leaving a white suspension. Finally, 3 mL of conc. HF (28.9 M) is added to acidify the solution and dissolve the white suspension.

The electrolysis cell consists of a 50-mL polyethylene beaker with two small holes drilled about 6 mm from the bottom and on opposite sides. Two platinum wires about 25 mm in length are used as electrodes. They are held in place with Apiezon N grease. The electrode-electrode separation is approximately 5 mm. The power source is a 1.5 V dry cell.

After about 4 hours of operation, the crystals on the bottom of the beaker are filtered in a plastic funnel containing filter paper. They are then washed with two approximately 6-mL portions of cold water and allowed to dry in air. Yield: 0.9-1.0 g (70-80%) based on $Cs_2[Pt(CN)_4] \cdot H_2O$.

Anal. Calcd. for $Cs_2[Pt(CN)_4](FHF)_{0.38}$: Cs, 44.46; Pt, 32.63; C, 8.04; N, 9.37; H, 0.40; F, 2.42. Found[9]: Cs, 43.69, 43.67; Pt. 33.95, 34.04; C, 8.19; N, 7.28; H, 0.23; F, 2.41, 2.39; halogen other than F, 0.0. Found[5]: C, 8.18; N, 9.32; H, 0.17; F, 2.80, 2.11; halogen other than F, 0.0. A platinum oxidation state of +2.35 was determined by iodine-thiosulfate titrations[10]; the value of +2.38 determined from a crystal-structure analysis is used in the formula in this synthesis.

Physical Properties[7]

The compound $Cs_2[Pt(CN)_4](FHF)_{0.38}$ forms body-centered tetragonal crystals that have a distinctive reddish-gold metallic luster. The cell constants deter-

---

*Associated Colleges of the Midwest research participant sponsored by the Argonne Center for Educational Affairs from Cornell College, Mount Vernon, IA.

[†] Chemistry Division, Argonne National Laboratory, Argonne, IL 60439.

[‡] Department of Chemistry, University of Kansas, Lawrence, KS 66045.

mined from a single-crystal X-ray diffraction analysis are as follows: $a = b = 13.057(2)$ Å, $c = 5.665(1)$ Å. The Pt—Pt spacing of 2.83 Å is the second shortest such repeat distance yet reported. The first 10 reflections of the powder pattern correspond to the following d-spacings[8]: 9.10 (m), 6.45 (vw), 4.60 (m), 4.08 (vs), 3.06 (vw), 2.91 (m), 2.83 (m), 2.763 (m), 2.761 (vw), 2.72 (vw) Å.

Although $Cs_2[Pt(CN)_4](FHF)_{0.38}$ is anhydrous, it appears to be relatively stable toward hydration and dehydration and may be stored safely over a saturated solution of NaCl that has a relative humidity of 75.1% at 25°. However, it should be stored in a plastic container and refrigerated to avoid possible loss of HF.

E. CESIUM TETRACYANOPLATINATE FLUORIDE (2:1:0.19): $Cs_2[Pt(CN)_4]F_{0.19}$

$$200Cs_2[Pt(CN)_4] \cdot H_2O + 38CsF \xrightarrow[1.5 \text{ V}]{CsOH, H_2O} 200Cs_2[Pt(CN)_4]F_{0.19} + 38CsOH + 162H_2O + 19H_2$$

Submitted by GRACE C. LEE,* KIM L. STEARLEY,* and JACK M. WILLIAMS†
Checked by KRISTIN BOWMAN MERTES‡

Procedure

Initially, 1.3 g (2.1×10^{-3} mole) of $Cs_2[Pt(CN)_4] \cdot H_2O$ is dissolved in 7 mL of water to produce a saturated solution (it may be necessary to warm the solution slightly to dissolve the $Cs_2[Pt(CN)_4] \cdot H_2O$). The solution is transferred to a 50-mL polyethylene beaker and 1.7 g (1.1×10^{-2} mole) of cesium fluoride is added. The white CsF appears to become yellow after it is added. It then dissolves, leaving a white suspension. Finally, 7.6 M CsOH is added to the solution dropwise to raise the pH from approximately 4 to 9 (±0.2). The pH of 9 must be carefully maintained throughout the reaction. The electrolysis cell consists of a 50-mL polyethylene beaker with two small holes drilled about 6 mm from the bottom and on opposite sides. Two platinum wires about 25 mm in length are used as electrodes. They are held in place with Apiezon N grease. The electrode–electrode separation is approximately 5 mm. The power source is a 1.5 V dry cell.

After 72 hours of operation at 1.5 V, the crystals are filtered in a plastic

---

*Undergraduate research participants sponsored by the Argonne Center for Educational Affairs: Grace C. Lee from Saint Mary-of-the-Woods College, Saint Mary-of-the-Woods, IN and Kim L. Stearley from Illinois State University, Normal, IL.
†Chemistry Division, Argonne National Laboratory, Argonne, IL 60439.
‡Department of Chemistry, University of Kansas, Lawrence, KS 66045.

TABLE I Chemical Analyses[5] of $Cs_2[Pt(CN)_4]F_{0.19}$

	Percentage Composition					
Elements	Sample 1	Sample 2	Sample 3	Sample 4	Average	Theoretical
C	8.88	8.11	8.61	8.59	8.55	8.19
N	10.14	11.70	9.51	11.60	10.73	9.55
F	0.69, 0.65, 0.67	0.47, 0.61, 0.63	0.60, 0.77, 0.58	0.67, 0.59, 0.62	0.63	0.63
Total halogen (Cl, Br, and I)	trace or none	trace or none	trace or none	trace of none	–	–

funnel containing filter paper. They are then washed with two approximately 1-mL portions of ice-cold water and are allowed to dry in air. After the crystals have been washed they should be examined under a microscope to ensure that no starting material remains. Yield: 0.94–0.99 g (81%), based on $Cs_2[Pt(CN)_4] \cdot H_2O$.

Chemical analyses[5] of $Cs_2[Pt(CN)_4]F_{0.19}$ are shown in Table I. Infrared spectroscopy[11] showed an absence of absorption bands in the hydrogen difluoride regions and confirmed that the salt is not a hydrogen difluoride. A platinum oxidation state of +2.25 was determined by iodine–thiosulfate[10] titrations. Thermogravimetric analysis[6] showed the $Cs_2[Pt(CN)_4]F_{0.19}$ to contain no water of hydration.

Physical Properties[7]

The compound $Cs_2[Pt(CN)_4]F_{0.19}$ forms body-centered orthorhombic crystals that have a distinctive reddish-gold metallic luster. The cell constants as determined from a preliminary single-crystal X-ray diffraction analysis are as follows: $a = 5.771(2)$ Å, $b = 12.770(5)$ Å, $c = 13.382(3)$ Å, $V = 986.2(5)$ Å^3, and Pt—Pt = 2.88 Å. The first 12 reflections of the powder pattern correspond to the following d-spacings:[8] 11.95 (vs), 6.82 (vs), 4.45 (vs), 4.00 (vs), 3.88 (vs), 2.85 (vs), 2.80 (vs), 2.73 (m), 2.66 (m), 2.38 (mw), 2.33 (m), 2.29 (m) Å.

Although $Cs_2[Pt(CN)_4]F_{0.19}$ is anhydrous, it appears to be relatively stable toward hydration and dehydration and may be stored safely over a saturated solution of NaCl that has a relative humidity of 75.1% at 25°. However, it should be stored in a plastic container to avoid reactions with glass.

References and Notes

1. H. Terry, *J. Chem. Soc.*, **1928**, 202.
2. L. A. Levy, *J. Chem. Soc.*, **101**, 1097 (1912).
3. R. L. Maffly, J. A. Abys, and J. M. Williams, *Inorg. Synth.*, **19**, 6 (1979).
4. T. R. Koch, J. A. Abys, and J. M. Williams, *Inorg. Synth.*, **19**, 9 (1979).

5. Midwest Microlabs, Indianapolis, Indiana. This laboratory does not perform Rb, Cs and Pt analyses.
6. We express our gratitude to A. J. Schultz and G. C. Lee for the thermal gravimetric analyses reported in this procedure.
7. We wish to thank Mr. J. P. Faris of Argonne National Laboratory, who checked impurity levels in all materials reported in this series using emission spectrographic analysis techniques. This also confirmed the presence of the indicated metals.
8. We wish to thank Ms. E. Sherry for obtaining the X-ray powder patterns.
9. Galbraith Laboratories, Knoxville, Tennessee.
10. Iodine–thiosulfate analyses were performed by Ms. E. Streets of Argonne National Laboratory.
11. Infrared spectroscopic analyses were performed by Dr. John Ferraro of Argonne National Laboratory.

Chapter Two

STOICHIOMETRICALLY SIMPLE COMPOUNDS

7. TELLURIUM TETRAFLUORIDE

Submitted by KONRAD SEPPELT*
Checked by LAWRENCE LAWLOR and JACK PASSMORE†

$$TeO_2 + 2SF_4 \longrightarrow TeF_4 + 2SOF_2$$

The preparation given here for TeF_4 is unusually convenient. None of the other methods, for example, preparation from Te and TeF_6[1] or from Te and F_2[2] has the simplicity of the procedure described here, nor the yield.[3] Tellurium dioxide (Merck Co.) and sulfur tetrafluoride are commercially available (Air Products, J. T. Baker Chemicals). The procedure requires a 200 mL or larger stainless steel pressure vessel in addition to a glass vacuum line and glass sublimator.

Procedure

■ **Caution.** *Sulfur tetrafluoride and SOF_2 are poisonous gases, although their intense smell is a good warning. Tellurium tetrafluoride is not so dangerous because of its low vapor pressure, but contact with the skin must be avoided. All tellurium compounds are highly toxic.*

*Anorganisch-Chemisches Institut der Universität, 6900 Heidelberg, Im Neuenheimer Feld 270, West Germany.
†Department of Chemistry, University of New Brunswick, P.O. Box 4400, Fredericton, N.B., Canada E3B 5A3.

Tellurium dioxide (16 g, 0.1 mole) is placed in the pressure vessel and 43 g (0.4 mole) of sulfur tetrafluoride is condensed onto it with help of the vacuum line and by cooling the vessel to liquid nitrogen temperature. The vessel is heated for 8 hours at 130°; the pressure reaches about 60 atm. After the vessel is cooled to room temperature, all gaseous products (excess SF_4, SOF_2) are removed by venting in a good hood. The crystalline residue in the pressure vessel is transferred under anhydrous conditions into the sublimator. Sublimation at $100°/10^{-2}$ torr affords pure TeF_4. Yield: 18 g (80%). *Anal.* Calcd. for TeF_4: Te, 62.7; F, 37.3. Found Te, 62.1; F 37.1.

Properties

Tellurium tetrafluoride is a colorless crystalline solid, mp 130° with no measurable vapor pressure at room temperature. It is readily characterized by its Raman spectrum.[4] The crystal structure of the compound shows TeF_4 to be a polymeric bridged species.[5] In HF solution it forms salts of the type $K^+[TeF_5]^-$.[5]

References

1. J. H. Jenkins, H. A. Bernhardt, and E. J. Barber, *J. Am. Chem. Soc.* **74**, 5749 (1952).
2. R. Campbell and P. L. Robinson, *J. Chem. Soc.*, **1956**, 785
3. D. Lentz, H. Pritzkow, and K. Seppelt, *Inorg. Chem.*, **17**, 1926 (1978).
4. D. J. Reynolds, *Advances in Fluorine Chemistry*, Vol. 7, Butterworths, London, 1973, pp. 1–68.
5. A. J. Edwards and F. J. Hewaidy, *J. Chem. Soc. (A)*, **1968**, 2977.
6. S. H. Martin, R. R. Ryan, and C. B. Asprey, *Inorg. Chem.*, **9**, 2100 (1970), A. J. Edwards and M. A. Monty, *J. Chem. Soc. (A)*, **1969**, 703.

8. SULFUR TETRAFLUORIDE OXIDE

Submitted by KONRAD SEPPELT*
Checked by F. TANZELLA and NEIL BARTLETT†

$$5SOF_2 + 2BrF_5 \longrightarrow 5SOF_4 + Br_2$$

Sulfur tetrafluoride oxide is the only stable sulfur fluoride or sulfur fluoride oxide that is not commercially available, although its value is increasing in synthetic chemistry. It was first made by burning SOF_2 in elemental fluorine,[1]

*Anorganisch-Chemisches Institut der Universität, Im Neuenheimer Feld 270, 6900 Heidelberg 1, West Germany.
†Chemistry Department, University of California, Berkley, CA 94720.

and later by catalytic oxidation of the expensive SF_4.[2] In neither case does the purity exceed 90%. The handling of elemental fluorine and the attack of stainless steel by the $SF_4/O_2/NO_2$ gas mixtures at elevated temperatures are other disadvantages of the alternate procedures.

The method presented here avoids all these problems. If the reaction is run at 300° with a small excess of BrF_5, the product contains only traces of SF_6, SO_2F_2, and SOF_2.[3]

Bromine pentafluoride, which is commercially available from Baker Chemicals, Phillipsburg, NJ, or Union Carbide Corp., may be used as such with the typical impurity of several percent bromine. It should be handled in a metal vacuum line.[4] Standard glass vacuum lines with Kel-F greased valves have also been used, but this technique results in a SiF_4 impurity. Sulfinyl difluoride, if not available, is best prepared from $SOCl_2$ and NaF in CH_3CN at 70° and by redistillation of the product at -43°. This method[5] can be used on the 100-500 g scale.

Procedure

■ **Caution.** *These materials must not be allowed to contact the skin. Bromine pentafluoride is an extremely reactive and very dangerous material. Silicone grease, or any other organic material, will cause explosions if contacted with BrF_5. Sulfur tetrafluoride oxide is very poisonous. Any exposure of the reagents to moist air should be avoided.*

Sulfinyl difluoride (64.5 g, 0.75 mole) and bromine pentafluoride (55.0 g, 0.31 mole) are condensed into a 300-mL stainless steel container. The pressure vessel is heated for 8-14 hr with the temperature held as close to 300° as possible. The pressure reaches about 70-90 atm (1000-1300 psi). The pressure vessel is cooled to -78°, and all volatile substances are pumped out in a dynamic vacuum into a glass trap cooled to -196°. The bulk of bromine remains in the autoclave with small amounts of bromine trifluoride. The volatile materials, essentially pure SOF_4, are purified from bromine by conventional trap-to-trap distillation. The impure sample is held at -100°, and the volatiles are collected at -196°.

Sulfur tetrafluoride oxide is best stored in a stainless steel pressure vessel. Addition of a few grams of dry mercury removes the last traces of bromine. Another way to separate the SOF_4/Br_2 mixture is by low-temperature distillation through a 70-cm column. Yield: 79 g (85%).

Properties

Sulfur tetrafluoride oxide is a colorless gas, mp -99.6°, bp -48.5°. Hydrolysis to SO_2F_2 is rapid. This compound has a trigonal-bipyramidal structure with the

36 *Stoichiometrically Simple Compounds*

oxygen atom in the equatorial position.[6] It readily forms $[SOF_3]^+$ cations and $[SOF_5]^-$ anions.[7]

The gas-phase infrared spectrum has strong absorption bands at 1379 ($\nu_{S=O}$), 927 and 820 (ν_{S-F}) cm^{-1}. Weak absorptions at 1501, 1269, 885, and 848 cm^{-1} would indicate an impurity of SO_2F_2 and those at 1333, 806, and 748 cm^{-1} indicate an impurity of SOF_2.

The ^{19}F nmr spectrum of the liquid at room temperature shows a single line, due to rapid exchange of the different pairs of fluorine atoms. This exchange cannot be frozen out, even at $-150°$.

References

1. H. Jonas, *Z. anor. allg. Chem.*, **265**, 273 (1951).
2. W. C. Smith and V. A. Englehardt, *J. Am. Chem. Soc.*, **82**, 3238 (1960).
3. K. Seppelt, *Z. anorg. allg. Chem.*, **386**, 229 (1971).
4. A description of a metal vacuum line is presented by C. L. Cernick, in *Noble Gas Compounds*, H. H. Hyman, Ed., Chicago University Press, Chicago, 1963, p. 158.
5. C. W. Tullock and D. D. Coffman, *J. Org. Chem.* **25**, 2026 (1960).
6. K. Kimura and S. H. Bauer, *J. Chem. Phys.*, **39**, 3172 (1963).
7. K. O. Christe, C. J. Schack, D. Pilipovich, E. C. Curtis, and W. Sawodny, *Inorg. Chem.*, **12**, 620 (1973). M. Brownstein, P. A. W. Dean, and R. J. Gillespie, *Chem. Commun.*, 1970, 9.

9. SELENONYL DIFLUORIDE

Submitted by KONRAD SEPPELT*
Checked by DARRYL D. DESMARTEAU†

$$H_2SeO_4 + 2HSO_3F \longrightarrow 2H_2SO_4 + SeO_2F_2$$

The method described below is simple and gives SeO_2F_2 in almost quantitative yields.[1,2] The hydrogen fluorotrioxosulfate(VI), HSO_3F, should be distilled at reduced pressure (65°/18 torr) before use. The selenic acid, H_2SeO_4, may best be prepared from SeO_2 and H_2O_2 in water. Care must be taken that it is as nearly anhydrous as possible. Crystallization of the H_2SeO_4 is a clear indicator of sufficient quality. The H_2SeO_4 may be substituted by $BaSeO_4$, which is as easily prepared as $BaSO_4$. However, the procedure using $BaSeO_4$ requires a much larger excess of HSO_3F, and the $BaSeO_4$ has to be dried carefully as well.

---

*Anorganisch-Chemisches Institut der Universität, Im Neuenheimer Feld 270, 6900 Heidelberg, West Germany.

†Department of Chemistry, Kansas State University, Manhattan, KS 66506.

In general, only Pyrex glass equipment is needed, although quartz apparatus will show less corrosion. Traces of SiF_4 are removed at the end of the procedure anyway.

Procedure

■ **Caution.** *Hydrogen fluorotrioxosulfate(VI) is very reactive towards all organic materials. Nevertheless, rubber gloves provide a reasonable protection. Selenonyl difluoride is poisonous and is a strong oxidizing agent. It immediately gives HF and H_2SeO_4 on hydrolysis.*

Preparation of Anhydrous H_2SeO_4

Selenuim dioxide (111 g, 1 mole) of colorless, crystalline quality is dissolved in 300 mL of distilled water. High-purity hydrogen peroxide (300 mL), 30% in water, is added dropwise. An exothermic reaction is observed. After complete addition the solution is heated to reflux for 2 days, while a slow stream of oxygen is bubbled through it. The water is pumped off under heating and high vacuum until the remaining acid reaches 140° for 5 hours. Heating should be provided by an oil bath to avoid local overheating. The vacuum has to be continued until the acid has cooled down. The yield is 150 g of oily, colorless H_2SeO_4, with about 2% water and 2% H_2SeO_3. Eventually the pure acid spontaneously crystallizes, but crystallization may be induced by cooling to $-20°$ and scratching the inner walls of the glass container with a glass rod. Like concentrated H_2SO_4, H_2SeO_4 is very hygroscopic.

To avoid the drying procedure for H_2SeO_4, the almost insoluble $BaSeO_4$ can be precipitated with dilute $BaCl_2$, or preferably, $Ba(NO_3)_2$. Use of $BaCl_2$ will give rise to a chlorine impurity in SeO_2F_2 as indicated by a yellow color. The $BaSeO_4$ is dried in vacuum at 50°.

Preparation of Selenonyl Difluoride

Anhydrous selenic acid (150 g, about 1 mole) or 280 g (1 mole) of $BaSeO_4$ is placed in a 1-L glass or quartz vessel, and 300 mL (or 500 mL, if $BaSeO_4$ is used) of HSO_3F is added. The vessel is then equipped with a water-cooled condenser (preferably a quartz one, if available). The upper opening of the condenser leads into a Teflon tube that is connected to two traps cooled to $-78°$. The mixture is stirred magnetically and heated slowly to reflux. The gas produced is trapped out completely in the first trap, while the second trap keeps away moisture of air. When the gas evolution has ceased (about 2 hr), the traps are disconnected and the excess HSO_3F is distilled immediately at reduced pressure for further use. However, this distilled HSO_3F can be used only for the same procedure, since it contains small amounts of tetravalent selenium compounds.

The contents of the second trap (few grams, mainly water) are discarded. The contents of the first trap are twice distilled from a trap at $-78°$ into a trap at $-196°$. The small residue is HSO_3F. Yield: 120 g (80%).* Silicon tetrafluoride can be pumped off at $-100°$.

Properties

Selenonyl difluoride is a colorless gas, mp $-99.5°$, bp $-8.4°$. The SeO_2F_2 from the $BaSeO_4$ procedure often contains traces of chlorine and is slightly yellow.

In contrast to SO_2F_2, SeO_2F_2 reacts immediately with water. In contrast to SO_2F_2, it slowly attacks glass and is a strong oxidizing agent. The IR spectrum of the gaseous compound shows absorptions at 1059, 971 ($\nu_{Se=O}$), and 702 (ν_{Se-F}) cm^{-1}.

The ^{19}F nmr spectrum has a single line at $\delta = 53$ ppm with small ^{77}Se isotope satellites, $J_{^{77}Se-^{19}F} = 1504$ Hz.

References

1. A. Engelbrecht and B. Stoll, *Z. anorg. allg. Chem.*, **292**, 20 (1957).
2. K. Seppelt, *Chem. Ber.*, **105**, 2431 (1972).

10. HYDROGEN PENTAFLUOROOXOSELENATE(VI)

Submitted by KONRAD SEPPELT[†]
Checked by DARRYL D. DESMARTEAU[‡]

$$3SeO_2F_2 + 4HF \rightleftharpoons H_2SeO_4 + 2HSeOF_5$$

Hydrogen pentafluorooxoselenate (VI) is the main source of the extremely electronegative ligand [SeOF$_5$]$^-$. This ligand and [TeOF$_5$]$^-$ are, for example, the only groups that form stable covalent derivatives of the types Xe(SeOF$_5$)$_2$, Xe(TeOF$_5$)$_4$, U(TeOF$_5$)$_6$, and others. In contrast, the chemistry of the SOF$_5$

---

*From 45 g of BaSeO$_4$, the Checkers obtained only 6 g of SeO$_2$F$_2$.
[†]Anorganisch-Chemisches Institut der Universität Heidelberg, Im Neuenheimer Feld 270, 6900 Heidelberg, West Germany.
[‡]Department of Chemistry, Kansas State University, Manhattan, KS 66506.

group is limited by the instability of hydrogen pentafluorooxosulfate (VI), $HSOF_5$.

During the synthesis of $HSeOF_5$, contact with glass or oxidizable materials must be avoided. The best materials for these substances are Kel-F [poly(trifluorochloroethylene)], FEP [poly(perfluoroethylenepropylene)], and stainless steel. To prepare $HSeOF_5$ completely free from HF, an all-Kel-F distillation column is necessary.

The procedure is based on a typical equilibrium reaction, and the equilibrium is slowly shifted by separating the volatiles HF, SeO_2F_2, and $HSeOF_5$ from the nonvolatile H_2SeO_4.

Procedure

■ **Caution.** *All Chemicals in this reaction are highly reactive and poisonous. A good hood is essential for protection against vapors of HF, SeO_2F_2, and $HSeOF_5$. Though selenium is known to be an essential trace element for mammals, larger amounts attack the liver.*

With the help of a metal vacuum line[1] selenonyl difluoride (120 g, 0.8 mole) and hydrogen fluoride (20 g, 1 mole) are condensed into a 300-mL stainless steel vessel, equipped with a stainless steel valve. The mixture is maintained at room temperature with occasional shaking for several days. Then all volatiles are condensed into another stainless steel vessel, cooled to −196°, in a static vacuum. Dynamic vacuum is even better as it reduces the distillation time considerably. However, when dynamic vacuum is used, the cooled trap must be equipped with two metal valves. This trap-to-trap distillation is finished when the weight of the reaction vessel stops decreasing. The residue in the reaction vessel is mainly H_2SeO_4 with some unstable $HSeO_3F$ and may be used for the preparation of more SeO_2F_2 (see p. 000). The volatile substances are again left at room temperature for several days under occasional shaking and then are distilled into a steel container. This procedure should be repeated until no more nonvolatile H_2SeO_4 is found. This is usually the case after the third distillation.

The separation of $HSeOF_5$ from HF and SeO_2F_2 is difficult. At −78° $HSeOF_5$ is almost nonvolatile, whereas SeO_2F_2 and HF can be removed in a good vacuum. Thus the mixture is held at −78° and is stirred magnetically, while SeO_2F_2 and HF are pumped through a −78° cold trap into a −196° cold trap in dynamic vacuum. When the weight loss of the trap holding the original mixture slows down, the temperature is raised, over several hours, to room temperature. Refined $HSeOF_5$ is collected in the −78° trap, whereas in SeO_2F_2/HF mixture is found in the −196° trap, along with some $HSeOF_5$. This separation is far from perfect and should be repeated at least once. Even then the resulting $HSeOF_5$ is not crystalline at room temperature because of an HF impurity. Yet this material can be used for preparations such as those of $Hg(OSeF_5)_2$,

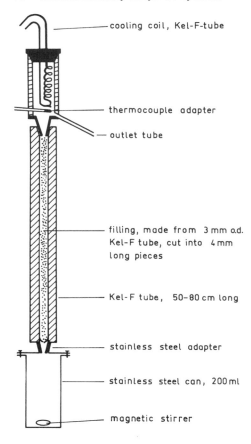

Fig. 1. Kel-F distillation column from 1-m-long, 4-cm-od stock. It is filled with pieces of 3-mm-od Kel-F tubing; the cooling coil and outlet are made from 5-mm tubing.

$KSeOF_5$, $Xe(SeOF_5)_2$, and $Br(SeOF_5)_3$. The yield is never better than 70 g (68%).*

For a complete separation of the H_2SeO_4 free mixture, fractional distillation through a 75-cm long, all-Kel-F column is necessary. This must be manufactured in a machine shop from Kel-F tubes and pieces, as it is not commercially available. A drawing of such a distillation column is given in Fig. 1. The coils of the distillation head must be cooled with a $-40°$ methanol or acetone solution.

During the slow distillation, continuous supervision is necessary. The temperature rises from $-10°$ (SeO_2F_2) slowly to $+47°$. A sufficient amount of reflux must be maintained for a good separation. Above $+30°$ the $HSeOF_5$ should be collected, though spectroscopically pure $HSeOF_5$ appears only at the end of

*The checkers obtained a 19% yield of $HSeOF_5$ starting with only 5 g of SeO_2F_2 and 1 g of HF.

the distillation. Then the material tends to crystallize in the outlet of the distillation head. The yield* of pure compound is about 78%. The lower boiling SeO_2F_2/HF mixture can be used again for preparation of $HSeOF_5$.

Properties

Hydrogen pentafluorooxoselenate(VI)[2,3] is a colorless solid with a melting point of 38° and a boiling point of 47°. It has a strong odor similar to that of rhubarb. Traces of HF lower the melting point below room temperature. It oxidizes chlorides to chlorine, but gives metathetical reactions with such fluorides as KF, RbF, CsF, HgF_2, XeF_2, and BrF_3 to form the corresponding pentafluorooxoselenate(VI) derivatives. Hydrogen pentafluorooxoselenate(VI) decomposes at 290°, forming SeF_4, O_2, and HF. The infrared spectrum of $HSeOF_5$ shows absorptions at 3609 (ν_{OH}), 1170 (δ_{OH}), 759 (ν_{SeF}), 443, 431 (δ_{SeF}) cm^{-1} in the gas-phase spectrum. Its ^{19}F nmr spectrum shows a characteristic AB_4 pattern, including small ^{77}Se isotope satellites. $\delta_A = -75.9$ ppm, $\delta_B = -66.1$ ppm, $J_{AB} = 227$ Hz.

References

1. A description of a metal vacuum line is presented by C. L. Cernick in *Noble Gas Compounds*, H. H. Hyman, Ed., Chicago University Press, Chicago, 1963, p. 158.
2. K. Seppelt, *Z. anorg. allg. Chem.*, **428**, 35 (1977).
3. K. Seppelt and D. Nöthe, *Inorg. Chem.*, **12**, 2727 (1973).

11. RHENIUM PENTACHLORIDE AND VOLATILE METAL CHLORIDES BY DIRECT CHLORINATION USING A VERTICAL-TUBE REACTOR

Submitted by ROGER LINCOLN* and GEOFFREY WILKINSON*
Checked by R. A. WALTON† and T. E. WOOD†

$$2Re + 5Cl_2 \longrightarrow 2ReCl_5$$

The chlorination of rhenium to rhenium pentachloride[1] and of other metals to volatile chlorides is conventionally done by passing chlorine over metal

*Chemistry Department, Imperial College of Science and Technology, London SW7 2AY, England.
†Department of Chemistry, Purdue University, West Lafayette, IN 47907.

placed, sometimes in a boat, in a horizontal tube.[2] High-pressure methods[3] have also been used, but yields are not as high as in this preparation. The use of a vertical tube, where the heavy metal chloride vapor is removed from the reaction zone in part by gravity, has proved more convenient. Chlorination of rhenium is given as an example, but we have made VCl_4, $NbCl_5$, $TaCl_5$ and, by use of bromine carried in a nitrogen stream, WBr_5. Runs using up to 70 g of rhenium metal with yields of over 99% have been achieved.

Procedure

■ **Caution.** *An efficient fume hood should be used for reactions involving chlorine or bromine.*

The chlorination apparatus, Fig. 1, consists of a 24-mm od quartz tube with a ground joint socket joined by way of a quartz-pyrex seal to a collection flask. For metal quantities of about 20 g a 250-mL flask should be used and for 5-g quantities a 100-mL flask is convenient.

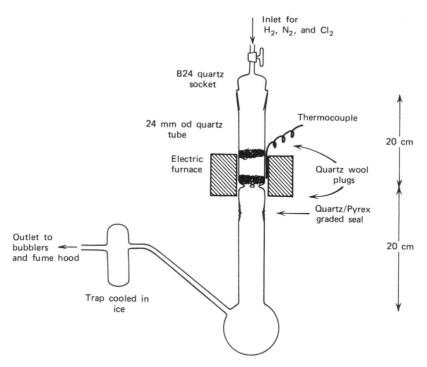

Fig. 1. *Vertical reactor for preparation of anhydrous metal halides.*

Since commercial rhenium (and other metals) usually contains surface oxide, prereduction by hydrogen at 250-300° for about 1 hr is advisable to avoid formation of chloride oxides.

The metal powder is placed on a quartz wool plug (supported on indentations in the tube) and topped by a second plug. Hydrogen is passed through the system and the metal is heated by an electric furnace to 250-300°. When no further water condenses in the flask or exit tubes, hydrogen is replaced by oxygen-free nitrogen and the water is removed by flaming with a Bunsen burner.

The furnace temperature is raised to 650-700° and chlorine is admitted at a rate sufficient to keep the metal burning. The temperature rises to about 800° and the chlorine flow is maintained until all the metal is consumed. The vapor of $ReCl_5$ descends and condenses in the flask. It is necessary to flame the tube below the furnace occasionally to prevent blockages. The small quantity of pentachloride escaping to the trap is also flamed into the flask. After the chlorine is replaced by a nitrogen stream and the system is allowed to cool, the flask is sealed off, taking care to ensure that there is no halide in the seal area as otherwise the seal will be impaired. The pentachloride can be transferred in a dry box to Schlenk tubes or ampules for storage if required. The yield on 20-g lots is over 99%: on smaller quantities small losses in handling may occur. *Anal.* Calcd. for $ReCl_5$: Cl, 48.76%; Found: Cl, 48.7%.

Properties

Rhenium pentachloride is obtained as black crystals. The solid melts at 261° with decomposition. It fumes in air, evolving hydrogen chloride and chlorine.

References

1. L. C. Hurd and E. Brimm, *Inorg. Synth.*, **1**, 182 (1939).
2. A. J. Leffler and R. Penque, *Inorg. Synth.*, **12**, 187 (1970).
3. E. R. Epperson, S. M. Horner, K. Knox, and S. Y. Tyree, Jr., *Inorg. Synth.*, **7**, 163 (1963).

12. TRIRHENIUM NONACHLORIDE

Submitted by ROGER LINCOLN* and GEOFFREY WILKINSON*
Checked by R. A. WALTON† and T. E. WOOD†

$$3\text{ReCl}_5 = \text{Re}_3\text{Cl}_9 + 3\text{Cl}_2$$

Previous preparations of Re_3Cl_9 have involved driving the rhenium halides horizontally along bulb trains.[1,2] The present procedure is a considerable improvement on these, because it involves decomposition of refluxing rhenium pentachloride in the reaction bulb. Yields are very high and, using the method for the preparation of rhenium pentachloride given in Sec. 11, yields of over 90% from rhenium metal are possible.

Procedure

The apparatus (Fig. 1) is flamed dry in a stream of dry nitrogen and transferred with the stopcock closed to a glove bag, previously flushed out with dry nitrogen. Rhenium pentachloride, 5-10 g, is placed in the bulb and the tube is removed from the glove bag and clamped in a vertical position with a moderately fast stream of nitrogen flowing.

The bulb is heated with a medium Bunsen flame until the rhenium pentachloride melts and is refluxing up the tube.‡ From time to time heating is stopped and Re_3Cl_9 formed in the tube is scraped back into the bulb. After about 4 hr, or when no more brown fumes are seen and only solid is observed in the bulb, the apparatus is allowed to cool. The bulb is sealed off and transferred to a glove bag and the dark-red crystalline solid is transferred to ampules. The product can be purified by grinding with a mortar and pestle under nitrogen and pumping off any volatile rhenium halide oxides under vacuum.

The method affords yields of 95% and above. *Anal.* Calcd. for Re_3Cl_9: Cl, 36.4%. Found: Cl, 36.3%.

Properties

The product Re_3Cl_9 is hygroscopic and on exposure to air forms a hydrate. This hydrate is more reactive than the anhydrous material, is soluble in polar solvents,

---

*Chemistry Department, Imperial College of Science and Technology, London SW7 2AY, England.
†Department of Chemistry, Purdue University, West Lafayette, IN 47907.
‡The checkers suggest that during reflux of ReCl_5 the scraper should be raised as high as possible to avoid decomposition of ReCl_5 upon it.

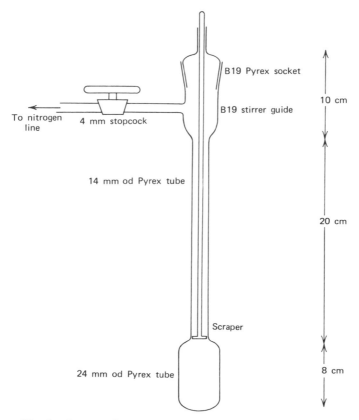

Fig. 1. Reactor for preparation of trirhenium nonachloride.

and is a useful starting material for the synthesis of rhenium(III) alkyls,[3] phosphine complexes, and other derivatives (see Section 13).

References

1. L. C. Hurd and E. Brimm, *Inorg. Synth.,* **1**, 182 (1939).
2. H. Gehrke, Jr. and D. Bue, *Inorg. Synth.,* **12**, 193 (1970).
3. A. F. Masters, K. Mertis, J. F. Gibson, and G. Wilkinson, *Nouv. J. Chim.,* **1**, 389 (1977).

13. TRIRHENIUM NONAHALIDES

Submitted by H. D. GLICKSMAN* and R. A. WALTON*
Checked by R. LINCOLN† and G. WILKINSON†

$$trans\text{-}ReOCl_3[P(C_6H_5)_3]_2 + (CH_3CO)_2O \longrightarrow Re_2(O_2CCH_3)_4Cl_2$$
$$3Re_2(O_2CCH_3)_4Cl_2 + 12HX(g) \longrightarrow Re_3X_9 + 12HO_2CCH_3$$

While several methods have been used in the past for the preparation of Re_3Cl_9,[1-3] Re_3Br_9,[4] and Re_3I_9,[5,6] no single preparative method has previously been developed that is applicable to all three halides. We now describe such a procedure[7,8] in which Re_3X_9, X = Cl, Br, or I, can be made from the same air-stable starting material, $Re_2(O_2CCH_3)_4Cl_2$. This experimental procedure is very simple, requiring no special equipment other than a tube furnace, and the product yields are very high.

The method we have used to prepare $Re_2(O_2CCH_3)_4Cl_2$, namely that involving the reaction of $trans\text{-}ReOCl_3[P(C_6H_5)_3]_2$ with acetic anhydride,[9] is preferred over the alternative procedure using octachlorodirhenate(III) salts and acetic acid.[10,11] Although the latter synthesis proceeds in high yield (about 95%),[11] it is less advantageous overall because it requires the synthesis of salts of the $[Re_2Cl_8]^{2-}$ anion, for which the best synthetic method[12] utilizes Re_3Cl_9 as the starting material.

Procedure

■ **Caution.** *An efficient fume hood should be used for reactions involving the hydrogen halides.*

A. TETRAACETATODICHLORODIRHENIUM

Following the procedure of Rouschias and Wilkinson,[9] a solution of 1.5 g (0.0018 mole) of $trans\text{-}ReOCl_3[P(C_6H_5)_3]_2$ (prepared from rhenium metal[13] or from sodium perrhenate[14]) in 150 mL of acetic anhydride is heated at reflux under nitrogen for 48 hr. The resulting orange precipitate of $Re_2(O_2CCH_3)_4Cl_2$ is filtered, washed with ethanol, benzene, and diethyl ether, and dried in vacuum. Yield: about 68% (0.42 g).

---

*Department of Chemistry, Purdue University, West Lafayette, IN 47907.
†Chemistry Department, Imperial College, South Kensington, London SW7 2AY, England.

B. TRIRHENIUM NONACHLORIDE

Finely ground $Re_2(O_2CCH_3)_4Cl_2$ (0.4 g, 0.00059 mole) is placed in a porcelain boat, which is then inserted into a dry open-ended glass tube. The tube is placed in a horizontal-tube furnace and after it is flushed with $N_2(g)$ for 15 minutes, a slow stream of hydrogen chloride gas is passed over the $Re_2(O_2CCH_3)_4Cl_2$ for 1 hr. (For this work all hydrogen halides were obtained from Matheson Gas Products and were of the following purities: HCl, 99.0%; HBr, 99.8%; HI, 98.0%.) The temperature of the furnace is then raised to 340° and the passage of HCl(g) is continued for 3 hrs. The furnace is then allowed to cool, the system is flushed out with $N_2(g)$, and the dark-red Re_3Cl_9 is collected. Yield: 95% (0.32 g). *Anal.* Calcd. for Re_3Cl_9: Cl, 36.35. Found: Cl, 36.2.

C. TRIRHENIUM NONABROMIDE AND TRIRHENIUM NONAIODIDE

The procedure used to prepare Re_3Br_9 and Re_3I_9 is the same as that used to make Re_3Cl_9. $Re_2(O_2CCH_3)_4Cl_2$ (0.21 g, 0.00031 mole) is allowed to react with HBr(g) at 340° for 3 hr to afford the dull-brown bromide Re_3Br_9. Yield 95% (0.22 g). *Anal.* Calcd. for Re_3Br_9: Br, 56.3. Found: Br, 56.5. In an analogous fashion, the reaction of $Re_2(O_2CCH_3)_4Cl_2$ (0.5 g, 0.00074 mole) with HI(g) for 5 hr at 320° afforded black Re_3I_9. Yield: 80% (0.67 g). *Anal.* Calcd. for Re_3I_9: I, 67.15. Found: I, 66.85.

Properties

Both Re_3Cl_9 and Re_3Br_9 are soluble in acetone, methanol, ethanol, and their corresponding hydrohalic acids. However, Re_3I_9 is insoluble in most solvents and slowly evolves iodine in vacuum. Crystallographic studies have shown that all three rhenium(III) halides possess a similar trimeric structure.[4,5,15]

Trirhenium nonachloride and trirhenium nonabromide combine with halide ion and many other Lewis bases to form complexes, such as $[(C_6H_5)_4As]_2Re_3X_{11}$ and $Re_3X_9[P(C_6H_5)_3]_3$.[4] They are readily reduced by the more basic heterocyclic tertiary amines such as pyridine, to afford the polymeric rhenium(II) species $[Re_3X_6L_3]_n$.[16,17] Monodentate tertiary phosphines such as tripropylphosphine also reduce Re_3Cl_9 and Re_3Br_9, as well as Re_3I_9, but in this case disruption of the trimeric structure occurs to afford metal-metal bonded dimers of the type $Re_2X_4[P(n-C_3H_7)_3]_4$.[8,17,18]

References

1. L. C. Hurd and E. Brimm, *Inorg. Synth.*, **1**, 182 (1939).
2. H. Gehrke, Jr. and D. Bue, *Inorg. Synth.*, **12**, 193 (1970).
3. R. Lincoln and G. Wilkinson, *Inorg. Synth.*, **20**, 44 (1980).
4. F. A. Cotton, S. J. Lippard, and J. T. Mague, *Inorg. Chem.*, **4**, 508 (1965).
5. M. J. Bennett, F. A. Cotton, and B. M. Foxman, *Inorg. Chem.*, **7**, 1563 (1968).
6. L. Malatesta, *Inorg. Synth.*, **7**, 185 (1963).
7. H. D. Glicksman, A. D. Hamer, T. J. Smith, and R. A. Walton, *Inorg. Chem.*, **15**, 2205 (1976).
8. H. D. Glicksman and R. A. Walton, *Inorg. Chem.*, **17**, 200 (1978).
9. G. Rouschias and G. Wilkinson, *J. Chem. Soc.(A)*, **1966**, 465.
10. F. A. Cotton, C. Oldham, and W. R. Robinson, *Inorg. Chem.*, **5**, 1798 (1966).
11. A. B. Brignole and F. A. Cotton, *Inorg. Synth.*, **13**, 85 (1972).
12. A. B. Brignole and F. A. Cotton, *Inorg. Synth.*, **13**, 83 (1972).
13. N. P. Johnson, C. J. L. Lock, and G. Wilkinson, *J. Chem. Soc.*, **1964**, 1054.
14. J. Chatt and G. A. Rowe, *J. Chem. Soc.*, **1962**, 4019.
15. F. A. Cotton and J. T. Mague, *Inorg. Chem.*, **3**, 1402 (1964).
16. D. G. Tisley and R. A. Walton, *Inorg. Chem.*, **12**, 373 (1973).
17. H. D. Glicksman and R. A. Walton, *Inorg. Chim. Acta*, **19**, 91 (1976).
18. J. R. Ebner and R. A. Walton, *Inorg. Chem.*, **14**, 1987 (1975).

14. β-PLATINUM(II) CHLORIDE

Submitted by GEORGE T. KERR* and ALBERT E. SCHWEIZER*
Checked by THEODORE DEL DONNO†

$$(H_3O)_2(PtCl_6) \cdot xH_2O \xrightarrow[350°]{\text{air}} PtCl_2 + 2HCl + Cl_2 + (2+x)H_2O$$

The preparation of platinum(II) chloride from dihydrogen hexachloroplatinate(IV) has been described in two previous volumes of *Inorganic Syntheses*. One method reduces the acid to dihydrogen tetrachloroplatinate(II) with hydrazine dihydrochloride. The dihydrogen tetrachloroplatinate(II) is then thermally decomposed at 150° to $PtCl_2$.[1] The other method consists of thermally decomposing the acid at 115° in a stream of chlorine to yield $PtCl_4$ followed by heating to 450° in flowing chlorine for 2 hours to yield the desired product.[2] Berzelius prepared $PtCl_2$ by evaporation of a solution of dihydrogen hexachloro-

---

*Mobil Research and Development Corporation, Central Research Division, PO Box 1025, Princeton, NJ 08540.
†Department of Chemistry, The Ohio State University, Columbus, OH 43210.

platinate(IV) and heating of the residue to about 300–350° with constant stirring.³ This is essentially the procedure described here, which utilizes only solid dihydrogen hexachloroplatinate(IV) and air. Modern thermogravimetric analysis instrumentation makes it possible to study accurately and easily thermal reactions that earlier workers found difficult to control and monitor.⁴

Procedure

Commercial dihydrogen hexachloroplatinate(IV) of high purity is now readily available (for a platinum assay of 40.0%, x is 2.32 in the above equation). The acid, 10 g, is spread into a thin layer less than 5-mm thick in a combustion boat and a steady air purge is maintained (200 mL/min) through a combustion tube of approximately 1 L in volume. The temperature of the furnace is raised from ~25 to 350° in 50° steps over a period of 3 hours. The temperature is held at 350° for 0.5 hours. The sample can then be removed from the furnace and stored in a desiccator. The yield is quantitative based on the Pt assay of the dihydrogen hexachloroplatinate(IV) used (>98%*).

A small portion of sample can be tested for purity by heating in a stream of hydrogen at 150° to yield platinum metal. *Anal.* Calcd. for $PtCl_2$: Pt, 73.3; Cl, 26.7. Found: Pt, 73.5, 73.2*; Cl, 26.8.*

Properties

Prepared in this manner, β-platinum(II) chloride is a chocolate-brown, nonhygroscopic solid with sufficient porosity to chemisorb several small molecules.[5-7] Although practically insoluble in water, it reacts with hydrochloric acid to form dihydrogen tetrachloroplatinate(II) and with ammonium hydroxide to form tetraammineplatinum(II) chloride. It is sensitive to high concentrations of ammonia, undergoing reduction to metallic platinum.[7]

References

1. W. E. Cooley and D. H. Busch, *Inorg. Synth.*, **5**, 208 (1957).
2. A. J. Cohen, *Inorg. Synth.*, **6**, 209 (1960).
3. J. J. Berzelius, *Schweigger's J.*, **7**, 55 (1813).
4. A. E. Schweizer and G. T. Kerr, *Inorg. Chem.*, **17**, 2326 (1978).
5. W. Peters, *Z. Anorg. Chem.*, **77**, 137 (1912).
6. M. F. Pilbrow, *J. Chem. Soc., Dalton Trans.*, **1975**, 2432.
7. A. E. Schweizer and G. T. Kerr, *Inorg. Chem.*, **17**, 2152 (1978).

*Values reported by the checker.

15. DIVALENT METAL CHROMIUM(III) OXIDES (CHROMITES)

Submitted by B. DURAND* and J. M. PÂRIS*
Checked by E. C. BEHRMAN,[†] A. HUANG[†] and A. WOLD[†]

$$2LiCrO_2 + K_2MCl_4 \longrightarrow MCr_2O_4 + 2LiCl + 2KCl$$

$$M = Co, Ni, Zn, Mg, Mn$$

Owing to the slow rates of diffusion of the cations, the direct solid-state reaction of the oxides Cr_2O_3 and MO at an elevated temperature is not a good preparation of divalent metal chromium(III) oxides. They can be prepared by more elaborate methods, such as controlled reduction of dichromates MCr_2O_7,[1] reaction of dichromium tungsten oxide Cr_2WO_6 with a molten divalent metal fluoride[2] at 1400°, pyrolysis of complexes,[3] and pulverization of slurries containing Cr_2O_3 and a divalent metal salt.[4]

Metal chromium(III) oxides are often used as catalysts, especially dehydrogenation catalysts; hence it is important to prepare them as powders.[5] The synthesis described here involves the double-decomposition reaction between lithium chromium(III) oxide and molten divalent metal chlorides, and produces finely divided powders.

A. LITHIUM CHROMIUM(III) OXIDE, LiCrO₂

$$2CrO_3 + Li_2CO_3 \longrightarrow Li_2Cr_2O_7 + CO_2$$

$$Li_2Cr_2O_7 + 7H_2C_2O_4 \longrightarrow 2Li[Cr(H_2O)_2(C_2O_4)_2] + 3H_2O + 6CO_2$$

$$Li[Cr(H_2O)_2(C_2O_4)_2] \xrightarrow{\Delta} LiCrO_2 + 2H_2C_2O_4$$

Because of the low reactivity of chromium(III) oxide, lithium chromium(III) oxide is prepared by pyrolysis of the oxalate $Li[Cr(H_2O)_2(C_2O_4)_2]$, which is prepared by the reduction of chromium(VI) oxide.[3]

Reagent grade chromium(VI) oxide (19.98 g, 0.2 mole) is dissolved in 500 mL of water in a 1-L beaker, and reagent grade lithium carbonate (7.39 g, 0.1 mole) is added to form an orange solution of lithium dichromate. To reduce the dichromate, reagent grade oxalic acid (88.25 g, 0.7 mole) is added slowly. When the reduction is finished, the dark red solution is placed on a heated magnetic

---

*University of Lyon I, 43 Boulevard du 11 Novembre 1918, 69621 Villeurbanne, France.
[†]Department of Chemistry, Brown University, Providence, RI 02912.

Divalent Metal Chromium (II) Oxides (Chromites) 51

TABLE I X-ray Powder Diffraction Data for Lithium Chromium Oxide, LiCrO$_2$[a]

d(Å)	I	d(Å)	I	d(Å)	I
4.85	vs	2.06	vs	1.455	m
2.47	s	1.895	w	1.39	w
2.42	vw	1.595	w		
2.37	w	1.47	m		

[a]vs, very strong; s, strong; m, medium; w, weak; vw, very weak.

stirrer and maintained at 60° to decrease the volume to about 150 mL. Finally, the complex is obtained by a complete evaporation of the solution at room temperature.

Lithium diaquabis(oxalato)chromate(III) is placed in a platinum crucible and pyrolysis is carried out in a hydrogen stream. The temperature is raised to 1000° at a rate of 150°/hr and is held for 5 hours. The crucible is then cooled to room temperature. *Anal.* Calcd. for LiCrO$_2$: Cr, 57.2. Found: 57.3. X-ray data are shown in Table I.

B. POTASSIUM METAL(II) CHLORIDES

$$2KCl + MCl_2 \cdot xH_2O \xrightarrow{\Delta} K_2MCl_4 + xH_2O$$

Reagent grade potassium chloride is intimately mixed with reagent grade metal-(II) chloride and the mixture is heated in a 250-mL pyrex beaker, in air at 200–250° for 3 hours. Proportions of mixed salts and compositions of the different double chlorides appear in Table II.

TABLE II Data for Double Chlorides

Double Chloride Composition[a]	Melting Point °	Proportions in the Mixture
K$_2$CoCl$_4$[b]	436	14.91 g KCl + 23.79 g CoCl$_2 \cdot$6H$_2$O
K$_2$NiCl$_4$	502	14.91 g KCl + 23.77 g NiCl$_2 \cdot$6H$_2$O
K$_2$ZnCl$_4$	446	14.91 g KCl + 13.63 g ZnCl$_2$
K$_2$MgCl$_4$[b]	433	14.91 g KCl + 20.33 g MgCl$_2 \cdot$6H$_2$O
MnCl$_2$–2.03 KCl	420	14.91 g KCl + 19.79 g MnCl$_2 \cdot$4H$_2$O

[a]For manganese, the proportions lead to a eutectic; for the other metals they give a double salt.
[b]These salts are especially hygroscopic. They must be kept dry to avoid the presence of unreacted LiCrO$_2$ in the product.

C. DIVALENT METAL CHROMIUM (III) OXIDES

$$K_2MCl_4 + LiCrO_2 \xrightarrow{\Delta} MCr_2O_4 + 2KCl + 2LiCl$$

Potassium metal (II) chloride K_2MCl_4 and lithium chromium (III) oxide $LiCrO_2$ are quickly mixed by milling in an agate mortar to avoid the hydration of the salt as much as possible. To get a rapid and complete reaction, an excess of salt is used: 0.1 mole of $LiCrO_2$ (9.09 g) for 0.1 mole of K_2MCl_4 (27.89 g K_2CoCl_4, 27.87 g K_2NiCl_4, 28.54 g K_2ZnCl_4, 24.43 g K_2MgCl_4, 27.72 g $MnCl_2$, 2.03 KCl). The reaction mixture is placed in a 70-mL platinum crucible, heated in a nonoxidizing atmosphere to 600° in a regulated furnace (heating rate: 150°/hr), and held there for 24 hours. The nonoxidizing atmosphere is produced by a stream of pure nitrogen that has been dried by passage over P_4O_{10}.

After the mixture is cooled, the solid is washed with water to remove the excess double chloride and the alkali salts formed. After filtration the divalent metal chromium(III) oxide is dried by heating in air at 200°.

TABLE III Analyses and X-ray Data for Divalent Metal Chromium(III) Oxides

	$M^{2+}Cr_2O_4$				
	$CoCr_2O_4$	$NiCr_2O_4$	$ZnCr_2O_4$	$MnCr_2O_4$	$MgCr_2O_4$
%M^{2+}					
Calc.	26.0	25.9	28.0	12.6	24.6
Found	26.1	25.7	27.9	12.7	24.8
%Cr^{3+}					
Calc.	45.8	45.7	44.5	54.2	46.8
Found	45.6	45.9	44.6	54.1	46.6
Lattice constant a (Å)	8.330 ± 0.004	8.310 ± 0.006	8.324 ± 0.004	8.432 ± 0.005	8.334 ± 0.005

			Diffraction Lines		
I	d (Å)	d (Å)	d (Å)	d (Å)	d (Å)
w	4.809	4.799	4.806	4.868	4.811
s	2.945	2.939	2.943	2.981	2.946
vs	2.512	2.506	2.510	2.542	2.513
vw	2.405	2.400	2.403	2.434	2.406
s	2.082	2.078	2.081	2.108	2.084
vw	1.700	1.697	1.699	1.721	1.701
s	1.603	1.600	1.602	1.623	1.604
s	1.472	1.470	1.471	1.491	1.473

Properties

The purity of metal(II) chromium(III) oxides was checked by chemical analysis. Chromium and divalent metal contents agree with the calculated values within less than 1%. Divalent metal chromium(III) oxides crystallize in the cubic spinel system. Chemical analysis data, cubic lattice constants and powder diffraction lines are given in Table III.

Divalent metal chromium(III) oxides produced by a double-decomposition reaction between $LiCrO_2$ and a molten metal(II) salt are fine powders. Electron microscopic examination shows that these powders are constituted of grains, the repartition size of which is heterogeneous (500–3000 Å), but each grain is made of smaller crystallites. The average diameter of the crystallites, determined by a radiocrystallographical method (measurement of the widening of X-rays 220 and 335), is about 250 Å.

References

1. E. Whipple and A. Wold, *J. Inorg. Nucl. Chem.*, **24**, 23 (1962).
2. W. Kunnmann, *Inorg. Synth.*, **14**, 134 (1973).
3. E. Vallet, *Thesis,* Lyon number 445 (1967) p. 78.
4. H. E. Manning, U. S. Pat. 4,056,490 (1977).
5. P. Scherrer, *Gottingen Nachr.*, **2**, 98 (1918).

16. COPPER(I) ACETATE

Submitted by S. J. KIRCHNER* and QUINTUS FERNANDO*
Checked by DANA DARBY† and J. A. DILTS†

$$Cu_2(CH_3COO)_4 \cdot 2H_2O + 2Cu^o \longrightarrow 2Cu_2(CH_3COO)_2 + 2H_2O$$

The thermal decomposition of copper(II) acetate has been proposed by several workers as a useful method for the synthesis of pure copper(I) acetate.[1-3] This method, however, suffers from several drawbacks. The major products of the decomposition are copper(I) oxide and acetic acid; in contrast, the copper(I) acetate is obtained in low yields (<5%) together with some acetone and carbon

---

*Department of Chemistry, The University of Arizona, Tucson, AZ 85721.
†Department of Chemistry, The University of North Carolina at Greensboro, Greensboro, NC 27412.

dioxide. Under oxygen-free conditions and in solvents that are capable of solvating copper(I) more effectively than water, copper(II) can be reduced to copper(I) with copper metal. Copper(I) acetate has been synthesized in this manner and has been isolated from both pyridine and acetonitrile solutions.[4,5] Copper(II) acetate has been reduced in a homogeneous system by use of hydrazine hydrate and hydroxylamine acetate as reducing agents.[3]

All the synthetic methods described above give products that require extensive purification. Moreover, the copper(I) acetate often undergoes rapid oxidation on exposure to the atmosphere because any solvent in contact with the copper(I) acetate absorbs water vapor, which promotes the conversion of copper(I) to copper(II) in air.

A simple heterogenous gas-phase reaction for the synthesis of extremely pure copper(I) acetate in high yields (50-65%) is described below. No solvents are required in the synthetic procedure and the product is relatively stable in air.

Procedure

Copper(II) acetate hydrate dimer, $[Cu_2(CH_3COO)_4 \cdot 2H_2O]$ (0.25 g, 0.0063 mole) is mixed with an excess of electrolytic copper dust (0.50 g, 0.0079 mole) (Fisher Cat. No. C-431) and is finely ground. The mixture is placed in a semimicro sublimation apparatus equipped with a cold finger through which tap water is passed. Sublimation can be carried out in this apparatus at reduced pressures and elevated temperatures. The apparatus is evacuated and its temperature is *slowly* increased to 100° with a paraffin wax bath. The copper(II) acetate is dehydrated in approximately 15 min. The temperature is then raised rapidly and maintained at 180-190° for 2-3 hours. Copper(I) acetate dimer, $[Cu_2(CH_3COO)_2]$, condenses on the cold finger of the sublimation apparatus and forms a white polymeric solid. The unreacted anhydrous copper(II) acetate, which is much less volatile than the copper(I) acetate, forms a blue ring just above the reaction mixture in the sublimation apparatus.

Properties

Copper(I) acetate is a white solid that slowly decomposes in air and is very unstable toward water. The green decomposition product has the molecular formula $Cu_2(CH_3COO)_2(OH)_2$. The far infrared spectrum of copper(I) acetate has bands at 230(s), 255(s), 375(m), and 419(m) cm^{-1} that are distinct from those of starting material and decomposition product. The Cu $2P_{1/2,3/2}$ band in the photoelectron spectrum of pure copper(I) acetate exhibits no secondary structure, in contrast to that of copper(II) acetate and the green decomposition product.

References

1. A. Angel and A. V. Harcourt, *J. Chem. Soc.*, **85**, 1385 (1902).
2. A. Perchard, *C. R.*, **131**, 504 (1903).
3. D. A. Edwards and R. Richards, *J. Chem. Soc., Dalton Trans.*, **1973**, 2463.
4. T. Ogura and Q. Fernando, *Inorg. Chem.*, **12**, 2611 (1973).
5. R. D. Mounts, T. Ogura, and Q. Fernando, *J. Am. Chem. Soc.*, **95**, 949 (1973).

Chapter Three

COORDINATION COMPOUNDS

17. COMPOUNDS CONTAINING THE cis-BIS(ETHYLENEDIAMINE)RHODIUM(III) SPECIES

Submitted by R. D. GILLARD*, JULIO PEDROSA DE JESUS* and P. S. SHERIDAN[†]
Checked by J. D. PETERSEN[‡] and STEPHEN F. CLARK[‡]

There have been several reports concerning isomers of bis(ethylenediamine) compounds of rhodium(III). A preparation of cis-$[Rh(en)_2Cl_2]^+$ salts appeared in 1939,[1] but the method has not yet been successfully repeated. A reproducible method of preparing cis-$[Rh(en)_2Cl_2]^+$ was given by Johnson and Basolo[2] and was amplified[3] in this series. The yields were low (about 11%). Bis(ethylenediamine)(oxalato)rhodium(III) was obtained[4] from the more accessible trans-$[Rh(en)_2Cl_2]^+$ with oxalate ion, but again in 6% yields only. By taking advantage of the ease of reduction of Rh(III) to stereolabile species using BH_4^-, better yields of salts of the cis-bis(ethylenediamine)rhodium(III) series may be obtained easily. We give details here of the convenient route[5] $RhCl_3 \rightarrow [Rh(en)_2Cl_2]^+$ (pure trans or mixture of cis/trans) $\rightarrow [Rh(en)_2(C_2O_4)]^+$ and describe preparations from the oxalato complex of cis-$[Rh(en)_2X_2]^+$, X = Cl^-, Br^-, NO_2^-. To make salts of such cations as $[Rh(en)_2(ab)]^{(3-m)+}$ where ab = malonate ($m = 2$) or glycinate ($m = 1$), the disodium oxalate in the procedures given is replaced by the corresponding amount of the appropriate sodium salt.

*University College, P.O. Box 78, Cardiff, Wales, U.K.
[†]Chemistry Department, State University of New York, Binghamton, NY.
[‡]Department of Chemistry, Kansas State University, Manhattan, KS 66506.

A. BIS(ETHYLENEDIAMINE)(OXALATO)RHODIUM(III) PERCHLORATE

(a) $RhCl_3 + 2\{C_2H_4(NH_2)_2 \cdot 2HCl\} + 4NaOH \longrightarrow$
$[Rh(en)_2Cl_2]Cl + 4NaCl + 4H_2O$

(b) $[Rh(en)_2Cl_2]Cl + Na_2C_2O_4 \longrightarrow$
$[Rh(en)_2(C_2O_4)]Cl + 2NaCl$

(c) $[Rh(en)_2(C_2O_4)]Cl + LiClO_4 \longrightarrow$
$[Rh(en)_2(C_2O_4)]ClO_4 + LiCl$

■ **Caution.** *Perchlorates are potentially explosive and must be treated with care.*

Method 1. Preparation from $RhCl_3 \cdot 3H_2O$

Rhodium trichloride trihydrate (1.0 g, 0.0038 mole supplied by Johnson Matthey, London, United Kingdom), dissolved in 30 mL of water, is placed in a 250-mL round-bottomed flask, which is fitted with a reflux condenser. After the solution is heated to boiling, 0.6 mL of ethylenediamine (0.01 mole) dissolved in 50 mL of water is added slowly through the condenser, in 5-mL increments, at 2-minute intervals. The resulting yellow solution is allowed to heat at reflux for 5 minutes more. Disodium oxalate (1.0 g, 0.0075 mole) dissolved in 25 mL of hot water is added through the condenser. The solution is maintained at reflux for 2 minutes, after which 1 speck (1 mg) of solid $NaBH_4$ (or 1 mL of 1 mg/mL $NaBH_4$ freshly made solution) is added (dropwise) to the boiling solution. Heating at reflux is continued for 15 minutes. The solution is allowed to cool slightly and while it is still hot (about 60°), activated charcoal is added. The mixture is filtered (paper) and concentrated to a volume of approximately 20 mL. (The checkers filtered again at this point.) Lithium perchlorate (3 g, 0.028 mole) is added to this solution, immediately precipitating pale-yellow, fine crystals of the product. Yield: 560 mg (37%). (The checkers report a yield of 60%.) *Anal.* Calcd. for $[Rh(en)_2(C_2O_4)]ClO_4 \cdot \frac{1}{2}H_2O$: C, 17.2; H, 4.1; N, 13.4. Found: C, 17.6; H, 4.5; N, 12.9.

Method 2. Preparation from trans-$[Rh(en)_2Cl_2]NO_3$

(a) $RhCl_3 + 2C_2H_4(NH_2)_2 \longrightarrow Rh(en)_2Cl_2^+ + Cl^-$

(b) $Rh(en)_2Cl_2^+ + HNO_3 \longrightarrow$ *trans*-$[Rh(en)_2Cl_2]NO_3 + H^+$

(c) *trans*-$[Rh(en)_2Cl_2]^+ + Na_2C_2O_4 \longrightarrow [Rh(en)_2(C_2O_4)]^+ + 2NaCl$

To make *trans*-$[Rh(en)_2Cl_2]NO_3$, the method described by Anderson and Basolo may be employed.[3] The following method involves slight modifications.

Rhodium trichloride trihydrate (1.0 g, 0.0038 mole) dissolved in 30 mL of water is placed in a 250-mL, round-bottomed flask, fitted with a condenser. The solution is heated to boiling, and 0.6 mL of ethylenediamine (0.01 mole) dissolved in 50 mL of H_2O is added through the top of the condenser, in 5-mL increments, at 2-minute intervals. The yellow solution is allowed to heat at reflux for an additional 15 minutes. Then it is allowed to cool slightly, but while it is still hot (about 60°) 0.02 g of activated charcoal is added. The mixture is filtered (paper) and the filtrate is concentrated under vacuum to a volume of 30 mL. Concentrated nitric acid (10 mL) is added to the solution, which is kept, with occasional shaking, for 3 hours. A crystalline golden yellow precipitate forms. This is collected, washed with ice-cold water, ethanol, acetone, and diethyl ether and air dried. Yield: 560 mg (40%). *Anal.* Calcd. C, 13.5; H, 4.5; N, 19.7. Found: C, 13.5; H, 4.2; N, 19.7%.

trans-Dichlorobis(ethylenediamine)rhodium(III) nitrate (900 mg, 0.0025 mole) is dissolved in 100 mL of H_2O, to which is added 0.50 g of $Na_2C_2O_4$ (0.0038 mole) dissolved in 25 mL of H_2O. The pH of this solution is adjusted to approximately 9 by adding 1 mL of 0.3 M NaOH (0.0003 mole).

The mixture is heated at reflux for 3 minutes in a 250-mL, round-bottomed flask. To the boiling solution, 1 mL of freshly made aqueous solution of $NaBH_4$ (1 mg/mL) is added dropwise. The golden-yellow solution becomes pale yellow. Heating at reflux is continued for another 15 minutes to ensure that the reaction is complete.

The solution is allowed to cool slightly, but while it is still hot (about 60°), 0.02 g of activated charcoal is added to help remove any rhodium metal that may have formed an adding tetrahydroborate. The mixture is filtered twice (paper) and the filtrate is concentrated under vacuum (at 50°) to a volume of 50 mL. Any solid formed is dissolved by heating the mixture. At this point, 4 g of $LiClO_4$ is added. A bright-yellow crystalline precipitate immediately starts to form. The solution is allowed to cool to room temperature and is then left in a refrigerator overnight. The crystalline solid is collected by filtration, washed with ice-cold water, ethanol, and diethyl ether, and air dried. Yield: 660 mg (60%). The checkers obtained only a 23% yield. *Anal.* Calcd. for [Rh(en)$_2C_2O_4$]-$ClO_4 \cdot \frac{1}{2} H_2O$: C, 17.29; H, 4.1; N, 13.4. Found: C, 17.3; H, 4.2; N, 13.5.

B. *cis*-BIS(ETHYLENEDIAMINE)DINITRORHODIUM(III) NITRATE

[Rh(en)$_2 C_2 O_4$]ClO$_4$ + 2NaNO$_2$ + 2HNO$_3$ ⟶
  cis-[Rh(en)$_2$(NO$_2$)$_2$]NO$_3$ + H$_2$C$_2$O$_4$ + NaNO$_3$ + NaClO$_4$

Bis(ethylenediamine)(oxalato)rhodium(III) perchlorate (990 mg, 0.0028 mole) is dissolved in 150 mL of H_2O, together with 3 g of $NaNO_2$ (0.043 mole). The pH of the solution is adjusted to approximately 4 by adding 4 mL of 0.3 M

HNO_3. The solution is then heated at reflux for about 1.5 hours. Ten milliliters of 0.3 M HNO_3 is added to the boiling solution and refluxing is continued for an additional 1.5 hr.

The solution is cooled to room temperature and 15 mL of concentrated HNO_3 is added (hood). Brown fumes of decomposition products of HNO_2 are liberated. When the evolution of fumes stops, the solution is light blue (because of HNO_2) and a white crystalline precipitate has formed. The mixture is allowed to stand in the hood for 3 hours and is then left in the cold overnight.

The solid is collected by filtration, washed with ice-cold water, ethanol, and diethyl ether, and air dried. Yield: 450 mg (50%). *Anal.* Calcd. for *cis*-[Rh(en)$_2$-(NO$_2$)$_2$]NO$_3$: C, 12.7; H, 4.2; N, 26.0. Found: C, 13.1; H, 4.4; N, 26.5.

C. *cis*-DICHLOROBIS(ETHYLENEDIAMINE)RHODIUM(III) CHLORIDE HYDRATE

Procedure

An aqueous solution (20 mL) of [Rh(en)$_2$(C$_2$O$_4$)]ClO$_4 \cdot \frac{1}{2}$H$_2$O (0.10 g, 0.25 mmole), acidified with 5 mL of concentrated HCl (12 *M*) is allowed to heat at reflux for 3 minutes. The solution, cooled and reduced in volume to about 5 mL, is kept at ice temperature for several hours. The yellow precipitate that forms is collected by filtration, washed carefully with cold water and methanol, and finally air dried. Yield: 0.05 g (64% based on Rh). The checkers obtained 71% (after removing a small amount of *trans*-isomer, as described below). Electronic and infrared spectra agree with published data.[6] *Anal.* Calcd. for [Rh(en)$_2$Cl$_2$]·Cl H$_2$O: C, 13.8; H, 5.2; N, 16.1. Found: C, 13.7; H, 5.5; N, 16.0.

When HBr is used instead of HCl, an identical procedure yields *cis*-[Rh(en)$_2$Br$_2$]Br. Yield: 0.058 g, (50%). The checkers report a 31% yield. However, they further report that, if after the reflux, the solution is not concentrated but merely kept in a refrigerator, the yield is 69%.

Properties

The ion [Rh(en)$_2$(C$_2$O$_4$)]$^+$ (λ, 325; ϵ, 270: see Reference 5) may be resolved readily into its optical enantiomers through the formation[7] of the less soluble diastereoisomeric salt {(+)[Rh(en)$_2$(C$_2$O$_4$)]} {(+)-[Co(edta)]}·xH$_2$O. The resolved cation then gives (+)-*cis*-[Rh(en)$_2$Cl$_2$]$^+$ by the treatment with HCl as given above.

cis-Dichlorobis(ethylenediamine)rhodium(III) salts (λ, 350 nm; ϵ, 195; lit. value[7] 203) are converted to the *trans*-isomers (λ, 406; ϵ, 75) rather readily at room temperature by treating them, in the presence of excess chloride in water, with a little sodium tetrahydroborate. The checkers report that samples of cis-[Rh(en)$_2$Cl$_2$]$^+$ salts often contain a little of the *trans*-isomer, which may be removed by dissolving the salt in the minimum amount of warm water and filtering into methanolic NaClO$_4$.

References

1. R. J. Meyer and H. Kienitz, *Z. Anorg. Chem.*, **242**, 281 (1939).
2. S. A. Johnson and F. Basolo, *Inorg. Chem.*, **1**, 925 (1962).
3. S. N. Anderson and F. Basolo, *Inorg. Synth.*, **7**, 217 (1963).
4. T. P. Dasgupta, R. M. Milburn, and L. Damrauer, *Inorg. Chem.*, **9**, 2789 (1970).
5. A. W. Addison, R. D. Gillard, P. S. Sheridan, and L. R. H. Tipping, *J. Chem. Soc., Dalton Trans.*, **1974**, 709.
6. F. P. Jakse and J. D. Petersen, *Inorg. Chim. Acta*, **27**, 225 (1978).
7. R. D. Gillard and L. R. H. Tipping, *J. Chem. Soc., Dalton Trans.*, **1977**, 1241.

18. POTASSIUM TETRAHYDROXODIOXOOSMATE(VI) AND *trans*-BIS(ETHYLENEDIAMINE)DIOXOOSMIUM(VI) CHLORIDE

Submitted by JOHN M. MALIN*
Checked by MICHAEL FLOOD† and ANN WHITMORE†

Osmium tetraoxide, an expensive and toxic[1] reagent, is reduced safely and quantitatively by ethanol in potassium hydroxide solution to form K$_2$[OsO$_2$(OH)$_4$].[2] Subsequent reaction in aqueous solution with excess ethylenediamine dihydrochloride affords, in very good yield, the salt *trans*-[OsO$_2$(en)$_2$]Cl$_2$.[3] Both these compounds are relatively stable and can be handled easily. Furthermore, they are useful starting materials for subsequent syntheses.[3,4]

A. POTASSIUM TETRAHYDROXODIOXOOSMATE(VI)

$$2OsO_4 + C_2H_5OH + 5KOH \longrightarrow CH_3COOK + 2K_2[OsO_2(OH)_4]$$

Procedure

■ **Caution.** *Osmium tetraoxide is extremely toxic and must be handled with utmost care. Residues containing osmium must be viewed as potential sources of this dangerous compound. The hazard is reduced by storage of osmium wastes in contact with alkaline aqueous ethanol.*

*Department of Chemistry, University of Missouri, Columbia, MO 65211.
†Chemistry Department, Beloit College, Beloit, WI 53511.

In a fume hood, four 1-g ampules of OsO_4 (0.016 mole) are opened by scoring them and touching with the molten tip of a glass rod. They are dropped into a 500-mL Erlenmeyer flask containing 40 g of KOH dissolved in 170 mL of water. The mixture is stirred gently until all the OsO_4 has dissolved (10-15 min).

With the addition of 40 mL of absolute ethanol, the dark-brown solution becomes opaque and violet in color. The solution then clears as the potassium osmate salt precipitates. After 5 minutes of gentle stirring, 75 mL of absolute ethanol is added. The precipitate is collected using a sintered-glass filter (medium frit) and fragments of the glass vials are removed with forceps and washed carefully to recover any adhering product. The yield, after washing with ethanol and diethyl ether and air drying, is 5.3 g of $K_2[OsO_2(OH)_4]$ (98%). This material can be stored for months under dry conditions and may be used directly in synthesis. However, it is unstable in acid solution with respect to disproportionation and moderately so in neutral solution. *See caution concerning OsO_4.*

B. *trans*-BIS(ETHYLENEDIAMINE)DIOXOOSMIUM(VI) CHLORIDE

$$K_2[OsO_2(OH)_4] + 2[C_2H_4(NH_2)_2 \cdot 2HCl] \longrightarrow$$

$$[OsO_2(en)_2]Cl_2 + 2KCl + 4H_2O$$

Procedure

In a fume hood, 5.7 g of $K_2[OsO_2(OH)_4]$ (0.016 mole) is dissolved in the minimum amount (about 125 mL) of water at room temperature. This solution is added dropwise with rapid stirring to one containing 20 g of ethylenediamine dihydrochloride (0.15 mole) and 75 mL of water. As the osmium solution is added, a yellow color appears, with subsequent formation of a yellow precipitate. After the addition, the mixture is cooled to 0° and 50 mL of cold 6 M HCl is added slowly with stirring. This is followed by 150 mL of 95% ethanol. The yellow precipitate is collected on a sintered-glass filter, washed with ethanol and diethyl ether, and dried *in vacuo*. Yield: 6.2 g (94%).

To recrystallize, 6.2 g of the crude product is dissolved in 135 mL of 10^{-5} M HCl, preheated to 65°. The solution is filtered *immediately* through a medium-grade sintered-glass filter and cooled to 0°. Yellow-orange platelets form in about 45 minutes. Afterward, 10 mL of 6 M HCl at 0° is added dropwise with slow stirring. When the crystals have been allowed to settle, most of the supernatant solution is decanted and 50 mL of methanol is added to the slurry that remains. The crystals are collected by filtration, washed with methanol and diethyl ether, and dried *in vacuo*. Yield: 5.3 g (80%). A second crop is obtained by treating the supernatant liquid with an equal volume of methanol and allowing it to stand overnight at 0°. The overall yield is 5.8 g

(87%). *Anal.* Calcd. for $OsO_2C_4N_4H_{16}Cl_2$: H, 3.87; N, 13.56; Cl, 17.19. Found: H, 3.90; N, 13.41; Cl, 17.26.

Properties

The yellow salt $[OsO_2(en)_2]Cl_2$ is diamagnetic. It has three main absorption peaks in the ultraviolet region, at 355, 310, and 240 nm, for which the log ϵ values are 2.31, 2.82, and 3.46, respectively. The peak at 310 nm is sharply pointed and is accompanied by a slightly smaller peak at 317 nm. In the Raman spectrum of the salt, a strong band due to the ν_1 symmetric stretch of the $O=Os=O$ structure is observed at 917 cm^{-1}. Salts of other anions can be prepared from this material by aqueous metathetical reactions with the appropriate sodium salts. An X-ray diffraction study of the hydrogen sulfate salt has confirmed the *trans* geometry for the ion.[5] After storage for long periods of time, the yellow salt becomes grayish. The ion $[OsO_2(en)_2]^{2+}$ is unstable in basic solution. Exchange of the dioxo oxygens takes place slowly in acid solution.[5]

References

1. W. P. Griffith, *The Chemistry of the Rarer Platinum Metals*, Interscience, New York, 1967, p. 68.
2. E. Fremy, *J. Prakt. Chem.*, **33**, 412 (1844).
3. J. Malin and H. Taube, *Inorg. Chem.*, **10**, 2403 (1971); also see: L. Gibbs, *Am. J. Chem.*, **3**, 233 (1881).
4. A. L. Coelho and J. M. Malin, *Inorg. Chim. Acta*, **14**, L41 (1975).
5. J. M. Malin, E. O. Schlemper, and R. K. Murmann, *Inorg. Chem.*, **16**, 615 (1977).

19. SODIUM BIS[2-ETHYL-2-HYDROXYBUTYRATO(2-)]-OXOCHROMATE(V)

$$Na_2Cr_2O_7 + 5(C_2H_5)_2C(OH)COOH \longrightarrow$$
$$2Na[((C_2H_5)_2COCO_2)_2CrO] + (C_2H_5)_2CO + CO_2 + 5H_2O$$

Submitted by M. KRUMPOLC* and J. ROČEK*
Checked by G. P. HAIGHT, JR.,† and PATRICK MERRILL†

Unlike most other known chromium(V) compounds,[1] the recently prepared chromium(V) complexes of α-hydroxy secondary acids are remarkably stable

---

*Department of Chemistry, University of Illinois at Chicago Circle, Chicago, IL 60680.
†School of Chemical Sciences, University of Illinois at Urbana-Champaign, Urbana, IL 61801.

even in aqueous solutions. Sodium bis[2-ethyl-2-hydroxybutyrato(2-)]oxochromate(V) is one of the most stable and easily accessible compounds of this class. The method of preparation described below represents a further simplification and improvement over our original procedure.[2] The entire synthesis can be easily completed within a 2-day period.

Chromium(V) has been widely used in the preparation of dynamically polarized proton targets in high-energy physics.[3] The availability of these types of compounds can also be expected to open up the investigation of the chemistry of chromium(V).

Procedure

To a solution of 19.8 g (0.150 mole) of 2-ethyl-2-hydroxybutyric acid (Aldrich Chemical Co., Milwaukee, WI 53233) in 125 mL of acetone in a 300-mL Erlenmeyer flask is added 6.5 g (0.025 mole) of anhydrous sodium dichromate (the dihydrate is dried *in vacuo* at 100° for about 30-40 min and finely pulverized), and the heterogeneous mixture is stirred magnetically until the dichromate is completely dissolved (about 10 min) (■ **Caution.** *Dichromates are highly toxic. Care should be taken to prevent skin contact, especially inhalation of such powders.*) The formation of the dark red-brown solution of the chromium(V) complex is soon apparent. The flask is fitted with a glass stopper and immersed in a temperature-controlled water bath at $25.00 \pm 0.1°$ for a period of 23-24 hours. The solution is poured into 375 mL of hexane, whereupon the chromium(V) complex precipitates as a dark red-violet solid. The crude product is collected, dried *in vacuo* at room temperature for about 30 minutes to remove water and volatile materials (solvents, 3-pentanone), dissolved in 125 mL of acetone, and reprecipitated by addition to 375 mL of hexane. The crystalline product is washed with 20 mL of hexane and dried *in vacuo* at room temperature to constant weight (about 30 min), giving 14.8-16.5 g (0.040-0.045 mole) of sodium bis[2-ethyl-2-hydroxybutyrato(2-)]oxochromate(V) monohydrate. The yield is between 80 and 90% based on sodium dichromate. *Anal.* Calcd. for $C_{12}H_{22}CrNaO_8$: C, 39.0; H, 6.01; Cr, 14.1; Na, 6.23; H_2O, 4.88. Found: C, 39.4; H, 6.00; Cr, 14.1; Na, 6.6; H_2O, 5.3. No Cr(VI) was detected.[4] The presence of one molecule of water may be established by the deuterium exchange technique and determined by nmr.[5]

Properties

Sodium bis[2-ethyl-2-hydroxybutyrato(2-)]oxochromate(V) monohydrate is a dark red-violet crystalline solid. It is very stable at room temperature; after exposure to air and light for several weeks no visible decomposition is observed. It does not show a melting point but slowly decomposes at ~170°. It is readily soluble in polar solvents (water, acetone, pyridine, dimethylformamide, dimethyl sulfoxide, acetic acid, liquid ammonia) but is insoluble in hydrocarbons, carbon tetrachloride, chloroform, and diethyl ether. The compound is relatively stable to hydrolysis: only about 28% of a 0.01 M solution of the complex is decomposed over a period of 24 hours at 25°. The stability can be substantially enhanced by the addition of a small amount of 2-ethyl-2-hydroxybutyric acid. Upon acidification the complex undergoes fast disproportionation to chromium(VI) and chromium(III); addition of sodium hydroxide results in instantaneous disproportionation.

The infrared spectrum (Nujol) contains major absorption bands at 3510(m, br), 1680(s,br), 1312(m), 1257(m), 1177(m), 1045(w), 998(m), 961(s), 888(w), 845(m), 815(w), and 712(m) cm^{-1}. The electronic absorption spectrum (in 0.1 M aqueous solution of 2-ethyl-2-hydroxybutyric acid) is [λ, nm (ϵ)] 250 (6510), 350 (1200), 485 *min* (160), 510 *max* (168), 633 *min* (28.6), 740 *max* (40.9), 750 (40.7), and 800 (39.4). The complex is paramagnetic (d^1 electron configuration).

References

1. For leading references, see Reference 2.
2. M. Krumpolc, B. G. DeBoer, and J. Roček, *J. Am. Chem. Soc.*, **100**, 145 (1978).
3. D. Hill, R. C. Miller, M. Krumpolc, and J. Roček, *Nucl. Instrum. Methods*, **150**, 331 (1978), and references cited therein.
4. M. Krumpolc and J. Roček, *J. Am. Chem. Soc.*, **99**, 137 (1977).
5. M. Krumpolc and J. Roček, *J. Am. Chem. Soc.*, **101**, 3206 (1978).

20. THE BIS(β-DIKETONATO)PLATINUM(II) COMPLEXES

Submitted by SEICHI OKEYA* and SHINICHI KAWAGUCHI†
Checked by ERVEN KUHLMANN‡ and MILTON ORCHIN‡

2,4-Pentanedione (acacH) is a common ligand, forming chelates with almost all transition metals.[1] Bis(2,4-pentanedionato)platinum(II) was first prepared by Werner,[2] but the yield reported later by Grinberg and Chapurskii[3] was only 35%. Furthermore, the platinum(II) chelates of other β-diketones have not been reported. The yield of Pt(acac)$_2$ has been increased to 75% by aquating the tetrachloroplatinate(II) anion by treatment with mercury(II) and silver(I) ions

---

*Faculty of Education, Wakayama University, Masago-cho, Wakayama 640, Japan.
†Faculty of Science, Osaka City University, Sumiyoshi-ku, Osaka 558, Japan.
‡Department of Chemistry, University of Cincinnati, Cincinnati, OH 45221.

prior to the reaction with acacH. The platinum(II) chelates of 1,1,1-trifluoro-2,4-pentanedione (tfacH) and 1,1,1,5,5,5-hexafluoro-2,4-pentanedione (hfacH) are prepared in a similar fashion.

A. PREPARATION OF STOCK SOLUTION

$$[PtCl_4]^{2-} + Hg^{2+}(Ag^+) + \text{excess } H_2O \longrightarrow [Pt(H_2O)_4]^{2+} + HgCl_2(AgCl)$$

Procedure

■ **Caution.** *Perchlorates are potentially explosive and must be treated with care.*

Pulverized K_2PtCl_4 (10.38 g, 25 mmole) is dissolved in 1 N $HClO_4$ (700 mL) and the mixture is stirred for about 30 minutes; this produces a red solution and a precipitate of $KClO_4$. Yellow mercury(II) oxide (16.25 g, 75 mmole) is dissolved in 1 N $HClO_4$ (200 mL), and the solution is added slowly (1-2 drops/sec) to the Pt(II) solution with stirring. Then a solution of silver perchlorate (5.18 g, 25 mmole) in 1 N $HClO_4$ (50 mL) is added to the reaction mixture. The precipitate is filtered promptly, and the filtrate is made up to 1 L by adding 1 N $HClO_4$. This is the stock solution. The Pt(II) seems to exist mainly as the tetraaqua ion in this orange-yellow solution[4] and is quite stable in this form if kept in a cool dark place.

B. BIS[2,4-PENTANEDIONATO(1-)]PLATINUM(II)

$$[Pt(H_2O)_4]^{2+} + 2[CH_2(COCH_3)_2] + 2OH^- \longrightarrow Pt(acac)_2 + 6H_2O$$

Procedure

The ligand 2,4-pentanedione (9 g, 90 mmole), dissolved in a 2.5 N aqueous NaOH solution (40 mL), is added dropwise to the Pt(II) stock solution (350 mL) by means of a pipette. The instant each drop of the ligand solution contacts the metal solution, a dark-grey precipitate is produced; this turns pale yellow quickly on agitation. After the ligand solution has been added, a 10 N NaOH solution is added slowly from a pipette until the pH of the mixture is 4.5. The amount of dark-grey precipitate formed by each drop of the alkali solution increases gradually, but it decreases remarkably when about 40 mL of the solution has been added. Stirring is continued for an additional 20 hours. A creamy-yellow precipitate is filtered, washed three times with water, and dried over silica gel *in vacuo*.

The crude product is dissolved in dichloromethane and the insoluble material is filtered and discarded. The filtrate is concentrated to about 10 mL and is then chromatographed over silica gel (100–200 mesh, 4 × 30 cm). The yellow dichloromethane eluate is evaporated under reduced pressure to obtain a yellow powder of Pt(acac)$_2$. The yield (2.58 g) is 75.0% on the basis of K$_2$PtCl$_4$. Recrystallization from benzene gives beautiful rodlike crystals with solvent of crystallization, which is lost gradually on standing in the air. (■ **Caution.** *Benzene is highly toxic. A hood should be used and gloves should be worn.*) *Anal.* Calcd. for C$_{10}$H$_{14}$O$_4$Pt: C, 30.54; H, 3.59. Found: C, 30.62; H, 3.55.

C. BIS[1,1,1-TRIFLUORO-2,4-PENTANEDIONATO(1-)]PLATINUM(II)

$$[Pt(H_2O)_4]^{2+} + 2CF_3COCH_2COCH_3 + 2OH^- \longrightarrow Pt(tfac)_2 + 6H_2O$$

Procedure

The ligand tfacH (4.6 g, 30 mmole) is dissolved in a 10 N NaOH solution (10 mL) and the mixture is added slowly from a pipette into the Pt(II) stock solution (300 mL) with stirring. This is followed by addition of 10 mL of an aqueous 10 N NaOH solution. The pH of the solution is about 1.3. After the solution is stirred for an additional 20 hours, a straw-yellow precipitate is filtered and washed with water. To the filtrate is added tfacH (1.2 g, 8 mmole) neutralized with NaOH solution. This solution is then stirred for 20 hours and the precipitate is filtered. This procedure is repeated twice, after which all the precipitates are combined, dried *in vacuo*, and extracted with dichloromethane. A concentrate containing the crude product is charged onto a column (5 × 60 cm) of silica gel and the column is developed with a mixture (7:10 by volume) of benzene and petroleum ether (boiling below 50°) to separate the product into two bands. After evaporation of the solvent, the first eluate gives tiny yellow crystals of *trans*-Pt(tfac)$_2$ (0.53 g, 14%) and the second eluate gives *cis*-Pt(tfac)$_2$ (0.60 g, 16%).[5] Both isomers are recrystallized from dichloromethane–petroleum ether to afford yellow needles. *Anal.* Calcd. for C$_{10}$H$_8$O$_4$F$_6$Pt: C, 23.96; H, 1.61. Found for the *trans*-isomer: C, 24.13; H, 1.63; for the *cis*-isomer: C, 24.37; H, 1.74.

D. BIS[1,1,1,5,5,5-HEXAFLUORO-2,4-PENTANEDIONATO(1-)]-PLATINUM(II)

$$[Pt(H_2O)_4]^{2+} + 2[CH_2(COF_3)_2] + 2OH^- \longrightarrow Pt(hfac)_2 + 6H_2O$$

Procedure

A solution of hfacH (7.3 g, 35 mmole) in 2 N NaOH aqueous solution (20 mL) is added to the Pt(II) stock solution (350 mL) with stirring by means of a pipette; this is followed by addition of a 10 N NaOH solution (22 mL). The pH of the solution is about 3. After the solution is stirred for 16 hours an orange precipitate is filtered and washed with water. The filtrate is stirred for 2 days more to produce an additional precipitate. The crude product is gathered, dried *in vacuo*, extracted with dichloromethane, and chromatographed through a column of silica gel (4 X 30 cm). The dichloromethane eluate is concentrated and kept in a refrigerator to afford orange needles, which are filtered and washed with a small quantity of *n*-hexane. The yield (2.95 g) is 55.3%. *Anal.* Calcd. for $C_{10}H_2O_4F_{12}Pt$: C, 19.71; H, 0.31. Found: C, 19.94; H, 0.47.

Properties

The infrared and ^1H nmr spectra of Pt(acac)$_2$ are essentially the same as reported in the literature.[6] The absorption spectrum in dichloromethane solution exhibits absorption maxima at 287 (ϵ = 6840) and 346 nm (ϵ = 3570).

Both *trans*- and *cis*-Pt(tfac)$_2$ are air stable and sublime at 97-101° and 108-111°, respectively. They are soluble in most organic solvents, such as benzene, dichloromethane, diethyl ether, and acetone, solubilities of the *cis*-isomer being much larger than those of the *trans*-isomer.

The infrared spectra in Nujol in the 4000-700 cm^{-1} region are similar for the two isomers, although the $\nu(C\doteq O)$ and $\nu(C\doteq C\doteq C)$ energies are a little lower for the *trans*-isomer [1580 (vs) broad and 1517 (vs) cm^{-1}] than for the *cis*-isomer [1590 (vs) broad and 1530 (vs) cm^{-1}]. In the lower frequency region the *trans*-isomer exhibits a single band at 458 cm^{-1}, while the *cis*-isomer has two bands at 461 and 446 cm^{-1}, which may be related to the $\nu(Pt-O)$ vibration.

Proton nmr signals in C$_6$D$_6$ with internal TMS: *trans*, CH$_3$ δ 1.10 ppm, $^4J(Pt-H) \simeq 3$ Hz and CH δ 5.40 ppm, $^4J(Pt-H) \simeq 10$ Hz; *cis*, CH$_3$ δ 1.17 ppm, $^4J(Pt-H) \simeq 3$ Hz and CH δ 5.44 ppm, $^4J(Pt-H) \simeq 11$ Hz. In CDCl$_3$, on the other hand, the isomers show identical signals: CH$_3$ δ 2.02 ppm, $^4J(Pt-H) \simeq 4$ Hz and CH δ 5.90 ppm, $^4J(Pt-H) \simeq 11$ Hz. The ^{19}F nmr signals in CDCl$_3$ are observed at 93.0 and 93.1 ppm downfield from external C$_6$F$_6$ for *trans*- and *cis*-Pt(tfac)$_2$, respectively, with $^4J(Pt-F)$ = 18 Hz in both cases.

Both of the isomers show an absorption maximum in dichloromethane at 309 nm accompanied by shoulders at 330 and 390 nm, but the molar extinction coefficients are different, $\epsilon(trans)$ being 5280 and $\epsilon(cis)$ 5570. Mass spectra of both isomers exhibit the parent peak at m/e = 501 in accordance with the calculated molecular weight of 501.

The crystals of Pt(hfac)$_2$ are air stable and sublime at around 65°, exhibiting the parent peak in the mass spectrum at *m/e* = 609 in accordance with the calculated molecular weight of 609. They are highly soluble in benzene, dichloromethane, diethyl ether, and acetone, and appreciably soluble in methanol, and hexane. The infrared spectra in Nujol show the $\nu(C\!=\!O)$ and $\nu(C\!=\!C\!=\!C)$ bands at 1585(vs), 1557(m), and 1533(m) cm^{-1}, and the lower frequency bands at 614(s), 536(m), and 352(vw) cm^{-1}. The nmr signal of the methine proton in CDCl$_3$ is observed at 6.39 ppm from TMS, ^4J(Pt—H) being 10.5 Hz. The ^{19}F nmr signal in CH$_2$Cl$_2$ is observed at 90.8 ppm downfield from external C$_6$F$_6$ with ^4J(Pt—F) = 17 Hz.

Absorption maxima are observed in dichloromethane at 329 (ϵ = 4380) and 426 (ϵ = 1870) nm together with shoulders at 345, 407, and 452 nm.

References and Notes

1. J. P. Fackler, Jr., *Prog. Inorg. Chem.*, **7**, 361 (1966).
2. A. Werner, *Chem. Ber.*, **34**, 2584 (1901).
3. A. A. Grinberg and I. N. Chapurskii, *Russ. J. Inorg. Chem.*, **4**, 137 (1959).
4. L. I. Elding, *Inorg. Chim. Acta*, **20**, 65 (1976). The Hg^{2+} ion alone does not remove chloride ion completely, but needs assistance by silver perchlorate. Use of the silver salt at the outset is not appropriate since Ag$_2$[PtCl$_4$] is precipitated.
5. The *trans* to *cis* ratio in the product varies with the preparative conditions, but the combined yield is almost constant at 30%. Assignment of the geometrical isomers was made by analogy to bis(benzoylacetonato)palladium(II). The *cis* isomer shows a smaller R$_f$ value on thin layer chromatography over silica gel with dichloromethane as a developing solvent; this was confirmed by single-crystal X-ray analysis [S. Okeya, H. Asai, S. Ooi, K. Matsumoto, S. Kawaguchi, and H. Kuroya, *Inorg. Nucl. Chem. Lett.*, **12**, 677 (1976)].
6. J. Lewis, R. F. Long, and C. Oldham, *J. Chem. Soc.*, **1965**, 6740.

21. TRIMETHYLPHOSPHINE IRON COMPLEXES

Submitted by HANS HEINZ KARSCH*
Checked by MICHELE ARESTA†

Trimethylphosphine complexes of transition metals are of considerable interest since the spectroscopic data from these complexes are usually simple and the steric requirements of the ligand are small compared with those of other tertiary phosphines. Because of the tedious preparative procedure of this phosphine, its application in this area has been restricted. However, recently laboratory[1] and industrial[2] procedures have been developed that make this ligand more easily obtainable.

*Technische Universität München, D- 8046 Garching, Lichtenbergstr. 4.
†University of Bari, Via G. Amendola 173, 70126 Bari, Italy.

A. DICHLOROBIS(TRIMETHYLPHOSPHINE)IRON(II)

$$FeCl_2 + 2P(CH_3)_3 \longrightarrow [(CH_3)_3P]_2FeCl_2$$

Many $P(CH_3)_3$ complexes of the first-row transition metals may be conveniently prepared by direct reaction with the appropriate anhydrous chromium(II), (III),[3] cobalt(II),[4] iron(II)[5] or hydrated iron(II),[5] nickel(II)[4a, 6, 7] salt. The compound $[(CH_3)_3P]_2FeCl_2$ is the starting material for a variety of phosphine iron complexes.[4, 8, 9]

Procedure

■ **Caution.** *Trimethylphosphine is toxic, very volatile, and flammable. All operations should be carried out in purified nitrogen and in a very well ventilated fume hood. The solvents should be free of oxygen and moisture.*

A bent, two-armed glass vessel with a sintered disc *e* resembling that previously described by Strohmeier[10] and Klein et al.[11] (Fig. 1) is fitted with flasks (100 mL) at *A* and at *B*, which is connected by a curved glass adapter *d*. The apparatus is attached to a high-vacuum system by means of the pivotal glass joint *c*.

Under a countercurrent of nitrogen, a flask containing $FeCl_2$[12] (2 g, 15.8 mmole) is substituted for the flask at *B*. The system is evacuated and after the flask is cooled at position *B* to $-78°$, 40 mL of a tetrahydrofuran solution of $P(CH_3)_3$ (2.9 g, 38.2 mmole) is condensed into it. [This solution is made by condensing 3.8 mL of $P(CH_3)_3$ from a flask, fitted with a stopcock, through a PVC tube into the tetrahydrofuran-containing flask, which is attached directly

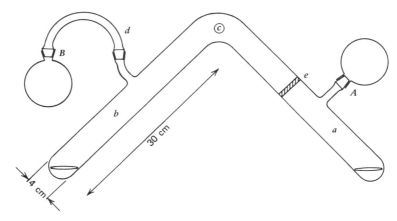

Fig. 1. Reactor for formation of complexes.

to the high-vacuum system]. The mixture is warmed to room temperature and after it is poured into arm b, it is stirred for approximately 1 hour. The solution is tipped onto the sintered disc (by turning the vessel) and is filtered into arm a by gently warming arm b and cooling arm a. The solvent is recondensed into arm b by cooling this arm; the procedure is repeated twice, thereby washing any remaining product from the solid onto the disc. The mixture (now in arm a) is tipped into flask A and cooled to $-78°$. Pentane (40 mL) is condensed into flask A and after it is warmed to room temperature and slowly recooled to $-78°$, white platelets with a faint blue tinge form, their size being increased by repeating the warming and cooling procedure. The system is refilled with nitrogen and the crystals are isolated by decanting the mother liquor back into arm a and maintaining it at $-78°$. The system is evacuated and the flask at A is allowed to warm to room temperature. The flask is removed under a countercurrent of nitrogen, and a second nitrogen-filled flask is attached quickly to the vessel at A. The system is evacuated again and the mother liquor is then warmed up and poured into the new flask. The solvent is evaporated under vacuum and a bluish-white solid is obtained, which is usually as pure as the crystal fraction. Total yield: 4.2 g (95%).

Properties

The properties of dichlorobis(trimethylphosphine)iron(II), melting at 115–117°, are described in detail elsewhere.[5] The solid may be handled in air for very short periods, but the solution immediately turns red on contact with air. After prolonged heating of the red solution, the original color may be restored. Pure samples show no ir absorption bands in the $\nu(OH)$ and $\nu(P=O)$ stretching-vibration regions. Besides the $\nu(Fe-Cl)$, 360 cm^{-1}, only bands due to the ligand are observed. *Anal.* Calcd. for $C_6H_{18}Cl_2FeP_2$: C, 25.84; H, 6.50. Found: C, 25.82; H, 6.52.

B. TETRAKIS(TRIMETHYLPHOSPHINE)IRON

cis-[(Dimethylphosphino)methyl-*C*, *P*] hydridotris(trimethylphosphine) iron

$$[(CH_3)_3P]_2FeCl_2 + 2P(CH_3)_3 + Mg \xrightarrow{THF}$$

$$MgCl_2 + [(CH_3)_3P]_4Fe \rightleftharpoons [(CH_3)_3P]_3HFe\begin{matrix}P(CH_3)_2\\|\\CH_2\end{matrix}$$

Phosphine complexes of metals in the oxidation state zero are versatile reagents. They offer access to a fascinating class of organometallic species, that is, those that are involved in or are related to catalytic processes. Tetrakis(trimethylphosphine)metal(0) complexes of nickel[13] and cobalt[14] have been made by reduction of the appropriate metal(II) complexes or anhydrous salts with sodium or magnesium in the presence of the ligand in tetrahydrofuran. The same method is applicable to the preparation of $[(CH_3)_3P]_4Fe$,[8,15] which is unique in its solvent and temperature-dependent equilibrium with a hydridoiron(II) isomer, incorporating a three-membered Fe—C—P ring. This behavior offers some remarkable features:

1. The compound may react as a typical metal(0) species, susceptible to substitution and oxidative addition reactions. Both are promoted by the unsaturated nature of the molecule, which is the first stable four-coordinate iron(0) species to be observed at ambient temperature.
2. The compound may also react by cleavage of the Fe—C and/or the Fe—H function of the iron(II) isomer.
3. A substrate (e.g., CO_2) may react with both the iron(0) and the iron(II) isomer simultaneously, the ratio of products being dependent on the temperature, solvents used, and the duration of the reaction.

Preparation

■ **Caution.** *Trimethylphosphine is toxic, very volatile, and flammable.*

For handling trimethylphosphine, solvent purity grades, and the reaction vessel, see the preparation of $[(CH_3)_3P]_2FeCl_2$ (Sec. A).

In Fig. 1 the bent tube *d* and flask *B* are removed from the two-armed glass vessel and 0.5 g (20.6 mmole) of magnesium is placed in arm *b*. The tube and the flask at *B* are replaced and the vessel is evacuated. The turnings are stirred overnight under vacuum to provide some fresh surface on the metal. The system is refilled with nitrogen and a flask containing 1.0 g (3.6 mmole) of $[(CH_3)_3P]_2FeCl_2$ (Sec. A) is substituted for the flask at *B*. The system is evacuated again, and the complex is tipped onto the magnesium. After arm *b* is cooled to -78°, 30 mL of tetrahydrofuran and 0.82 g (10.8 mmole; 50% excess) of $P(CH_3)_3$ are condensed onto the magnesium. While it is stirred vigorously, the mixture is warmed to about 40°. As soon as the pale-blue color of the solution begins to show a trace of green, the arm is quickly immersed in an ice-water bath. With continued stirring, the color turns deep yellow-brown. The reaction is usually complete after 1 hour. The ice bath is removed and the solvent is evaporated, while stirring is continued (without pumping!), by condensing it into a trap on the high-vacuum line, which is cooled to -78°. The last traces of the solvent sometimes are difficult to remove, and short pumping while warming to about 40° may be helpful at the end. However, very small amounts of tetrahydrofuran do not affect the isolation. (It should be noted that heating and pumping promote decomposition of the product and such procedures should be kept to a minimum.) The yellow solid obtained is cooled to $-78°$, and pentane (50 mL), along with 0.5 mL of $P(CH_3)_3$, is condensed onto the yellow solid. The solution is then rewarmed with stirring. The filtering and washing procedure described in the preparation of $[(CH_3)_3P]_2FeCl_2$ (Sec. A) is used. Then the filtered solution, now in arm a, is poured into the flask at A. The solvent is evaporated into a cooled trap ($-78°$) (some stirring at the end is effective in the prevention of bumping) until the substance comes to dryness. The orange solid obtained (1.17 g, 91%) is pure enough for most practical purposes, but sometimes traces of pentane are retained and their presence reduces the thermal stability of the compound at room temperature. Yellow crystals may be obtained by concentrating the pentane solution to 5-10 mL and cooling to $-78°$ for at least 10 hours, but the yield is low and the isolation is tedious.

Properties

The solid compound slowly decomposes at room temperature and quickly at $70°$, while at $-30°$ it may be stored for several months. In solution, it slowly decomposes at room temperature, yielding a black, ferromagnetic powder and $P(CH_3)_3$. However, the stability in solution may be enhanced by addition of $P(CH_3)_3$. Like $[(CH_3)_3P]_4Co$[14] and $[(CH_3)_3P]_4Ni$,[13] the iron compound is very soluble in hydrocarbons and ethers. These solutions are extremely air sensitive, the solid being pyrophoric. The spectroscopic and analytical data and some reactions are reported elsewhere[8,9,15,16]. The purity is tested by means of the ir spectrum[8]. Indicative are bands at 1822 $[\nu(Fe-H)]$, 895, and 455 cm^{-1}. *Anal.* Calcd. for $C_{12}H_{36}FeP_4$: C, 40.02; H, 10.08; Fe, 15.51. Found: C, 39.61; H, 9.86; Fe, 15.52.

C. (CARBON DIOXIDE—C,O)TETRAKIS(TRIMETHYLPHOSPHINE)IRON

$$\text{``}[(CH_3)_3P]_4Fe\text{''} + CO_2 \xrightarrow[20°]{\text{pentane}} [(CH_3)_3P]_4Fe(CO_2)$$

Recently, carbon dioxide complexes have attracted increasing attention for preparative and catalytic application.[17] However, the number of known simple and reasonably stable CO_2 adducts is limited.[16a, 17, 18] Because of the isomerization behavior of the previously described complex, $[(CH_3)_3P]_4Fe$ (Sec. B), a variety of very different compounds is obtained on reaction with CO_2, depending on the reaction conditions. For example, insertion into the iron-carbon and iron-hydrogen bonds may occur, as well as disproportionation to CO_3^{2-} and

74 Coordination Compounds

CO. In addition, simple adduct formation is also possible.[16a] Somewhat related behavior seems to occur with an iridium(I) complex.[18g]

Procedure

- **Caution.** *Trimethylphosphine is toxic, very volatile, and flammable.*

For handling P(CH$_3$)$_3$ and air-sensitive compounds, solvent purity grades, and apparatus, see the preparation of [(CH$_3$)$_3$P]$_2$FeCl$_2$ (Sec. A). A flask containing 1.17 g (3.25 mmole) of [(CH$_3$)$_3$P]$_4$Fe (Sec. B) is attached to the nitrogen-filled two-armed glass vessel at *B* in Fig. 1 under a countercurrent of nitrogen. The system is evacuated and the flask is cooled to $-78°$. Pentane (50 mL) and P(CH$_3$)$_3$ (0.5 mL) are condensed into the flask. The mixture is warmed to room temperature and poured (by turning the bent tube and/or the reaction vessel) into arm *b*. The system, still under reduced pressure, is filled with pure carbon dioxide[19] (1 atm). When the color of the solution begins to turn from orange to brown-red (about 5 min), the assembly is turned and the solution is poured quickly onto the sintered disc. Short cooling at arm *a* and gentle warming at arm *b* of the vessel assist the filtering process. The solution is immediately transferred into the flask at *a* and in approximately 3 hours, dark-red crystals separate along with some fine yellow powder. The solvent is swirled by quickly shaking the flask (turning one arm up and down). Most of the light-yellow particles, consisting of [(CH$_3$)$_3$P]$_3$(CO)Fe(CO$_3$)[16a], which remain suspended in the solution longer then the heavier red crystals, are decanted from the flask into arm *a* with the solvent. The red crystals are dried by cooling the mother liquor in arm *a* and evacuating the system for a short period. The flask containing the crystals is removed under a countercurrent of nitrogen. When the crystals are gently loosened from the glass wall of the flask in a glove box and transferred to another nitrogen-filled flask, they are essentially pure. Any traces of yellow precipitate remain more strongly fastened to the glass walls of the original flask. The yield of the deep-red crystals varies from 0.83 to 1.09 g (63–83%). The compound may be recrystallized from acetone at low temperature, but for most purposes, the purity of the original crystals is sufficient. *Anal.* Calcd. for C$_{13}$H$_{36}$FeO$_2$P$_4$: C, 38.63; H, 8.98; Fe, 13.82. Found: C, 38.50; H, 8.99; Fe, 14.12.

Properties

The dark-red, octahedral crystals decompose very slowly at room temperature and quickly at $80°$.[16a] The purity of the compound is checked by means of ir spectroscopy,[16a] common impurities giving rise to bands in the $\nu(C\equiv O)$ and in the $\nu(P=O)$ stretching frequency regions.

Cryoscopic molecular weight determinations are not fully conclusive because of ligand dissociation in solution. A dimeric structure with two bridging CO$_2$

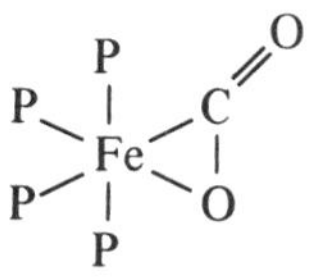

```
  P       O
  |      ‖
P\ |  ,C
  Fe
P/ |  \
  |    O
  P
```

Fig. 2. *Proposed structure of the CO_2 adduct.*

ligands (see Reference 18f) seems unlikely considering steric and bonding arguments.

Bands at 1620, 1108, and 612 cm^{-1} in the ir spectrum (Nujol)[16a] are due to the CO_2 ligand. The stereochemistry of the four phosphine ligands is indicated by the ^{31}P nmr spectrum, which exhibits characteristic features of a second-order A_2BC spin system.[16a]

Two doublets and one "pseudotriplet," observed in the ^{1}H nmr spectrum for the PCH_3 groups,[16a] are characteristic for two chemically different methylphosphine ligands that are linked to the metal in *cis* position and two methylphosphine ligands that are linked in *trans* positions.[20] The structure shown in Fig. 2 best fits these findings. The X-ray structure of a closely related CS_2 iron complex has been determined.[21]

References

1. W. Wolfsberger and H. Schmidbaur, *Synth. Inorg. Met. Org. Chem.*, **4**, 149 (1974).
2. Knapsack AG, D.O.S. 2116439; *c.f.*, *Chem. Abstr.* **78**, 16309d (1973); D.O.S. 211635; *c.f.*, *Chem. Abstr.* **78**, 43704 k (1973); D.O.S. 2522021; *c.f.*, *Chem. Abstr.* **86**, 72879 (1977).
3. H. H. Karsch, *Angew. Chem.*, **89**, 57 (1977); *Angew. Chem. Int. Ed., Engl.*, **16**, 56 (1977).
4(a). K. A. Jensen, P. H. Nielsen, and C. T. Petersen, *Acta Chem. Scand.*, **17**, 1115 (1963). (b) H.-F. Klein and H. H. Karsch, *Chem. Ber.*, **109**, 1453 (1976). (c) M. Zinoune, M. Dartiguenave, and Y. Dartiguenave, *C. R. Ser. C.*, **278**, 849 (1974). (d) M. Bressan and P. Rigo, *Inorg. Chem.*, **14**, 38 (1975).
5. H. H. Karsch, *Chem. Ber.*, **110**, 2222 (1977).
6(a). O. Dahl, *Acta Chem. Scand.*, **23**, 2342 (1969). (b) M. A. A. Beg and H. C. Clark, *Can. J. Chem.*, **39**, 595 (1961).
7. H.-F. Klein and H. H. Karsch, unpublished results.
8. H. H. Karsch, H.-F. Klein, and H. Schmidbaur, *Chem. Ber.*, **110**, 2200 (1977).
9. H. H. Karsch, *Chem. Ber.*, **110**, 2699, 2712 (1977).
10. W. Strohmeier, *Chem. Ber.*, **88**, 1218 (1955).
11. H.-F. Klein and H. H. Karsch, *Inorg. Chem.*, **14**, 473 (1975).
12. P. Kovacic and M. C. Brace, *Inorg. Synth.*, **6**, 172 (1960).
13. H.-F. Klein and H. H. Karsch, *Chem. Ber.*, **109**, 2515 (1976).
14. H.-F. Klein and H. H. Karsch, *Chem. Ber.*, **108**, 944 (1975).
15(a). H. H. Karsch, H.-F. Klein, and H. Schmidbaur, *Angew. Chem.*, **87**, 630 (1975); *Angew. Chem. Int. Ed., Engl.*, **14**, 637 (1975); (b) J. W. Rathke and E. L. Muetterties, *J. Am. Chem. Soc.*, **97**, 3272 (1975).
16(a). H. H. Karsch, *Chem. Ber.*, **110**, 2213 (1977). (b) H. H. Karsch, *Chem. Ber.*, **111**, 1650 (1978).
17(a). M. E. Vol'pin, *Z. Chem.*, **12**, 361 (1972). (b) M. E. Vol'pin and I. S. Kolmnikov,

Pure Appl. Chem., 33, 567 (1973). (c) M. E. Vol'pin and I. S. Kolomnikov, Organomet. React., 5, 313 (1975).

18(a). B. R. Flynn and L. Vaska, Chem. Commun., 1974, 703. (b) M. Aresta and C. F. Nobile, Inorg. Chim. Acta, 24, L 49 (1977). (c) M. Aresta and C. F. Nobile, J. Chem. Soc. Dalton Trans., 1977, 708. (d) I. Ito, H. Tsuchiya, and A. Yamamoto, Chem. Lett., 1976, 851. (e) A. Miyashita and A. Yamamoto, J. Organomet. Chem., 113, 187 (1976). (f) P. W. Jolly, K. Jonas, C. Krüger, and Y. H. Tsay, J. Organomet. Chem., 33, 109 (1971). (g) T. Herskowitz, J. Am. Chem. Soc., 99, 2391 (1977).

19. Gmelin Handbuch der Anorganischen Chemie, 8th ed., Vol. 14, Part C, Verlag Chemie, Weinheim 1970.

20(a). R. K. Harris, Can. J. Chem., 42, 2275 (1964). (A) (b) J. M. Jenkins and B. L. Shaw, J. Chem. Soc., (A) 1966, 770.

21. H. le Bozec, P. Dixneuf, N. J. Taylor, and A. J. Carter, J. Organomet. Chem., 135, C 29 (1977).

22. PENTAKIS(TRIMETHYL PHOSPHITE) COMPLEXES OF THE d^8 TRANSITION METALS

Submitted by J. P. JESSON,* M. A. CUSHING,* and S. D. ITTEL*
Checked by L. W. YARBROUGH II† and J. G. VERKADE†

The physical aspects of five-coordination have intrigued a great number of workers. The numerous crystallographic investigations have been reviewed by Frenz and Ibers.[1] Detailed nmr line-shape analyses have been performed for a number of systems.[2,3] The pentakis(trimethyl phosphite) complexes of the d^8 transition metals are found to have trigonal bipyramidal coordination and to be stereochemically nonrigid on the nmr time scale at room temperature.[4] They represent the first class of ML_5 complexes for which limiting slow exchange nmr spectra, corresponding to D_{3h} symmetry, have been observed.

A. PENTAKIS(TRIMETHYL PHOSPHITE)NICKEL(II) BIS(TETRAPHENYLBORATE)

$$Ni[BF_4]_2 \cdot 6H_2O + 5P(OMe)_3 + 2Na[BPh_4] \longrightarrow$$
$$[Ni[P(OCH_3)_3]_5][BPh_4]_2 + 2Na[BF_4] + 6H_2O$$

Procedure

The reaction is carried out at room temperature in a hood. A 500-mL filter flask is equipped with a stirring bar coated with Teflon, and a slow stream of

*Central Research and Development Department, E. I. du Pont de Nemours and Company, Wilmington, DE 19898, Contribution No. 2369.
†Department of Chemistry, Iowa State University, Ames, IA 50010.

nitrogen is blown through the side arm. The flask is charged with 200 mL of methanol and hydrated nickel bis(tetrafluoroborate)* (5.85 g, 17.2 mmole). The solution is stirred and trimethyl phosphite (25 g, 0.20 mole) is added dropwise. The color of the solution changes from green to red-orange and the solution is filtered after several minutes. A solution of sodium tetraphenylborate (11.8 g, 34.4 mmole) in methanol (50 mL) is added slowly, causing the precipitation of a fine yellow powder. The powder is collected on a medium fritted funnel and dried *in vacuo* at room temperature overnight. The yield is 20.4 g, 90% of theory, and the melting point is 137°. *Anal.* Calcd. for $C_{63}H_{85}B_2NiO_{15}P_5$: C, 57.4; H, 6.50; Ni, 4.5; P, 11.8. Found: C, 57.3; H, 6.43; Ni, 4.0; P, 12.2.

Properties

The yellow powder is fairly stable in air, but solutions are less stable. The compound is soluble in acetone and dichloromethane, insoluble in hydrocarbons, tetrahydrofuran and alcohols. The complex displays a $^{31}P\{^1H\}$ nmr spectrum that is a single line in fast exchange and an A_2B_3 spin system in slow exchange.[4a]

B. PENTAKIS(TRIMETHYL PHOSPHITE)PALLADIUM(II) BIS(TETRAPHENYLBORATE)

$$PdCl_2 + 5P(OCH_3)_3 + 2Na[BPh_4] \longrightarrow [Pd[P(OCH_3)_3]_5][BPh_4]_2 + 2\,NaCl$$

Procedure

The reaction is carried out in a hood at room temperature. A 125-mL filter flask is equipped with a stirring bar coated with Teflon, and a slow stream of nitrogen is blown through the side arm. The flask is charged with 50 mL of methanol. Palladium dichloride (0.88 g, 5.0 mmole) is added, followed by trimethyl phosphite (6.2 g, 5.0 mmole). Stirring is continued until all the solids have dissolved. The yellow solution is then filtered and a methanol (20 mL) solution of sodium tetraphenylborate (3.42 g, 10.0 mmole) is added dropwise, precipitating a fine white powder. Stirring is continued for an additional 10 minutes before the product is collected on a medium fritted funnel. The product is then dried *in vacuo* overnight. The yield is 6.4 g, 90% of theory. The melting point is 162°. *Anal.* Calcd. for $C_{63}H_{85}B_2P_5O_{15}Pd$: C, 55.43; H, 6.28; P, 11.30; Pd, 7.8. Found: C, 55.17; H, 6.09; P, 10.8; Pd, 7.9. The same procedure can be

---

*Available from Alfa Products, Ventron Corp., P.O. Box 299, Danvers, MA 01923. The more readily available compound $Ni(NO_3)_2 \cdot 6H_2O$ (5.0 g) can be used, but the results are not as reliable.

used to prepare the platinum analogue, starting with platinum dichloride (1.32 g, 5.0 mmole) in place of the palladium dichloride. All other quantities are held constant. The yield is 6.6 g, 90% of theory. The melting point is 199°. *Anal.* Calcd. for $C_{63}H_{85}B_2P_5O_{15}Pt$: C, 52.04; H, 5.87; P, 10.7; Pt, 13.4. Found: C, 52.43; H, 6.09; P, 10.9; Pt, 13.1.

Properties

The platinum complex is reasonably air stable, but the palladium complex is moderately air sensitive, turning to a black powder overnight if stored in air. Their solubilities are similar to those of the nickel analogue. There is a tendency for both the palladium and platinum complexes to lose a phosphite ligand on recrystallization.

$$[ML_5]^{2+} \rightleftharpoons [ML_4]^{2+} + L \quad [M = Pd, Pt; L = P(OMe)_3]$$

Recrystallization in the presence of a slight excess of phosphite can make purification more straightforward.

The complex displays a $^{31}P\{^1H\}$ nmr spectrum that is a single line in fast exchange and an A_2B_3 spin system in slow exchange.[4a]

C. PENTAKIS(TRIMETHYL PHOSPHITE)RHODIUM(I) TETRAPHENYLBORATE

$$[(C_2H_4)_2RhCl]_2 + 10P(OCH_3)_3 + 2Na[BPh_4] \longrightarrow$$
$$2[Rh[P(OCH_3)_3]_5][BPh_4] + 2NaCl + 4C_2H_4$$

Procedure

The reaction is carried out under an inert atmosphere, using standard techniques.[5] A 125-mL flask equipped with a magnetic stirring bar coated with Teflon is charged with 50 mL of methanol and di-μ-chloro-tetrakis(ethylene)-dirhodium(0)[6] (1.16 g; 3.6 mmole). Trimethyl phosphite (5.00 g, 40 mmole) is added dropwise over a period of 5 minutes. Upon completion of the addition, stirring is continued for 30 minutes until all the starting material is in solution.

Sodium tetraphenylborate (2.50 g, 7.0 mmole) is dissolved in 20 mL of methanol, and this solution is added dropwise to the yellow solution to precipitate white solids. The solids are collected on a medium-porosity frit and dried *in vacuo.*

The crude product is dissolved in a minimum amount of dry dichloromethane and filtered through a medium frit. Methanol is added slowly to the stirred solution to precipitate white crystals. The final product is collected on a medium frit, washed with 60 mL of methanol, and dried *in vacuo* for 3 hours at room temperature. The yield of recrystallized product is 4.00 g, 64% of theory. The melting point is 199–200°. *Anal.* Calcd. for $C_{39}H_{65}BP_5O_{15}Rh$: C, 44.93; H, 6.28; P, 14.85. Found: C, 44.27; H, 6.10; P, 15.33. The iridium analog can be prepared using a similar technique but starting with di-μ-chlorochlorotetrakis-(cyclooctene)diiridium[7] (2.80 g, 3.00 mmole). It is seldom necessary to recrystallize this material. The yield is 5.77 g, 85% of theory. The melting point is 214–216°. *Anal.* Calcd. for $C_{39}H_{65}BP_5O_{15}Ir$: C, 41.39; H, 5.79; O, 21.19. Found: C, 41.72; H, 5.78; O, 21.06.

Properties

The complexes are soluble in tetrahydrofuran, chlorinated solvents, acetone, and dimethylformamide. They are insoluble in diethyl ether, hydrocarbons, and alcohols.

The rhodium complex displays a $^{31}P\{^1H\}$ nmr spectrum that is a doublet in fast exchange and an A_2B_3X spin system in slow exchange. The iridium complex displays a single line in fast exchange and an A_2B_3 spin system in slow exchange.[4a,4b]

D. PENTAKIS(TRIMETHYL PHOSPHITE)IRON(0)

$$FeCl_2 + 5P(OCH_3)_3 + 2Na \xrightarrow[Hg]{THF} Fe[P(OCH_3)_3]_5 + 2NaCl$$

Procedure

The entire procedure is carried out in an inert atmosphere using standard techniques.[5] A 1-L, three-necked flask equipped with mechanical stirrer and nitrogen flush is charged with anhydrous $FeCl_2$ (7.61g; 60 mmole), tetrahydrofuran (500 mL), and $P(OCH_3)_3$ (62 g, 500 mmole). Sodium amalgam is prepared in an inert atmosphere by adding chunks of sodium, the size of split peas, one at a time to the mercury under a 3-cm layer of pentane. The pentane, which moderates the highly exothermic reaction by boiling, is then decanted before the amalgam is used. The sodium amalgam (Na, 2.8 g, 120 mmole in 60 mL Hg) is added after the mixture has been cooled to −40°. The yellow-orange mixture quickly turns brown. Stirring is continued at −40 to −10° for 4 hours. The spent amalgam is removed by means of a separatory funnel. The brown suspension is filtered through Celite filter aid, and the filtrate is stripped to dryness

80 Coordination Compounds

on a vacuum line at room temperature. The dark solids are extracted with several 50-mL portions of pentane. The brown pentane solution is then chromatographed on a short neutral grade alumina column (5-cm column, 6-cm diameter). The column is eluted with pentane until the solution is colorless or the dark band on the column approaches the bottom. The yellow solution is evaporated to dryness, yielding Fe[P(OCH$_3$)$_3$]$_5$ (yields as high as 40%). Recrystallization is difficult because of the high solubility of the complex in all solvents. However, the unrecrystallized material is pure by ^{31}P nmr spectroscopy. The material can be sublimed onto a liquid nitrogen cold finger under high vacuum with loss in yield. The decomposition temperature is dependent on the rate of heating but may be as high as 160°. *Anal.* Calcd. for C$_{15}$H$_{45}$P$_5$O$_{15}$Fe: Fe, 8.3; P, 22.9; O, 35.5; C, 26.7; H, 6.7. Found: Fe, 9.0; P, 24.3; O, 33.1; C, 27.0; H. 6.8.

Properties

The yellow crystalline material is very air sensitive. When stored for long periods of time, it should be kept cold, or it will darken and become tacky. The compound is very soluble in hydrocarbon and ether solvents and is slightly protonated by alcohols. It decomposes when exposed to halocarbon solvents.

The complex displays a ^{31}P{^{1}H} nmr spectrum that is a single line in fast exchange and an A$_2$B$_3$ spin system in slow exchange.[4]

E. PENTAKIS(TRIMETHYL PHOSPHITE)RUTHENIUM(O)

$$\text{RuCl}_2[\text{P(OCH}_3)_3]_4 + \text{P(OCH}_3)_3 + 2\text{Na} \xrightarrow[\text{Hg}]{\text{THF}} \text{Ru}[\text{P(OCH}_3)_3]_5 + 2\text{NaCl}$$

The entire procedure is carried out under an inert atmosphere using standard techniques.[5] A three-necked, 100-mL flask is equipped with a stirring bar coated with Teflon and a nitrogen flush. The flask is charged with RuCl$_2$-[P(OCH$_3$)$_3$]$_4$[8] (2.0 g, 3.0 mmole), THF (40 mL), and P(OCH$_3$)$_3$ (4 g, 32 mmole). It is then cooled to −30° and an amalgam (Na, 0.14 g; 6 mmole in 25 mL Hg) is added. The clear yellow solution becomes turbid as stirring is continued for 3 hours. The spent amalgam is removed and the suspension is stripped under vacuum to a milky, glasslike solid. The solids are extracted with pentane and the solution is filtered through 0.5 cm of neutral grade alumina before it is evaporated to dryness yielding a white crystalline solid. The yield is 1.84 g, 85% of theory. The melting point is 190°. *Anal.* Calcd. for C$_{15}$H$_{45}$P$_5$O$_{15}$Ru: C, 25.0; H, 6.29; P, 21.5; O, 33.3. Found: C, 25.2; H, 6.13; P, 21.0; O, 32.1.

Properties

The white cyrstalline solid is very air sensitive and should be stored in sealed glass tubes. It is soluble in most solvents and is protonated by alcohols, giving [HRu[P(OCH$_3$)$_3$]$_5$]$^+$, which can be isolated as the tetraphenylborate salt.

The complex displays a ^{31}P{^{1}H} nmr spectrum that is a single line in fast exchange and an A$_2$B$_3$ spin system in slow exchange.[4c]

F. PENTAKIS(TRIMETHYL PHOSPHITE)COBALT(I) TETRAPHENYLBORATE[9]

$$2Co(BF_4)_2 \cdot 6H_2O + 11P(OCH_3)_3 \longrightarrow [Co[P(OCH_3)_3]_5][BF_4]$$
$$+ [Co[P(OCH_3)_3]_6][BF_4]_3$$

$$[Co[P(OCH_3)_3]_5]BF_4 + Na[BPh_4] \longrightarrow [Co[P(OCH_3)_3]_5][BPh_4]$$
$$+ Na[BF_4]$$

Submitted by J. G. VERKADE* and L. W. YARBROUGH II*
Checked by S. D. ITTEL† and M. A. CUSHING†

Procedure

In an inert atmosphere, a solution of Co[BF$_4$]$_2 \cdot$6H$_2$O (3.4 g, 10 mmole) in 2,2-dimethoxypropane (25 mL) and acetone (25 mL) is added to stirred trimethyl phosphite (80 mL) in a 500-ml flask. The reaction is slightly exothermic. After 2-hr, the suspension is filtered to remove the less soluble [Co[P(OCH$_3$)$_3$]$_6$]-[BF$_4$]$_3$. The yellow filtrate is slowly added to a dry diethyl ether (400 mL), precipitating crude [Co[P(OCH$_3$)$_3$]$_5$][BF$_4$]. This crude material is collected by vacuum filtration and dried under vacuum. It is then redissolved in dichloromethanel (20 mL), filtered, and reprecipitated by addition of diethyl ether (200 mL). The filtered and dried product is taken up in methanol (20 mL) and precipitated as the tetraphenylborate salt by dropwise addition of a solution of Na[BPh$_4$] (1.4 g, 4.0 mmole) in methanol (10 mL). The yield is 3.59 g, 70% of theory based on conversion to Co(I). The melting point is 202–203° (dec.). *Anal.* Calcd. for C$_{39}$H$_{65}$BP$_5$O$_{15}$Co: C, 46.92; H, 6.56; Co, 5.90; P, 15.51. Found: C, 47.50; H, 6.52; Co, 5.61; P, 15.66.

*Department of Chemistry, Iowa State University, Ames, IA 50010.
†Central Research and Development Department, E. I. du Pont de Nemours and Company, Wilmington, DE 19898, Contribution No. 2369.

Properties

The complex is soluble in tetrahydrofuran, chlorinated solvents, acetone, and dimethylformamide. It is insoluble in diethyl ether, hydrocarbons, and alcohols.

The complex displays a $^{31}P\{^1H\}$ nmr spectrum that is a single line in fast exchange and an A_2B_3 spin system in slow exchange.[4a]

References

1. B. A. Frenz and J. A. Ibers, *M.T.P. Int. Rev. Sci. Phys. Chem.*, Ser. *1*, **11**, 33 (1972).
2. E. L. Muetterties and J. P. Jesson in *Dynamic Nuclear Magnetic Resonance Spectroscopy*, L. M. Jackman and F. A. Colton, Eds., Academic, New York, 1975, Chap. 8, p. 253.
3. J. P. Jesson, *M.T.P. Int. Rev. Sci., Inorg. Ser. 2*, **9**, 227 (1974).
4(a). P. Meakin and J. P. Jesson, *J. Am. Chem. Soc.*, **96**, 5751 (1974). (b) J. P. Jesson and P. Meakin, *J. Am. Chem. Soc.*, **96**, 5760 (1974). (c) A. D. English, S. D. Ittel, C. A. Tolman, P. Meakin, and J. P. Jesson, *J. Am. Chem. Soc.*, **99**, 117 (1977).
5. D. F. Shriver, *The Manipulation of Air-Sensitive Compounds,* McGraw-Hill, New York, 1969.
6. R. Cramer, *Inorg. Synth.*, **15**, 14 (1974).
7. J. L. Herde, J. C. Lambert, and C. V. Senoff, *Inorg. Synth.*, **15**, 18 (1974).
8. W. G. Peet and D. H. Gerlach, *Inorg. Synth.*, **15**, 40 (1974) reports the preparation of $Ru[P(OEt)_3]_4Cl_2$. The trimethyl phosphite analogue used here can be prepared by the same technique.
9. J. C. Verkade, *Coord. Chem. Rev.*, **9**, 1 (1972). Originally reported as the perchlorate and nitrate salts.

23. COMPLEXES OF ALUMINUM IODIDE WITH PYRIDINE AND RELATED BASES

Submitted by F. J. ARNÁIZ GARCÍA*
Checked by G. D. ELLIS† and E. P. SCHRAM†

Complexes of aluminum iodide with pyridine (1:1 and 1:3) have been described as being formed by direct reaction of the two compounds.[1] However, the limited availability of high-purity AlI_3 has hampered the development of the coordination chemistry of this compound. The synthesis described here makes use of an AlI_3 ethereal solution, prepared *in situ*, as the starting material, and it is based on the reaction:

$$AlI_3 \cdot B + 3B' \longrightarrow AlI_3 \cdot B'_3 + B$$

*Departmento de Química Inorgánica, Colegio Universitario, Burgos, Spain.
†Department of Chemistry, The Ohio State University, Columbus, OH 43210.

Similar compounds of 3- and 4-methylpyridine can be prepared with the same stoichiometry and yield.

A. PREPARATION OF TRIIODOTRIS(PYRIDINE)ALUMINUM

$$Al + \tfrac{3}{2}I_2 \xrightarrow{(C_2H_5)_2O} AlI_3 \cdot (C_2H_5)_2O$$

$$AlI_3 \cdot (C_2H_5)_2O + 3C_5H_5N \xrightarrow{(C_2H_5)_2O} AlI_3 \cdot 3C_5H_5N + (C_2H_5)_2O$$

Procedure

■ **Caution.** *If very finely powdered aluminum is used, the reaction may be uncontrollable. If the diethyl ether or the pyridine are not quite dry, the white product will slowly turn yellow in vacuo. Pyridine is toxic and its odor is unpleasant; all operations involving it should be conducted in a fume hood.*

A 500-mL, two-necked flask, fitted with a nitrogen inlet and outlet and a reflux condenser, is flushed with nitrogen and then charged with anhydrous diethyl ether (200 mL), pure aluminum turnings (3 g, 0.11 mole), and dry iodine (20 g, 0.078 mole). The mixture is boiled under reflux until the solution turns colorless (1–2 hr), while a slow stream of nitrogen is passed through it. The mixture is then filtered under nitrogen.

The solution is stirred magnetically while 13 mL of freshly distilled pyridine is added dropwise. Heat is evolved and precipitation occurs quickly. The white precipitate is separated by filtration, washed twice with diethyl ether, and dried *in vacuo*. All operations are carried out in an inert atmosphere.

Yield: 32 g, based on iodine (95%). *Anal.* Calcd. for $AlI_3 \cdot 3C_5H_5N$: Al, 4.18; I, 59.04; C_5H_5N, 36.78. Found: Al, 4.1; I, 58.7; C_5H_5N, 36.5.

Properties

The complex is obtained as a white microcrystalline solid that melts at 195–198° with decomposition. It is insoluble in most organic solvents but rather soluble in methanol. The product is moisture sensitive but remains unchanged after 2 years when stored in a tightly stoppered container in darkness. The infrared spectrum, taken as CsI pellet, has bands at 1630 (s), 1598 (vs), 1524 (s), 1478 (s), 1364 (w), 1330 (m), 1246 (m), 1238 (s), 1186 (m), 1152 (m), 1050 (s), 1026 (m), 988 (m), 740 (s), and 672 (vs) cm^{-1}.

Reference

1. J. W. Wilson and I. J. Worrall, *Inorg. Nucl. Chem. Lett.*, **3**, 57 (1967).

Chapter Four

COMPLEXES WITH COMPLICATED CHELATE LIGANDS

24. CAGED METAL IONS: COBALT SEPULCHRATES

Submitted by J. MacB. HARROWFIELD,* A. J. HERLT,* and A. M. SARGESON*
Checked by THEODORE Del DONNO†

The organic synthesis of cryptate-like molecules is usually a fairly complex matter.[1,2] The problem arises in part from the size of the ring systems and the large unfavorable entropy terms that limit the ring closure reactions. The problem can be remedied by reducing the syntheses to coupled small ring closures, that is, five- or six-membered systems. This synthesis is an example of how a large cage 1,3,6,8,10,13,16,19-octaazabicyclo[6.6.6]-eicosane (bicyclo[6.6.6]-

*Research School of Chemistry, The Australian National University, Canberra, 2600, Australia.
†Chemistry Department, The Ohio State University, Columbus, OH 43210.

ane-1,3,6,8,10,13,16,19-N_8) can be made around a metal ion in a rather simple manner using tris(ethylenediamine)cobalt(III) ion, ammonia, and formaldehyde in a basic medium.[2]

A. [Co(bicyclo[6.6.6]ane-1,3,6,8,10,13,16,19-N_8)]Cl_3

Procedure

To a stirred solution of Li_2CO_3 (25 g) in a solution of (±)-[Co(en)$_3$]Cl_3 (9.1 g, 0.025 mole) in water (125 mL) is added aqueous ammonia (166 mL, 2.5 mole) diluted to 502 mL and aqueous formaldehyde (592 mL, 38%, 7.5 mole). The solutions are added separately and dropwise over 2 hours using a peristaltic pump. The mixture is stirred for another 30 minutes, the Li_2CO_3 is filtered off, and the pH of the filtrate is adjusted to ~3 with 12 M HCl. The solution is diluted to ~8 L and sorbed on an ion exchange column (Dowex 50-WX2, 200-400 mesh, H^+ form, 5 × 10 cm). The column is eluted with Na_3citrate (5 L of 0.2 M) to remove a pink species, which is discarded. The resin bed is then washed with H_2O (1 L) and 1 M HCl (1 L) to remove Na^+ and the orange species is removed from the column by eluting with HCl (3 M, ~3 L). The sepulchrate ion elutes before Co(en)$_3^{3+}$ on SP-Sephadex C-25 using 0.3 M NaCl or $NaClO_4$ and on Dowex 50W-X2, 200-400 mesh using 0.5 M K_2HPO_4. The eluate is taken to dryness on a vacuum evaporator at 50°. The isolated compound is recrystallized by dissolving it in water (~90°) and adding acetone dropwise, while it is cooled in an ice bath with stirring. Yield: 16.7 g, 74%. *Anal.* Calcd. for $CoC_{12}H_{30}N_8Cl_3$: Co, 13.05; C, 31.91; H, 6.69; N, 24.18. Found: Co, 12.85; C, 31.8; H, 7.2; N, 24.6.

Properties

For the cobalt(III) sepulchrate, $\epsilon_{max}^{472} = 109\ M^{-1}\ cm^{-1}$, $\epsilon_{max}^{340} = 116\ M^{-1}\ cm^{-1}$. ^{1}H nmr spectra ($D_2O$): δ4.0 ppm, 12 protons, AB doublet pair, *J* about 12 Hz; δ3.2 ppm, complex AA'BB' pattern, 12 protons. Also the ^{13}C nmr spectrum (D_2O) shows two equal signals at δ -0.389 and +13.245 ppm versus dioxane. The optically active forms of the cobalt(III) sepulchrate have been prepared similarly starting with Δ-[Co(en)$_3$]Cl_3 and Λ-[Co(en)$_3$]Cl_3, respectively. The cobalt(III) sepulchrate was reduced to the cobalt(II) derivative with Zn dust in H_2O under N_2 and isolated as [Co(bicyclo[6.6.6]ane-1,3,6,8,10,13,16,19-N_8)]$ZnCl_4 \cdot H_2O$.

References

1. J. M. Lehn, *Struct. Bonding (Berl.)*, **16**, 1 (1973).
2. I. I. Creaser, J. MacB. Harrowfield, A. J. Herlt, A. M. Sargeson, J. Springborg, R. J. Geue, and M. R. Snow, *J. Am. Chem. Soc.*, **99**, 3181 (1977).

25. CLATHROCHELATES FORMED BY THE REACTION OF TRIS(2,3-BUTANEDIONE DIHYDRAZONE)-METAL COMPLEXES WITH FORMALDEHYDE: [5,6,14,15,20,21-HEXAMETHYL-1,3,4,7,8,10, 12,13,16,17,19,22-DODECAAZATETRACYCLO-[8.8.4.1^{3,17}.1^{8,12}]TETRACOSA-4,6,13,15,19,21-HEXAENE-N^4, N^7, N^{13}, N^{16}, N^{19}, N^{22}]METAL(n+) TETRAFLUOROBORATE

$$3C_4H_{10}N_4 + M[BF_4]_2 \cdot 6H_2O \longrightarrow [M(C_4H_{10}N_4)_3][BF_4]_2$$

$$[M(C_4H_{10}N_4)_3][BF_4]_2 + 6H_2CO \xrightarrow{CH_3CN} [M(C_{18}H_{30}N_{12})][BF_4]_2$$

$$M = Fe(II), Ni(II), Co(II)$$

Submitted by V. L. GOEDKEN*
Checked by ERIC V. DOSE†

The possibility of completely encapsulating a metal ion in a three-dimensional ligand was first alluded to by Busch.[1] In recent years, a number of such metal complexes, dubbed clathrochelates, cryptates, or sepulchrates, have been synthesized and characterized.[2-5] Generally, with transition metals, encapsulation can be accomplished by means of metal-directed template syntheses. Once formed, the metal is functionally inert toward donor-atom displacement, since this can be accomplished only by breaking N—C or C—C bonds, an energetically prohibitive process. Encapsulation of the metal leads to complete insulation of the metal from the immediate environment of the solvent and enables studies involving electron transfer by means of outer-sphere mechanisms of metal complexes that are normally substitutionally labile; such studies are not possible with analogous noncyclic structures. The clathrochelates (Fig. 1) formed by the condensation reaction of formaldehyde with tris(2,3-butanedione dihydrazone) metal complexes are easily prepared, are exceedingly robust when the metals are in their most stable oxidation state, and are suitable for a variety of studies involving unusual coordination geometries or electrochemical studies.

Procedure

2,3-Butanedione dihydrazone was synthesized by a modification of the procedure described by Busch and Bailar.[6] A solution of hydrazine hydrate, 27 g

---

*Department of Chemistry, Florida State University, Tallahassee, FL 32306.
†Department of Chemistry, The Ohio State University, Columbus, OH 43210.

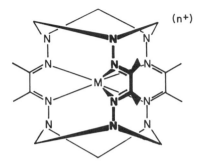

Fig. 1. *Structure of the clathrochelates.*

(0.8 mole, 2.7 equiv) in 300 mL of ethanol is brought to gentle boil with stirring in a large-necked Erlenmeyer flask in a hood. 2,3-Butanedione, 26 g (0.3 mole) is added dropwise over 1 hour. The solution turns dark yellow or orange. Boiling is continued for 10 minutes after addition is complete. The flask is then removed from the hot plate-stirrer, wrapped in a towel, and allowed to cool slowly to room temperature. Large, pure white crystals of 2,3-butanedione dihydrazone are filtered off, washed with ethanol and vacuum dried. The filtrate is refrigerated and a second crop of crystals is recovered. The combined yield is 85%.

Tris(2,3-butanedione dihydrazone)metal bis(tetrafluoroborate) complexes, $[M(C_4H_{10}N_4)_3][BF_4]_2$, are prepared by dissolving 2.00 g (0.0175 mole) of 2,3-butanedione dihydrazone in 100 mL of hot absolute ethanol and adding a solution of 2.15 g (0.00586 mole) of the appropriate metal(II) tetrafluoroborate hexahydrate salt (or 41% aqueous solution in the case of commercially available $Fe(BF_4)_2$) in 30 mL of absolute ethanol. After the solution is cooled, the product is filtered on a Büchner funnel, washed with absolute ethanol, and dried *in vacuo*. Yields range from 70 to 95%. All these complexes, except those of iron(II), slowly decompose when stored under ordinary conditions (i.e., at room temperature, bottled in air) and should be used within a week of preparation.

The clathrochelate complexes, $[M(C_{18}H_{30}N_{12})][BF_4]_2$, are prepared by adding 5.0 g of 37% aqueous formaldehyde solution to a suspension of 5 g of the appropriate tris(2,3-butanedione dihydrazone)metal bis(tetrafluoroborate) complex in acetonitrile. The condensation reaction is catalyzed by the addition of 0.5 mL of concentrated tetrafluoroboric acid. An immediate darkening of the solutions is observed as the condensation reaction proceeds. The iron(II)-containing solution becomes a very intense purple; the nickel(II) solution becomes very dark olive-brown. Precipitation of the clathrochelates requires the addition of approximately 30% by volume of diethyl ether and refrigeration. The products are filtered from the solution, washed with methanol, and air dried. The iron(II) complex is somewhat more soluble in acetonitrile than the other complexes, and increased yields can be obtained by the dropwise addition of more diethyl ether. Yields are 34% for the nickel(II) complex and 37% for the iron(II) complex. *Anal.* Calcd. for [Fe($C_{18}H_{30}N_{12}$)][BF_4]$_2$: C, 33.6; H, 4.69; N, 26.10. Found: C, 33.4; H, 4.75; N, 25.5. Calcd. for [Ni($C_{18}H_{30}N_{12}$)][BF_4]$_2$: C, 33.4; H, 4.66; N, 26.1. Found: C, 33.1; H, 4.50; N, 25.6.

The cobalt(III) complex, [Co($C_{18}H_{30}N_{12}$)][BF_4]$_3$, is obtained in 25% yield by a procedure similar to that described above, even though cobalt(II) is used as the starting material. Oxidation of cobalt(II) to cobalt(III) occurs even when the reaction is carried out under nitrogen. The Co(II) complex, [Co($C_{18}H_{30}N_{12}$)][BF_4]$_2$, is obtained by preparing a dispersion of 1.0 g of the Co(III) complex in 15 mL of acetonitrile and adding 20 mg (0.66 mole) of fresh, anhydrous hydrazine. More hydrazine may be slowly added, as necessary, to dissolve unreacted Co(III) complex. An excess of hydrazine must be avoided to prevent further reduction of the cobalt(II) to cobalt(I). Nitrogen is evolved and the solid Co(III) complex goes into solution as the dark-green Co(II) complex. The solution is filtered to remove any insoluble residues. Diethyl ether is added dropwise to the filtrate until a slight turbidity is observed. The solution is then refrigerated for 4 hr. The product is filtered from the solution, washed with diethyl ether and dried *in vacuo*; the yield is about 40%. *Anal.* Calcd. for [Co($C_{18}H_{30}N_{12}$)][BF_4]$_3$: C, 29.5; H, 4.11; N, 22.9. Found: C, 29.1; H, 4.15; N, 22.7. Calcd. for [Co($C_{18}H_{30}N_{12}$)][BF_4]$_2$: C, 33.4; H, 4.66; N, 26.1. Found: C, 33.7; H, 4.70; N, 25.8.

Properties

The iron(II), nickel(II), and cobalt(III) complexes may be stored indefinitely under normal atmospheric conditions. The complexes are soluble in acetonitrile, but only sparingly soluble to insoluble in other solvents. The complexes are very robust toward attack by protic acids; strong acids such as hydrogen tetrafluoroborate may be added to solutions without complex decomposition. The complexes exhibit very simple nmr spectra, with the methyl groups appearing as singlets near 2.50 ppm and the methylene groups appearing as *ab* patterns centered near 3.71 ppm with J = 13 Hz. The nickel(II) complex has a triplet ground state, μ_{eff} = 3.15 BM. The cobalt(II) complex is low spin, μ_{eff} = 2.13 BM. The electrochemical behavior is interesting; each complex displays a number of reversible oxidation and reduction waves. It has been shown by esr investigations that these redox processes may be either metal- or ligand-centered processes.[7]

The perchlorate salts of these clathrochelates may be prepared according to the same procedure outlined for the tetrafluoroborate salts by substituting the appropriate metal perchlorate salts and perchloric acid. They have similar reactivity properties but are less soluble than their BF_4 counterparts. They must be regarded as treacherous and must be treated with the same respect as all other perchlorate salts containing oxidizable matter, that is, they should be prepared only in small quantities using safety shields. Neither concentrated solutions nor the solids should be heated.

References

1. D. H. Busch, *Rec. Chem. Prog.*, **25**, 107 (1964).
2. D. R. Boston and N. J. Rose, *J. Am. Chem. Soc.*, **95**, 4163 (1973).
3. J. E. Parks, B. E. Wagner, and R. H. Holm, *J. Am. Chem. Soc.*, **92**, 3500 (1970).
4. V. L. Goedkin and S. M. Peng, *Chem. Commun.*, **1973**, 62.
5. I. I. Creager, J. MacB. Harrowfield, A. J. Herlt, A. M. Sargeson, J. Springborg, R. J. Gene, and M. R. Snow, *J. Am. Chem. Soc.*, **99**, 3181 (1977).
6. D. H. Busch and J. Bailar, *J. Am. Chem. Soc.*, **78**, 1137 (1956).
7. J. Palmer and V. L. Goedkin, to be published.

26. COMPARTMENTAL LIGANDS

Submitted by DAVID E. FENTON,* STEPHEN E. GAYDA,* and CATRIONA M. REGAN*
Checked by M. A. CUSHING, JR.† and S. D. ITTEL†

There has been considerable interest in the development of ligands derived from triketones and α,ω-diamines.[1-5] Such ligands are capable of binucleation by provision of compartments consisting of $-N_2O_2$ and $-O_2O_2$ donor sets. The availability of such ligand compartments affords the opportunity to study the chemical and spectral properties of metals held in close proximity. These metals may be different (heterobinuclear) or similar (homobinuclear). In the latter case, it is possible to have present two ions of the same metal in different geometries, spin states, and oxidation numbers. Both types of complex have been shown to be of value in studying magnetic interactions between adjacent metal ions.[6-8] The dicopper complexes are of interest as potential models for copper oxidases.

The macrocyclic Schiff base 5,9,14,18-tetramethyl-1,4,10,13-tetraazacyclo-octadeca-5,8,14,17-tetraene-7,16-dione (I) (H_4daen) can serve as a precursor to the formation of either mononuclear macrocyclic complexes or mononuclear acyclic complexes, depending on the reaction conditions. The mononuclear complexes may then be used in the synthesis of homo- or heterobinuclear complexes. The synthetic procedures described have been extended to include Schiff bases derived from a range of symmetrical and nonsymmetrical tri-

*Department of Chemistry, The University, Sheffield S3 7HF, United Kingdom.
†Central Research and Development Department, Experimental Station, E.I. du Pont de Nemours and Company, Inc., Wilmington, DE 19898.

ketones.[7,9] Mononuclear nickel(II) and cobalt(II) derivatives of H_4daen have been synthesized[2,5,9,10] and ring-closure reactions have been effected on acyclic Schiff base complexes, such that Cu(H_2daen) and related macrocyclic species may be isolated.[8] Numerous homo- and heterobinuclear complexes of the acyclic Schiff bases have been reported (see References 3, 4, 7–10).

A. 5,9,14,18-TETRAMETHYL-1,4,10,13-TETRAAZACYCLOOCTADECA-5,8,14,17-TETRAENE-7,16-DIONE, H_4daen (I)

$$2\ \text{CH}_3\text{-CO-CH}_2\text{-CO-CH}_2\text{-CO-CH}_3 + 2\text{NH}_2\text{CH}_2\text{CH}_2\text{NH}_2 \longrightarrow$$

(I)

Procedure

2,4,6-Heptanetrione[11] (7.1 g, 0.05 mole) dissolved in hot 95% ethanol (60 mL) is mixed with anhydrous ethylenediamine (3 g, 3.3 mL, 0.05 mole) also dissolved in 95% ethanol (10 mL). The solution is warmed on a steam bath for 5 minutes, after which pale-yellow crystals are precipitated. After the solution is cooled, the crystals are collected by filtration and dried under vacuum. Recrystallization may be effected from a methanol–dichloromethane mixture. The yield is 3.3 g (40%). Anal. Calcd. for $C_{18}H_{28}N_4O_2$: C, 65.0; H, 8.4; N, 16.8. Found: C, 64.7; H, 8.3; N, 16.8.

Properties

Compound (I), H_4daen, forms cream colored needles that are soluble in dichloromethane and chloroform and insoluble in most other organic solvents. The infrared spectrum contains *inter alia* bands at 1621 ($\nu_{C=O}$, hydrogen bonded), 1580, and 1520 cm^{-1} ($\nu_{C=C}$ and $\nu_{C=O}$). The ^{1}H nmr spectrum in chloroform-d has a CH_3 singlet at 1.84 ppm, a CH singlet at 4.73 ppm, a broad N—H····O resonance at 10.80 ppm, and a complex multiplet at 3.30 ppm for the CH_2 group. The parent peak P$^+$ at m/e = 332 amu is observed in the mass spectrum.

B. [5,9,14,18-TETRAMETHYL-1,4,10,13-TETRAAZACYCLOOCTADECA-5,8,14,17-TETRAENE-7,16-DIONATO(2-)-N^1, N^4, O^7, O^{16}]COPPER(II), Cu(H$_2$daen)(II)

Procedure

The ligand H$_4$daen (3.32 g, 0.01 mole) is dissolved in the minimum volume of hot dichloromethane and added to a solution of copper(II) acetate monohydrate (2.00 g, 0.01 mole) dissolved in the minimum volume of hot 95% ethanol. The resulting solution immediately becomes dark green, and after it is warmed on a steam bath a fine, shiny-brown microcrystalline solid precipitates. The mixture is allowed to cool and the product is collected by filtration. The complex is purified by suspending it in hot toluene and then filtering. This procedure is carried out three times. The pure complex is dried under vacuum. The yield is 2.3 g (60%). *Anal.* Calcd. for C$_{18}$H$_{26}$N$_4$O$_2$Cu: C, 54.8; H, 6.6; N, 14.2. Found: C, 54.6; H, 6.7; N, 14.3.

Properties

This brown microcrystalline solid has a magnetic moment μ_{eff} = 1.70 BM at 296 K. The infrared spectrum contains absorption bands at 3160 (ν_{NH}), 1630 (ν_{CO}), 1590 ($\nu_{CO} + \nu_{CC}$), and 1508 cm^{-1} ($\nu_{C=N}$). The mass spectrum exhibits a peak at m/e = 393 amu, corresponding to the parent peak P$^+$ based on ^{63}Cu.

The diffuse reflectance spectrum* gives bands at 623 (m), 446 (sh), 427 (s), and 378 (s) nm.

C. [[6,6'-(ETHYLENEDINITRILO)BIS(2,4-HEPTANEDIONATO)] (2-)-$N,N',O^4,O^{4'}$]COPPER(II), Cu(H$_2$daaen-N$_2$O$_2$) (III)

$$\text{(I)} + 2\text{Cu(CH}_3\text{COO)}_2 \cdot \text{H}_2\text{O} \xrightarrow[\text{CH}_2\text{Cl}_2]{\text{H}_2\text{O}} \text{(III)} + 2\text{CH}_3\text{COOH} + \text{NH}_2\text{CH}_2\text{CH}_2\text{NH}_2$$

Procedure

Copper(II) acetate monohydrate (0.6 g, 0.003 mole) dissolved in water (50 mL) is added to a solution of H$_4$daen (1.0 g, 0.003 mole) in dichloromethane (50 mL), and the two phases are intimately mixed by stirring for 1.5 hours at room temperature. The dichloromethane layer first becomes deep green and then purple. The mixture is poured into a separatory funnel and the lower dichloromethane layer is collected and evaporated under reduced pressure to leave a purple gum. This gum is dissolved in a small volume of chloroform and the resulting solution is eluted with chloroform down a column of neutral activated alumina (50 g). (■ **Caution**. *Chloroform is a suspected carcinogen. It should be handled in a well-ventilated hood, with inert gloves to avoid inhalation and skin contact*.) The purple fraction is collected, and removal of the solvent by evaporation under reduced pressure leaves a purple solid that is recrystallized

*m = medium, sh = shoulder, s = strong.

from a hexane-dichloromethane mixture to give a deep-purple microcrystalline solid in a yield of 0.3 g (20%). *Anal.* Calcd. for $C_{16}H_{22}N_2O_4Cu$: C, 52.0; H, 5.9; N, 7.3. Found: C, 51.8; H, 6.2; N, 7.5.

The aqueous layer in the above preparation may contain small quantities of macrocyclic $Cu(H_2daen)$ and the dicopper derivative of 2,4,6-heptanetrione, $Cu_2(C_7H_8O_3)_2$.

Properties

Compound (III), $Cu(H_2daaen-N_2O_2)$ is a purple solid giving bands in the diffuse reflectance spectrum at 549 (m), 379 (sh), and 345 (s) nm. It has a magnetic moment μ_{eff} = 1.76 BM at 296 K. The parent peak, P^+ m/e = 369 amu based on ^{63}Cu is observed in the mass spectrum. The infrared absorptions include bands at 1721, 1705 ($\nu_{noncoordinated\ C=O}$), 1591 ($\nu_{CO} + \nu_{CC}$), and 1518 cm^{-1} ($\nu_{C=N}$). The presence of infrared bands corresponding to free carbonyl groups affirms the compartmental occupancy.

D. μ-{[6,6'-(ETHYLENEDINITRILO)BIS(2,4-HEPTANEDIONATO)] (4-)-$N,N',O^4,O^{4'}:O^2,O^{2'},O^4,O^{4'}$} DICOPPER (II), Cu_2(daaen) (IV)

Procedure

Compound (III), $Cu(H_2daaen-N_2O_2)$ (1.0 g, 0.0027 mole), is dissolved in hot 95% ethanol (10 mL), and an ethanolic solution (10 mL) of copper(II) acetate monohydrate (0.54 g, 0.0027 mole) is added. The reaction mixture becomes dark green and a silky precipitate is deposited. After the mixture has cooled, the product is collected by filtration and dried under vacuum. The yield is 1.0 g (90%). *Anal.* Calcd. for $C_{16}H_{20}N_2O_4Cu_2$: C, 44.5; H, 4.6; N, 6.4. Found: C, 44.3; H, 4.8; N, 6.2.

Properties

Compound (IV), Cu_2(daaen), is a dark-green solid with bands in the diffuse reflectance spectrum at 667 (m) and 448 (s) nm. It has a reduced magnetic moment μ_{eff} = 0.62 BM per metal atom at 293 K. The highest peak in the mass spectrum at m/e = 430 amu corresponds to the parent peak P^+ based on 2 × ^{63}Cu. The infrared spectrum contains bands at 1564 ($\nu_{CO} + \nu_{CC}$) and 1509 cm^{-1} (ν_{CN}).

E. AQUA-μ-[[6,6'-(ETHYLENEDINITRILO)BIS-(2,4-HEPTANEDIONATO)](4-)-$N, N', O^4, O^{4'}: O^2, O^{2'}, O^4, O^{4'}$]-COPPER(II)OXOVANADIUM(IV), CuVO(daaen)(H_2O), (V)

Procedure

Compound (III), Cu(H_2daaen-N_2O_2) (1.0 g, 0.0027 mole), is dissolved in hot 95% ethanol (10 mL) and added to a stirred suspension of bis(acetato)oxovanadium(IV)[12] (0.5 g, 0.0027 mole) in hot 95% ethanol (100 mL). The mixture is heated at reflux with stirring for 2 hours. The deep-green solution is filtered hot to remove unreacted bis(acetato)oxovanadium(IV). On cooling the filtrate deposits a deep-green microcrystalline solid, which is collected by filtration and recrystallized from dichloromethane. The product is dried under vacuum. The yield is 0.59 g (50%). *Anal.* Calcd. for $C_{16}H_{20}N_2O_5CuVH_2O$: C, 42.4; H, 4.9; N, 6.2. Found: C, 42.8; H, 4.7; N, 6.1.

Properties

Compound (V) is a dark-green crystalline solid with a reduced magnetic moment μ_{eff} = 0.88 BM per metal ion at 296 K. The mass spectrum has a parent ion peak at m/e = 434 amu (^{63}Cu, ^{51}V). The infrared spectrum has absorbances at 3420 (H_2O), 1580 (ν_{CO}), 1560 ($\nu_{CO} + \nu_{CC}$), 1480 ($\nu_{C=N}$), and 988 cm^{-1} ($\nu_{V=O}$). The diffuse reflectance spectrum has bands at 621 (m) and 440 (s) nm. The assignment of compartmental occupancy of this compound has been confirmed by an X-ray crystal-structure determination.[13]

References

1. T. Yano, T. Ushijama, M. Sasaki, H. Kobayashi, and K. Ueno, *Bull. Chem. Soc. Jap.*, **42**, 2452, (1972).
2. D. E. Fenton and S. E. Gayda, *Chem. Commun.*, **1974**, 960.
3. D. E. Fenton and S. E. Gayda, *Inorg. Chim. Acta.*, **14**, L11, (1975).
4. B. Tomlonovic, R. L. Hough, M. D. Glick, and R. L. Lintvedt, *J. Am. Chem. Soc.*, **97**, 2925, (1975).
5. P. A. Vigato, M. Vidali, U. Casellato, R. Graziani, and F. Benetollo, *Inorg. Nucl. Chem. Lett.*, **11**, 595, (1975).
6. R. L. Lintvedt, M. D. Glick, B. K. Tomlonovic, D. P. Gavel, and J. M. Kuszaj, *Inorg. Chem.*, **15**, 1633, (1976).
7. R. L. Lintvedt, M. D. Glick, B. K. Tomlonovic, and D. P. Gavel, *Inorg. Chem.*, **15**, 1646, 1654, (1976).
8. M. D. Glick, R. L. Lintvedt, T. J. Anderson, and J. L. Mack, *Inorg. Chem.*, **15**, 2258, (1976).
9. D. E. Fenton and S. E. Gayda, *J. Chem. Soc., Dalton Trans.*, **1977**, 2095, 2101, 2109.
10. M. Vidali, U. Casellato, P. A. Vigato, and R. Graziani, *J. Inorg. Nucl. Chem.*, **38**, 1455 (1976).
11. J. R. Bethell and P. Maitland, *J. Chem. Soc.*, **1962**, 3751.
12. R. C. Paul, S. Bhatia, and A. Kumar, *Inorg. Synth.*, **13**, 181, (1972).
13. N. A. Bailey and C. A. Phillips, personal communication, 1977.

27. [[7,12:21,26-DIIMINO-19,14:28,33:35,5-TRINITRILO-5H-PENTABENZO[c, h, m, r, w] [1,6,11,16,21]-PENTAAZACYCLOPENTACOSINATO](2-)]-DIOXOURANIUM(VI)(URANYL SUPERPHTHALOCYANINE)[1]

Submitted by EDWARD A. CUELLAR,* DJORDJE R. STOJAKOVIC,* and TOBIN J. MARKS*
Checked by ALAN D. ADLER† and RONALD S. GIORDANO†

Template reactions[2] represent an extensive and important class of chemical transformations in which a metal ion serves as the hub for the coordinative cyclization of organic ligands. The method can lead either to enhanced yields over direct cyclizations or to the formation of otherwise inaccessible macrocycles. It is in this latter context that the uranyl ion acts, as a result of its large ionic radius and propensity toward pentagonal bipyramidal coordination, to form expanded macrocycles.

It has been known for some time that the reaction of uranyl dichloride with phthalonitrile (typical phthalocyanine condensation methodology[3]) yields an unusual phthalocyanine-like material;[4] however, it has been only recently that this product was unambiguously shown to contain an expanded, five-subunit analogue of the phthalocyanine ligand[5] (Eq. 1).

$$5 \begin{array}{c} \text{CN} \\ \text{CN} \end{array} + UO_2Cl_2 \longrightarrow \text{[macrocycle]} + \text{``}Cl_2\text{''} \quad (1)$$

● = UO_2

This "superphthalocyanine" (SPc) complex possesses a number of interesting chemical and physicochemical properties; these are discussed in detail in Reference 6. The synthesis of uranyl superphthalocyanine, $U(spc)O_2$, can be

*Department of Chemistry, Northwestern University, Evanston, IL 60201.
†Chemistry Department, Western Connecticut State College, Danbury, CT 06810.

98 *Complexes with Complicated Chelate Ligands*

conducted in a number of solvents and with a variety of uranyl salts; however, the following procedure using dry quinoline and anhydrous uranyl dichloride was found to be the most reproducible in terms of product yield and purity. This general procedure can also be applied to the synthesis of alkyl-substituted uranyl superphthalocyanines by employing alkyl-substituted phthalonitriles.[6a, 7]

Procedure

■ **Caution.** *Phthalonitrile, quinoline, UO_2Cl_2, and dibutyl phthalate are considered hazardous. Avoid skin contact and inhalation.*

In a glove box, a 25-mL Schenk tube is charged with 2.0 g (15.6 mmole) of phthalonitrile (Aldrich Chemical Co., Milwaukee, WI 53233, vacuum dried overnight at 10^{-3} torr prior to use) and 0.50 g (1.47 mmole) of anhydrous UO_2Cl_2[8] and fitted with a magnetic stirring bar. It is advantageous to mix the UO_2Cl_2 and phthalonitrile well at this point to prevent caking of the UO_2Cl_2 during the reaction. The reaction vessel is removed to a fume hood and connected to a nitrogen source so that 1.8 mL of dry quinoline can be introduced by syringe under a vigorous nitrogen flow. The quinoline is predried over KOH and vacuum distilled from BaO. A small forerun is discarded and the solvent is stored under a positive nitrogen pressure over Davison 4A molecular sieves in the absence of light.* Next, the reaction vessel is lowered into a stirred dibutyl phthalate bath whose oil temperature has been brought to 170° by a stirrer-hot plate or nichrome heating coil.† The reaction mixture is stirred vigorously. The upper parts of the Schlenk tube are sufficiently cool to condense the quinoline vapors without the use of a reflux condenser. The initial yellow suspension begins to darken 3-4 minutes after immersion and is a deep black-green within 5 minutes. The UO_2Cl_2 is soon completely dissolved and dark-violet crystals are apparent. The reaction is halted after 45 minutes, and the reaction mixture is filtered in a Büchner funnel while hot. There is no longer any need for a nitrogen atmosphere, the purpose of which is to exclude moisture. Water, larger volumes of quinoline, higher reaction temperatures, or longer reaction times decrease the product yield and increase the yield of phthalocyanine, PcH_2. Any material remaining in the reaction Schlenk tube is scraped out and washed into the Büchner funnel with methanol or acetone.

The combined black-purple crude product is next washed with methanol until the washings are essentially colorless. This procedure is repeated with acetone. Approximately 0.4 g of bright blue-violet crystalline product remains. This consists of a mixture of $U(spc)O_2$ and PcH_2 in a mole ratio of approxi-

*The checkers found it adequate to pass the quinoline over silica gel and 4A molecular sieves just prior to use.
†We found the reaction proceeds satisfactorily at 152° with essentially the same yield of $U(spc)O_2$ but less PcH_2 contamination. The checkers preferred a temperature of 170°.

mately 5:1 (by electronic spectrophotometry). The mixture is next transferred to a Soxhlet extractor (with a diatomaceous earth layer if the sintered-glass type is used) and is extracted for 18 hours with absolute ethanol to remove any remaining phthalonitrile and the thermal trimer of phthalonitrile, which forms as a side-product. Extraction is continued with benzene for 4 days, yielding 0.29 g (20%) of a dark-blue, microcrystalline product that is ⩾94% U(spc)O$_2$ by spectrophotometry (⩽6% PcH$_2$). A more nearly pure product can be obtained by extracting with benzene for shorter periods of time. Sublimation at 400° (10^{-3} torr) also gives a more nearly pure product; however, the yield is severely diminished. *Anal.* Calcd. for (C$_8$H$_4$N$_2$)$_5$UO$_2$: C, 52.72; H, 2.22; N, 15.37. Found: C, 52.54; H, 2.16; N, 15.21.

Properties

Uranyl superphthalocyanine is a dark-blue microcrystalline solid that becomes bright green when finely ground. It is practically insoluble in most common solvents and sparingly soluble in aromatic solvents (e.g., 1-chloronapthalene, benzene, toluene), giving bright-green solutions. Large crystals can be grown by slow evaporation of 1,2,4-trichlorobenzene solutions on a hot plate. The ^{1}H nmr spectrum of U(spc)O$_2$ in benzene-d_6 exhibits an AA'BB' pattern of two mirror-image multiplets (pseudoquartets) at δ 9.06 and δ 7.68. The infrared spectrum (Nujol mull or KBr pellet) exhibits a strong antisymmetric ν_{OUO} stretching transition at 925 cm^{-1}. Other infrared absorptions are observed at 1505 (ms), 1495 (sh), 1460 (m), 1410 (m), 1370 (w), 1330 (s), 1280 (m), 1180 (vw), 1165 (w), 1110 (m), 1070 (s), 1025 (s), 1013 (s), 940 (w), 897 (w), 865 (m), 765 (m), 715 (s), 700 (s), 660 (w), and 625 (w) cm^{-1}. The electronic spectrum of U(spc)O$_2$ in 1-chloronapthalene shows a strong, broad absorption at 914 nm (ε = 6.67 × 10^4) with a pronounced shoulder at about 810 nm, and a strong absorption at 424 nm (ε = 5.02 × 10^4). These provide both ready identification and an indication of purity, PcH$_2$ having intense absorptions in 1-chloronapthalene at 663 nm (ε = 1.51 × 10^5)[9] and 698 nm (ε = 1.62 × 10^5).[9] Attempts to dislodge the uranyl ion with acids or metal salts invariably result in demetalation accompanied by ring contraction to produce the corresponding phthalocyanine or metallophthalocyanine.[6]

References and Notes

1. The authors thank the Editor for correcting a mistake made previously by *Chemical Abstracts* in the formal nomenclature for this complex.
2. (a) L. F. Lindoy, *Chem. Soc. Rev.*, **4**, 421 (1975). (b) J. J. Christensen, D. J. Eatough, and R. M. Izatt, *Chem. Rev.*, **74**, 351 (1974). (c) D. St. C. Black and A. J. Hartshorn, *Coord. Chem. Rev.*, **9**, 219 (1973). (d) D. H. Busch, K. Farmery, V. Katovic, A. C. Melnyk, C. R. Sperati, and N. E. Tokel, *Adv. Chem. Ser.*, **100**, 44 (1971). (e) L. F.

Lindoy and D. H. Busch, *Prep. Inorg. React.*, **6**, 1 (1971). (f) N. F. Curtis, *Coord. Chem. Rev.*, **3**, 3 (1968). (g) D. H. Busch, *Helv. Chim. Acta. Fasc. Extraordinarius*, Alfred Werner Commemoration Volume, 1967.
3. (a) A. B. P. Lever, *Adv. Inorg. Chem. Radiochem.*, **7**, 27 (1965). (b) F. A. Moser and A. L. Thomas, *Phthalocyanine Compounds*, Reinhold, New York, 1963, pp. 104-141.
4. (a) J. E. Bloor, C. C. Walden, A. Demerdache, and J. Schlabitz, *Can. J. Chem.*, **42**, 2201 (1964). (b) F. Lux, *Proc., 10th Rare Earth Res. Conf., Carefree, Arizona, May 1973*, p. 871.
5. V. W. Day, T. J. Marks, and W. A. Wachter, *J. Am. Chem. Soc.*, **97**, 4519 (1975).
6. (a) T. J. Marks and D. R. Stojakovic, *J. Am. Chem. Soc.*, **100**, 1695 (1978). (b) T. J. Marks and D. R. Stojakovic, *Chem. Commun.*, **1975**, 28.
7. E. A. Cuellar and T. J. Marks, unpublished results.
8. Prepared after J. A. Leary and J. F. Suttle, *Inorg. Synth.*, **5**, 148 (1957). UO_2Cl_2 [Gallard-Schlesinger Chemical Corp., 548 Mineola Ave., Carle Place, New York, 11514, "minimum assay (ex U) 90%"] has been used successfully. We caution, though, that variable amounts of water present in most commercial UO_2Cl_2 may affect the reproducibility of this procedure.
9. F. A. Moser and A. L. Thomas, *Phthalocyanine Compounds*, Reinhold, New York, 1963, p. 33.

Chapter Five

COMPOUNDS OF BIOLOGICAL INTEREST

28. METALLOINTERCALATION REAGENTS: THIOLATO COMPLEXES OF (2, 2':6', 2"-TERPYRIDINE)PLATINUM(II)

Submitted by MARY HOWE-GRANT* AND STEPHEN J. LIPPARD*
Checked by PURUSH CHALILPOYIL† AND LUIGI G. MARZILLI†

The metallointercalation reagents are a class of heavy metal derivatives that bind to double-stranded polynucleotides by inserting between adjacent base pairs in the helix.[1,2] Prototype members of this class of intercalators are (2,2':6',2"-terpyridine)(thiolato)platinum(II) complexes.[3] These may be synthesized from chloro(2,2':6',2"- terpyridine)platinum(II), which can both intercalate and bind covalently by losing chloride ion. Covalent binding of the thiolato complexes is much slower owing to the more inert character of the Pt—S bond. Metallointercalation reagents also have the potential to bind to proteins that have natural receptor sites for nucleic acid bases. They may therefore also be used to provide isomorphous heavy atom derivatives for X-ray analysis.

The synthesis of chloro(2,2':6',2"-terpyridine)platinum(II) was first reported by Morgan and Burstall in 1934.[4] Better yields for this compound can be obtained by the method of Intille,[5] which is given below. The (2,2':6',2"-terpyridine)thiolato compounds were reported in 1974,[1] and a general synthetic procedure for these complexes was recently described.[3] In this method silver ions are introduced during an intermediate step. Since silver is known to bind

---

*Department of Chemistry, Columbia University, New York, NY 10027.
†Department of Chemistry, The Johns Hopkins University, Baltimore, MD 21218.

strongly to polynucleotides, a synthesis that avoids its use was developed. As is described here, this new procedure not only avoids the introduction of a possibly contaminating heavy metal, but also results in consistently higher yields and affords pure microcrystalline nitrate salts as products.

Synthesis of the chloro compound can take a few days, whereas the thiolato complexes can be prepared in a few hours.

- **Caution.** *2,2′:6′,2″-Terpyridine and its complexes can be absorbed through the skin. Since these compounds are toxic, gloves should be worn.*

A. CHLORO(2,2′:6′,2″-TERPYRIDINE)PLATINUM(II) CHLORIDE DIHYDRATE, [Pt(terpy)Cl]Cl·2H$_2$O

$$[\text{(terpy)PtCl structure}] \text{Cl·2H}_2\text{O}$$

$$K_2PtCl_4 + C_{15}H_{11}N_3 + 2H_2O \longrightarrow [(\text{terpy})PtCl]Cl \cdot 2H_2O + 2KCl$$

Procedure

Potassium tetrachloroplatinate(II) (4.15 g, 0.010 mole) is dissolved in 200 mL of deionized water in a 250-mL, round-bottomed flask equipped with a magnetic stirrer and condenser. 2,2′:6′,2″-Terpyridine (terpy), from the Sigma Chemical Company (2.77 g, 0.012 mole), is added and the stirred suspension is heated at reflux until a clear red solution is evident, a process that takes anywhere from 20 to 100 hours. Although refluxing may be terminated any time after the first 24 hours with no sacrifice of product purity, halting the reaction before the mixture clarifies significantly lowers the product yield. Prolonged heating of the clear solution, however, may result in product disproportionation, as evidenced by metallic platinum plating out onto the sides of the flask. The solution is filtered, any solid residue is set aside, and the filtrate is evaporated on a steam bath to a volume of around 20 mL. The red-orange trihydrate salt of the product precipitates upon cooling. The solid is collected by vacuum filtration and may be washed with dilute 0.1 N HCl and acetone.

The solid [Pt(terpy)Cl]Cl·3H$_2$O may be recrystallized from a minimum amount of hot 1:1 water–ethanol, giving long needlelike crystals. Drying in a vacuum desiccator gives the slightly more orange dihydrate [Pt(terpy)Cl]Cl·2H$_2$O. The typical yield is 65% (3.48 g) and is dependent on the purity of the starting materials, as well as the time allowed for refluxing. Higher yields may be obtained by combining any solid residue from the first filtration with the filtrate from the second, diluting with water, and refluxing until the solid material dissolves. Nonaqueous washings or the ethanol–water filtrate from the recrystallization must *not* be introduced into the refluxing medium. *Anal.* Calcd. for $C_{15}H_{15}N_3Cl_2O_2Pt$: C, 33.66; H, 2.82; N, 7.85; Cl, 13.25. Found: C, 33.95; H, 2.73; N, 7.74; Cl, 13.17.

Properties

Chloro(2,2′:6′,2″-terpyridine)platinum(II) chloride is a red-orange crystalline compound. It is very soluble in water, but only slightly soluble in polar organic solvents, such as acetone. The electronic absorption spectrum is dependent on the chloride ion concentration and may be used to determine purity. Spectral bands follow Beer's law up to a concentration of 15 μM. At neutral pH in 0.1 M NaCl the compound has the following absorption maxima and molar extinction coefficients (given in parentheses): 343 (11,300), 327 (12,600), 278 (25,100), and 248 (28,800) nm.

B. (2-MERCAPTOETHANOLATO-S)(2,2′:6′,2″-TERPYRIDINE)-PLATINUM(II) NITRATE, [Pt(terpy)(SCH$_2$CH$_2$OH)]NO$_3$

[Pt(terpy)Cl]$^+$ + HS(CH$_2$)$_2$OH + NO$_3^-$ + OH$^-$ ⟶
[Pt(terpy)(SCH$_2$CH$_2$OH)]NO$_3$ + H$_2$O + Cl$^-$

Procedure

A 0.535-g (1 mmole) portion of [Pt(terpy)Cl]Cl·2H$_2$O is dissolved in about 20 mL of deionized water in a 50-mL beaker, which is then covered. The red-orange solution is stirred and continually flushed with N$_2$ at room temperature. Pure* 2-mercaptoethanol (80 μL, 1 mmole) is added slowly, and the dark-red

---

*Formation of a brown precipitate or a color change in the solution from dark red to brown at any time after the addition of the 2-mercaptoethanol indicates decomposition products that presumably result from impure thiol. Unless the brown component can be successfully separated from the desired dark-red product before the addition of the saturated nitrate solution, the synthesis should be abandoned and the 2-mercaptoethanol should be purified by vacuum distillation and stored under N$_2$, where it may be kept in the cold indefinitely.

color of the product immediately becomes apparent. The pH of the solution is adjusted to ~6 with ~1 mL of 1 N NaOH and the solution is filtered in air through a medium glass frit. To the filtered dark-red solution is added an equal volume of a saturated aqueous $NaNO_3$ or KNO_3 solution plus a few milligrams of the solid nitrate salt. This solution is cooled in an ice bath and allowed to stand several hours until dark-red crystals form. These are removed by filtration and the filtrate is set aside. The product is washed once with very cold 0.5 N HNO_3, once with cold ethanol, and then once with diethyl ether. The yield is ~80% but can be increased to >90% (0.51 g) by treating the filtrate with additional nitrate salt to saturate the solution again. *Anal.* Calcd. for $C_{17}H_{16}N_4O_4SPt$: C, 35.98; H, 2.84; N, 9.87; S, 5.65. Found: C, 35.41; H, 2.97; N, 9.83; S, 6.22.

Properties

The compound [Pt(terpy)(SCH_2CH_2OH)]NO_3 is a dark red-purple microcrystalline material that is extremely water soluble. The complex is stable in cold aqueous solution for periods up to several weeks. High pH or heating ($T > 60°$) leads to decomposition, however. The electronic absorption spectra of the terpyridine thiolato compounds are characteristically definitive in the 300-350 nm region and may be used to determine product purity. Solutions of [Pt(terpy)-(SCH_2CH_2OH)]NO_3 of less than 15 μM obey Beer's law and exhibit the following absorption maxima and molar extinction coefficients: 475 (890), 342 (12,900), 327 (10,700), 311 (10,300), 277 (20,300), and 242 (28,700) nm.

C. (2-AMMONIOETHANETHIOLATO-S)(2,2':6',2"-TERPYRIDINE)-PLATINUM(II) NITRATE, [Pt(terpy)($SCH_2CH_2NH_3$)](NO_3)$_2$ AND OTHER (2,2':6',2"-TERPYRIDINE)(THIOLATO)PLATINUM(II) COMPLEXES

[Pt(terpy)Cl]$^+$ + HS(CH_2)$_2$$NH_3^+$ + OH$^-$ + 2NO_3^- $\longrightarrow$
[Pt(terpy)($SCH_2CH_2NH_3$)](NO_3)$_2$ + H_2O + Cl$^-$

Procedure

[Pt(terpy)(SCH$_2$CH$_2$NH$_3$)](NO$_3$)$_2$ and other (2,2′:6′,2″-terpyridine)thiolato complexes can be prepared by the method described for [Pt(terpy)(SCH$_2$CH$_2$OH)]NO$_3$ with the following provisos. The 2-aminoethanethiol hydrochloride, or other solid starting materials, should be mixed in an absolute minimum amount of deionized water and the solution should be flushed with N$_2$ before it is slowly added to the aqueous [Pt(terpy)Cl]$^+$. The more soluble products such as [Pt(terpy)(SCH$_2$CH$_2$NH$_3$)]$^{2+}$ require the use of less water in the reaction mixture and/or the addition of more solid nitrate salt to ensure a high yield. An orange precipitate is observed after the addition of the 2-aminoethanethiol hydrochloride solution to the aqueous [Pt(terpy)Cl]$^+$. This material will redissolve as the pH of the solution is adjusted to about 6 to give the characteristic dark-red color.

The purity of the [Pt(terpy)(SCH$_2$CH$_2$NH$_3$)](NO$_3$)$_2$ may be determined by its electronic absorption spectrum. Solutions of less than 15 μM obey Beer's law and exhibit the following absorption maxima and molar extinction coefficients: 341 (13,100), 326 (11,700), 311 (10,800), 277 (18,900) and 242 (29,700) nm. Upon standing for prolonged periods in neutral or basic solutions, the complex may decompose to form the [{Pt(terpy)(SCH$_2$CH$_2$NH$_2$)$_2$}Pt]$^{4+}$ cation.[6]

References

1. K. W. Jennette, S. J. Lippard, G. A. Vassiliades, and W. R. Bauer, *Proc. Natl. Acad. Sci. U.S.A.*, 71, 3839 (1974).
2. P. J. Bond, R. Langridge, K. W. Jennette, and S. J. Lippard, *Proc. Natl. Acad. Sci. U.S.A.*, 72, 4825 (1975).
3. K. W. Jennette, J. T. Gill, J. A. Sadownick, and S. J. Lippard, *J. Am. Chem. Soc.*, 98, 6159 (1976).
4. G. T. Morgan and F. H. Burstall, *J. Chem. Soc.*, 1934, 1498.
5. G. M. Intille, Ph.D. Dissertation, Syracuse University, Syracuse NY, 1967.
6. J. C. Dewan, S. J. Lippard, and W. R. Bauer, *J. Am. Chem. Soc.*, 102, (1980) in press.

29. SATURATED, UNSUBSTITUTED TETRAAZAMACROCYCLIC LIGANDS AND THEIR COBALT(III) COMPLEXES

A large number of tetraazamacrocyclic metal complexes are known. The ring sizes of these macrocycles vary from 12 to 16 members. The simplest of the ligands are the unsubstituted fully saturated compounds. They are ideal for

106 Compounds of Biological Interest

systematic studies on ring size effects among complexes of macrocyclic ligands. 1,4,8,11-Tetraazacyclotetradecane (cyclam, [14]aneN$_4$) has long been known,[1] and a simple high-yield synthesis has been reported for this ligand.[2] The preparations of 1,4,7,10-tetraazacyclotridecane ([13]aneN$_4$)[4,5], 1,4,8,12-tetraazacyclopentadecane ([15]aneN$_4$)[3-5] and 1,5,9,13-tetraazacyclohexadecane ([16]aneN$_4$)[4-6] have been reported more recently. The complexes of these ligands with many metal ions have been studied.[4,5,7,8] The preparations of the free ligands [13]aneN$_4$, [15]aneN$_4$, and [16]aneN$_4$ and their cobalt(III) complexes are described here.

A. 1,4,7,10-TETRAAZACYCLOTRIDECANE ([13]aneN$_4$)

Submitted by YANN HUNG*
Checked by DANIEL R. ENGLISH† and THEODORE A. DEL DONNO†

Procedure

1. *N,N′,N″,N‴-Tetrakis(p-toluenesulfonyl)triethylenetetramine).* In an 800-mL beaker, sodium hydroxide (16 g, 0.4 mole) is dissolved in 100 mL of water and triethylenetetramine (2,2,2) (14.6 g of 2,2,2; 0.1 mole) is added. A solution of *p*-toluenesulfonyl chloride (76.2 g, 0.4 mole) in 400 mL of diethyl ether is added dropwise with stirring. The mixture is stirred for an hour at room tem-

---
*Stanford University, Stanford, CA 94305.
†Chemistry Department, The Ohio State University, Columbus, OH 43210.

perature. The tosylate separates as an off-white solid. It is recrystallized from a large volume of methanol (84 mL/g) or by dissolving in acetone (30 mL/g), followed by the addition of an equal volume of ethanol. Yield: 80-90%.

2. *N,N′,N″,N‴-Tetra(p-toluenesulfonyl) Derivative of [13]aneN$_4$*. Two equivalents of sodium ethoxide (3.02 g in 66 mL of ethanol) are added to 50 g of the tosylated linear tetraamine in 200 mL of boiling ethanol. After the solution is boiled for 20 minutes, the ethanol is removed by rotary evaporating to dryness to yield the disodium salt of the tosylated linear tetraamine. The sodium salt is dissolved in 660 mL of dimethylformamide (0.1 M) and transferred to a 2-L, three-necked flask equipped with an addition funnel and a thermometer. The mixture is heated to 110°, and 1 equiv of 1,3-dibromopropane (14 g in 330 mL of DMF, 0.2 M) is added dropwise over a period of 1 hour while the solution is stirred vigorously. The volume of DMF is reduced to one-fourth the initial volume. The solution is slowly added to a volume of water equal to $1\frac{1}{2}$ times that of the initial DMF volume. This yields a tacky off-white precipitate. The product is recrystallized from hot benzene. A white product is precipitated by reducing the volume of the benzene to about 150-mL, adding ethanol to the solution, and then letting it stand at room temperature for 1 hour. Yield: 40%.

3. *Hydrolysis of Tosyl Groups*. The tosylated [13]aneN$_4$ is hydrolyzed by heating it in 30% hydrobromic-acetic acid (50 mL/g), which is prepared by adding 9 volumes of glacial acetic acid to 16 volumes of 47% hydrobromic acid. The mixture is refluxed for 2 days, after which time the solution is reduced to one-tenth its initial volume. After it is cooled, the solution is added to a threefold volume of diethyl ether–ethanol(1:1). The solid that separates is filtered and washed repeatedly with diethyl ether and ethanol. The product is air dried on the frit. Yield: 90%.

4. *Free Ligand Extraction*. Ten grams (about 0.02 mole) of the tetrahydrobromide salt of [13]aneN$_4$ is dissolved in 40 mL of water and is neutralized with a slight excess of sodium hydroxide (about 4 g, 0.1 mole). Several extractions (about five) with chloroform (20 mL) are then carried out. The chloroform extract is dried over sodium sulfate and then rotary evaporated. Diethyl ether is added and then rotary evaporated, leaving behind the solid ligand. Sometimes, refrigeration is required to bring about solidification of the ligand. The solid is recrystallized from diethyl ether; white needles are obtained. Yield: 60-80%.

Properties

The free ligand, [13]aneN$_4$, melts at 40-41°. It is soluble in most organic solvents but slightly soluble in water. Nmr spectral data: δ 1.68 ppm (quintet, 2H, $J \simeq 6$ Hz). δ 2.20 ppm (s, 4H), δ 2.78 ppm (complex multiplet, 16H).

B. 1,4,8,12-TETRAAZACYCLOPENTADECANE ([15]aneN$_4$)

Submitted by E. KENT BAREFIELD* and GARY FREEMAN*
Checked by DANIEL R. ENGLISH[†]

Procedure

Nickel(II) chloride hydrate, 10.2 g (0.043 mole), is dissolved in 150 mL of water in a 250-mL Erlenmeyer flask. With stirring, 8 g (0.043 mole) of N,N'-bis(3-aminopropyl)-1,3-propanediamine[‡] is added followed by 10 mL of 30-40% aqueous glyoxal solution. The blue solution is then heated at 60-80° for 3 hours during which time the color becomes red-brown. The solution is transferred to a 500-mL Parr hydrogenation pressure bottle along with 20 g of freshly prepared W-2 Raney nickel or an equal amount of *fresh*[§] Grace Chemical Co. Grade 28 nickel catalyst. The bottle is placed on the hydrogenation apparatus and heated at 60-80° under 50 psi of hydrogen for 16-24 hours (until hydrogen uptake ceases). The bottle is removed from the apparatus and the catalyst is separated by gravity filtration. (■ **Caution.** *The spent catalyst is pyrophoric when dry and must be treated with care.*) The blue-violet solution is cooled in an ice bath and 5.2 g (0.065 mole) of NaNCS dissolved in 25 mL of water is added to precipitate [Ni(C$_{11}$H$_{26}$N$_4$)(NCS)$_2$]. The lavender precipitate is collected by fil-

---

*School of Chemistry, Georgia Institute of Technology, Atlanta, GA 30332.
[†]Chemistry Department, The Ohio State University, Columbus, OH 43210.
[‡]Prepared according to the procedure for N,N'-bis(2-aminoethyl)-1,3-propanediamine (*Inorg. Syn.*, **16**, 222 (1976), substituting 202 g (1 mole) of 1,3-dibromopropane for the dibromoethane. The yield of tetramine is about 60% with a boiling point of 102° at 1 torr.
[§]Catalyst that has been opened for some time may not be active enough for the hydrogenation.

tration and transferred with 75 mL of water to a 200-mL, round-bottomed flask; 5 g of NaCN (0.1 mole) is added and the solution is heated at reflux for 1 hour. The orange solution is cooled and 5 g of NaOH is added. The solution is then extracted six times with 30-mL portions of $CHCl_3$. The $CHCl_3$ extracts are combined and dried with anhydrous Na_2SO_4. After removal of the drying agent, the $CHCl_3$ is removed on a rotary evaporator to leave 1,4,8,12-tetraazacyclopentadecane ([15]aneN$_4$) as a nearly white solid that is satisfactory for preparation of metal complexes. This product may be further purified by recrystallization from hexane to give fine white needles. Yield (recrystallized): 4.2-4.5 g, 45-48%. *Anal.* Calcd. for $C_{11}H_{26}N_4$: C, 61.63; H, 12.22; N, 26.18. Found: C, 61.36; H, 12.32; N, 26.18.

Properties

1,4,8,12-Tetraazacyclopentadecane is a hygroscopic solid melting at 99-100° when pure. The free base is very soluble in polar solvents, including water. The infrared spectrum (Nujol mull) contains several strong N—H stretching absorptions in the region 3200-3300 cm^{-1}. Its nmr spectrum (60 MHz, CDCl$_3$, TMS) consists of an overlapping singlet and triplet ($J \approx 5$ Hz) at τ 7.27 and an overlapping singlet and quintet ($J \approx 5$ Hz) at τ 8.2 and 8.3, respectively.

C. 1,5,9,13-TETRAAZACYCLOHEXADECANE ([16]aneN$_4$)

Submitted by WILLIAM L. SMITH* and KENNETH N. RAYMOND*
Checked by DANIEL R. ENGLISH† and THEODORE A. DEL DONNO†

*Department of Chemistry and Materials and Molecular Research Division, Lawrence Berkeley Laboratory, University of California, Berkeley, CA 94720.
†Chemistry Department, The Ohio State University, Columbus, OH 43210.

Procedure

The starting linear amine, N,N'-bis(3-aminopropyl)-1,3-propanediamine, is prepared using the procedure for N,N'-bis(2-aminoethyl)-1,3-propanediamine,[2] which consists of the condensation of 1,3-dibromopropane with 1,3-diaminopropane. The resulting amine is purified by vacuum distillation from barium oxide, bp 135-136° at 1 torr, 65-70% yield. The tetraamine forms a solid hydrate when exposed to moist air; its tetrahydrochloride decomposes at 297°.

The tetratosylate of the linear amine is prepared by the dropwise addition of 477 g (2.5 mole) of p-toluenesulfonyl chloride dissolved in 2.5 L of diethyl ether to a solution of 80 g (2 mole) of sodium hydroxide and 94.2 g (0.5 mole) of the linear tetraamine in 200 mL of water. The mixture is vigorously stirred by a mechanical stirrer during the addition and for an hour after its completion. The resulting sticky paste is chilled in an ice bath and then removed by filtration. It is washed well with diethyl ether and water and thoroughly dried in a vacuum at 50° to obtain a clumpy white powder. After recrystallization from methanol, 300-350 g (70-90%) of the tetratosylate is obtained, mp 126-127°.

Small portions of sodium hydride (as a suspension in oil) are added to a solution of 80.5 g (0.1 mole) of the tetratosylated linear amine in 1 L of DMF until the evolution of hydrogen ceases. The DMF is purified by vacuum distillation from CaH_2 (bp ~29° at 0.01 torr). The excess sodium hydride is removed by filtration and the filtrate is heated to 110° in a flask equipped with an efficient stirrer. A solution of 39.4 g (0.1 mole) of the ditosylate of 1,3-propanediol[9] in 500 mL of DMF is added dropwise with stirring, after which the stirring is continued for 2 hours and the temperature is maintained at 110°. The volume is reduced to 250-300 mL by vacuum distillation and the solution is added to 2 L of water to cause precipitation. The solid is removed by filtration, washed well with water, and dried at 60° overnight in a vacuum. The tetratosylated [16]aneN_4 is used in the next step without purification. Recrystallization from $CHCl_3$-EtOH yields a white powder decomposing at 252-255°.

The free amine is formed by reaction of the tetratosylated [16]aneN_4 with conc. H_2SO_4 (2 mL H_2SO_4/g tetratosylate) at 100° for 48 hours. After the solution is cooled in an ice bath, it is added dropwise into a solution made up of 2 volumes of ethanol. This is followed by the addition of 5 volumes of diethyl ether. If two phases form, enough ethanol is added to make one phase.

The precipitate is removed by filtration, washed with diethyl ether, and dried thoroughly. It is then dissolved in the minimum amount of water and brought to pH ≥ 10 (Hydrion paper) with sodium hydroxide. Any precipitate is removed by filtration and the filtrate is extracted four times with equal volumes of dichloromethane. The dichloromethane extracts are dried with anhydrous sodium sulfate and evaporated to dryness leaving a light-yellow solid. The crude product is sublimed (~60°, 0.01 torr) to yield 12-14 g (55-60% on the basis of tetratosylated linear amine) of white crystals, mp 84° (lit.[8] 82-83°).

Properties

The white, waxy, slightly hygroscopic crystals of [16]aneN$_4$ are soluble in water, alcohol, chloroform, and THF. They can be recrystallized from chlorobenzene or diethyl ether and are insoluble in pentane. The ^{1}H nmr spectrum[8] in CDCl$_3$ versus TMS consists of a quintet δ 1.70 ppm, 8H, a singlet δ 1.82 ppm, 4H, and a triplet δ 2.72 ppm, 16H. The IR spectrum shows a strong N—H stretching absorption at 3285 cm^{-1} and C—H stretches at 2927, 2870, and 2805 cm^{-1}. Other strong absorptions occur at 1480, 1468, 1137, 1070, 853, and 779 cm^{-1}. In the solid state,[6] the conformation of [16]aneN$_4$ consists of a square-planar array of nitrogen atoms with adjacent trimethylene chains folded toward opposite sides of the nitrogen plane. There is weak internal hydrogen bonding between *cis* nitrogen atoms, with the *trans* orientation of these hydrogen atoms limiting the molecular point symmetry to S_4.

D. *trans*-DICHLORO([13]aneN$_4$)COBALT(III) CHLORIDE, (I)- and (II)-*trans*-DICHLORO([15]aneN$_4$)COBALT(III) CHLORIDE, (I)- and (II)-*trans*-DICHLORO([16]aneN$_4$)COBALT(III) PERCHLORATE

Submitted by YANN HUNG*
Checked by DANIEL R. ENGLISH† and THEODORE A. DEL DONNO†

The saturated, unsubstituted tetraazamacrocycles [13–16]aneN$_4$ all form typical pseudooctahedral *trans*-dichlorotetraamine complexes with cobalt(III).[7] These have been chosen as typical complexes of these macrocyclic ligands.

1. *trans*-[Co([13]aneN$_4$)Cl$_2$]Cl

$$CoCl_2 \cdot 6H_2O + [13]aneN_4 \xrightarrow[HCl, O_2]{CH_3OH} trans\text{-}[Co([13]aneN_4)Cl_2]Cl$$

Procedure

Cobalt(II) chloride·6-hydrate (1.0 g, 4 mmole) and a slight excess of [13]aneN$_4$ (0.8 g, 4.3 mmole) are warmed (∼60°) in 100 mL of methanol for 15 minutes. Concentrated HCl is then added dropwise to the solution until the red-brown solution turns green-brown (about 1 mL). Air is bubbled through the solution for 2 hours and its volume is reduced. At this point, red crystals of the *cis*-isomer separate (yield: 15%). Reduction of the volume of the filtrate results in isolation

---

*Chemistry Department, Stanford University, Palo Alto, CA 94305.
†Chemistry Department, The Ohio State University, Columbus, OH 43210.

112 *Compounds of Biological Interest*

of the green *trans*-compound; it is recrystallized from methanol (yield: 35%). *Anal.* Calcd. for $CoC_9H_{22}N_4Cl_3$: Co, 16.76; C, 30.75; H, 6.31; N, 15.94; Cl, 30.75. Found: Co, 16.53; C, 30.52; H, 6.24; N, 15.66; Cl, 30.99.

2. *trans*-[Co([15]aneN$_4$)Cl$_2$]Cl (Isomers I and II)

$$CoCl_2 \cdot 6H_2O + [15]aneN_4 \xrightarrow[HCl, O_2]{CH_3OH} \xrightarrow{HClO_4} (I)\text{-}trans\text{-}[Co([15]aneN_4)Cl_2]ClO_4$$

$$(I)\text{-}trans\text{-}[Co([15]aneN_4)Cl_2]ClO_4 \xrightarrow[CH_3CN]{LiCO_3} \xrightarrow{HCl}$$
$$(II)\text{-}trans\text{-}[Co([15]aneN_4)Cl_2]ClO_4$$

$$trans\text{-}[Co([15]aneN_4)Cl_2]ClO_4 \xrightarrow[Cl^- \text{ form}]{Dowex\ IX8} trans\text{-}[Co([15]aneN_4)Cl_2]Cl$$

Procedure

■ **Caution.** *Perchlorate salts of complexes are hazardous. Heating of the solids can lead to explosions.*

Because of the cobalt(II) contamination, the perchlorate salt is synthesized and then converted to other salts. Cobalt(II) chloride 6-hydrate (1.0 g, 4 mmole) and free [15]aneN$_4$ (0.9 g, 4.2 mmole) are warmed in 100 mL of methanol for 15 minutes. Concentrated hydrochloric acid is added dropwise (1 mL) and air is bubbled through the solution for at least 2 hours. Several drops of perchloric acid are added to precipitate the products as the perchlorates. The yield is 72%. The product is brown and is about 90% pure isomer I. The brown perchlorate salt is dissolved in acetonitrile and passed through a Dowex IX8, 200-400 mesh anion exchange column (Cl$^-$ form) at a rate of 10 sec/drop. The eluent is concentrated by rotary evaporation. Some green solid (isomer II) precipitates. This is filtered and the volume of the filtrate is further reduced to precipitate the tan isomer (I) as the chloride salt. Yield: 60% *Anal.* Calcd. for $CoC_{11} \cdot H_{26}N_4Cl_3$: C, 34.80; H, 6.90; N, 14.76. Found: C, 34.35; H, 7.09; N, 14.50.

The perchlorate salt of *trans*-[Co([15]aneN$_4$)Cl$_2$]$^+$ is dissolved in 40 mL of acetonitrile containing 5 mL of water and warmed with excess lithium carbonate with stirring for 2 hours. The solution turns violet. The excess lithium carbonate is removed by filtration. Concentrated HCl is added dropwise (2 mL) and the solution is warmed for 30 minutes. By this time, the color has changed to green. The product is obtained by concentrating the solution. The solid product, which is about 90% isomerically pure green isomer, is dissolved in acetonitrile and passed through a Dowex IX8, 200-400 mesh, anion exchange column (Cl$^-$ form). The chloride salt of the green isomer II is obtained by reducing the volume of the eluent. Yield: 50%. *Anal.* Calcd. for $CoC_{11}H_{26}N_4Cl_3$. C, 34.80; H, 6.90; N, 14.76. Found: C, 34.38; H, 7.24; N, 14.41.

3. *trans*-[Co([16]aneN$_4$)Cl$_2$]ClO$_4$

$$CoCl_2 \cdot 6H_2O + [16]aneN_4 \xrightarrow[LiClO_4]{CH_3OH} Co([16]aneN_4)(ClO_4)_2$$

$$Co([16]aneN_4)(ClO_4)_2 + Br_2 \xrightarrow{CHCl_3} [Co([16]aneN_4)Br_2]ClO_4$$

$$[Co([16]aneN_4)Br_2]ClO_4 + 2LiCl \longrightarrow$$

$$\text{(I) and (II)-}trans\text{-}[Co([16]aneN_4)Cl_2]ClO_4 + 2LiBr$$

Procedure

■ **Caution.** *Perchlorate salts of metal complexes can be explosive and must be handled with care. Compounds should not be heated as solids.*

One gram of [16]aneN$_4$ (4 mmole) is dissolved in 100 mL of methanol to which is added cobalt(II) chloride 6-hydrate (1.04 g, 4 mmole). The resulting magenta solution is warmed for 20 minutes, a slight excess of lithium perchlorate (0.6 g, 5.6 mmole) is added, and the solution is warmed (60°) for another 10 minutes. The methanol solution is rotary evaporated to dryness and the solid residue is redissolved in chloroform and filtered. Evaporation followed by dissolution in chloroform is repeated once more. Bromine is added dropwise to the chloroform solution until the magenta color is completely gone (several drops) and brown solids separate from the solution. The product is filtered and washed with chloroform. The brown product is dissolved in hot methanol to which a slight excess of lithium chloride (0.2 g, 4.7 mmole) is added. Brown isomer I is obtained in successive fractions as the solution is concentrated and cooled. Yield: 40%. Green isomer II is obtained by continuing to reduce the volume of the filtrate. Yield: 10%. Both are recrystallized from acetonitrile. *Anal.* Calcd. for $CoC_{12}H_{26}$·$N_4Cl_3O_4$; C, 31,49; H, 6.17; N, 12.24. Found: C, 31.28; H, 5.92; N, 12.04 (isomer I). C, 31.40; H, 6.20; N, 12.21 (isomer II).

Properties

The cobalt complexes are all diamagnetic and behave as 1:1 electrolytes in methanol.[7] The infrared spectra of the complexes contain sharp, strong N—H stretching absorptions (Table I). The electronic spectral data are also shown in Table I. Isomers I and II differ in the chiralities of nitrogen atoms in the macrocyclic ligands.[7] Both chemical and physical properties are dependent on ring size.[7,10]

TABLE I Properties of the Cobalt(III) Complexes

Complex	$\nu_{N-H}(cm^{-1})^a$	Electronic Spectral Data (ϵ_M) $(m\mu^{-1})$
trans-[Co([13]aneN$_4$)Cl$_2$]$^+$	3155 (s, sp)	1.667(33), 2.370(144)
(I)-trans-[Co([15]aneN$_4$)Cl$_2$]$^+$	3200 (s, sp)	1.527(40), 1.923(69), 2.370(106)
(II)-trans-[Co([15]aneN$_4$)Cl$_2$]$^+$	3160 (s, sp)	1.563(27), 2.041(38), 2.410(62)
(I)-trans-[Co([16]aneN$_4$)Cl$_2$]$^{+b}$	3236 (s,), 3175 (s, sp)	1.473(57), 1.869(34), 2.326(99)
(II)-trans-[Co([16]aneN$_4$)Cl$_2$]$^{+b}$	3247 (s), 3174 (s, sp)	1.511(41), 1.961(35), 2.387(76)

^{a}Nujol mull, s = strong, sp = sharp.
^{b}Electronic spectral data obtained in acetonitrile solutions, all others in methanol.

References

1. (a) B. Bosnich, C. K. Poon, and M. L. Tobe, *Inorg. Chem.*, **4**, 1102 (1976); (b) H. Stetter and K. H. Mayer, *Chem. Ber.*, **14**, 1410 (1961).
2. E. K. Barefield, F. Wagner, A. W. Herlinger, and A. R. Dahl, *Inorg. Synth.*, **16**, 220 (1976).
3. E. K. Barefield, F. Wagner, and K. D. Hodges, *Inorg. Chem.*, **15**, 1370 (1976).
4. D. D. Watkins, Jr., D. P. Riley, J. A. Stone, and D. H. Busch, *Inorg. Chem.*, **15**, 387 (1976).
5. L. Y. Martin, C. R. Sperati, and D. H. Busch, *J. Am. Chem. Soc.*, **99**, 2968 (1977).
6. W. L. Smith, J. D. Ekstrand, and K. N. Raymond, *J. Am. Chem. Soc.*, **100**, 3539 (1978).
7. Y. Hung, L. Y. Martin, S. C. Jackels, A. M. Tait, and D. H. Busch, *J. Am. Chem. Soc.*, **99**, 4029 (1977).
8. L. Y. Martin, L. J. DeHayes, L. J. Zompa, and D. H. Busch, *J. Am. Chem. Soc.*, **96**, 4046 (1974).
9. E. R. Nelson, M. Maienthal, L. A. Lane, and A. A. Benderly, *J. Am. Chem. Soc.*, **79**, 3467 (1957).
10. Y. Hung and D. H. Busch, *J. Am. Chem. Soc.*, **99**, 4977 (1977).

30. [7,16-DIHYDRO-6,8,15,17-TETRAMETHYLDIBENZO[b,i]- [1,4,8,11]TETRAAZACYCLOTETRADECINATO(2-)]- NICKEL(II) ([5,7,12,14-Me$_4$-2,3: 9,10-Bzo$_2$[14]- HEXAENATO(2-)N$_4$]Ni(II) AND THE NEUTRAL MACROCYCLIC LIGAND

Submitted by V. L. GOEDKEN* and M. C. WEISS*
Checked by D. PLACE† and J. DABROWIAK†

The [5,7,12,14-Me$_4$-2,3: 9,10-Bzo$_2$[14]hexaenato(2-)N$_4$] nickel(II) complex‡ (I) of the title ligand is easily synthesized and the free ligand is readily obtained

(I) (Numbering is for abbreviation)

from this complex.[1] Other transition metal complexes, not obtainable by direct metal-template methods, can then be prepared from the free ligand by reaction with appropriate metal acetate salts.[2] The Me$_4$Bzo$_2$[14]hexaene-N$_4$ macrocycle complexes are more readily handled than their Bzo$_2$[14]hexaene-N$_4$ counterparts. This behavioral difference results from the interaction of the four methyl substituents with the benzenoid rings to yield a ligand having a pronounced saddle shape.[3] The marked nonplanarity of the ligand, while not profoundly affecting the coordinating ability of the ligand, inhibits aggregation that leads to marginal solubility, characteristic of planar, totally conjugated macrocycles, such as porphyrin, phthalocyanine, and the simple dibenzo macrocyclic ligand. These steric interactions have important consequences that are reflected in the stoichiometry and reactivity of the complexes. The saddle shape of the ligand directs the metal to one side of the N$_4$ donor plane, resulting in a preference for five-coordinate complexes. In some instances, the distortions produced in the 2,4-pentanediiminato chelate rings result in unusual addition reactions.[4] These

---

*Department of Chemistry, Florida State University, Tallahassee, FL 32306.
†Department of Chemistry, Syracuse University, Syracuse, NY 13210.
‡For numbering of abbreviations, see *Inorg. Synth.*, 18, 45 (1968).

include 1,4-addition of acetylenes and nitriles to the Co(III) complex and reactions with oxygen to produce a carbonyl function at the γ-carbon atom. In the absence of good donor solvents, cofacial metal–metal bonded complexes are formed with rhodium and ruthenium complexes of this ligand.[5]

A. [5,7,12,14-Me$_4$-2,3: 9,10-Bzo$_2$[14]HEXAENATO(2-)N$_4$]Ni(II)

$$2 \begin{array}{c} \text{o-C}_6\text{H}_4(\text{NH}_2)_2 \end{array} + 2 \begin{array}{c} \text{CH}_3\text{COCH}_2\text{COCH}_3 \end{array} + \text{Ni}(\text{C}_2\text{H}_3\text{O}_2)_2 \cdot 4\text{H}_2\text{O} \xrightarrow{\text{CH}_3\text{OH}} \text{Ni}(\text{C}_{22}\text{H}_{22}\text{N}_4) + 8\text{H}_2\text{O}$$

Procedure

Nickel diacetate tetrahydrate (50.0 g, 0.201 mole), o-phenylenediamine (43.47 g, 0.402 mole), and 2,4-pentanedione (40.23 g, 0.402 mole) are added to 500 mL of anhydrous methanol in a liter flask and refluxed under nitrogen for 48 hours. The very pale green o-phenylenediamine complex of nickel forms upon mixing the reagents. Before refluxing begins, a deep-purple color develops in the reaction vessel. This species has been identified as the 2,4-dimethyl-1H-1,5-benzodiazapinium cation[6] by isolation as the tetrafluoroborate salt. Within the first hour of refluxing, the purple color gradually turns to an intense blue-green. This intermediate has been isolated as the BF$_4^{1-}$ salt and characterized as the open-chain quadridentate species II. The long period of refluxing is necessary for

(II)

the rate-limiting ring closure to occur. After it is refluxed for 48 hours, the solution is cooled and the finely divided dark-violet (deep blue-green when crushed) product is filtered and washed with generous portions of methanol until the effluent is light green to colorless. The product may be either air dried or dried *in vacuo*. Yield of the crude product is about 45%. The crude product may be used directly for the isolation of the free ligand. The pure, highly crystalline product may be obtained by recrystallization. This is done by dissolving 10 g of the crude reaction product in 150 mL of hot toluene and filtering while hot

[5,7,12,14-Me₄-2,3: 9,10-Bzo₂[14]Hexaenato(2-)N₄]Ni(II)   117

to remove any insoluble residues. Then 50 mL of anhydrous methanol is added slowly to the toluene solution; the flask is stoppered and placed in a refrigerator for several hours. The product is filtered, washed with methanol, and dried. Yield is about 3.5 g. *Anal.* Calcd. for $Ni(C_{22}H_{22}N_4)$: C, 65.9; H, 5.49; N, 14.0. Found: C, 65.7; H, 5.40; N, 13.8.

Properties

The $Ni(C_{22}H_{22}N_4)$ is stable to the atmosphere, but an inert atmosphere is recommended for extended storage because a perceptible odor develops with long-term standing under atmospheric conditions. It is very soluble in most nonpolar to slightly polar solvents, such as benzene, toluene, carbon tetrachloride, and chloroform, to give very intense blue-green solutions. It is sparingly soluble to insoluble in polar solvents, such as alcohols and acetonitrile. The stability of the nickel(II) complex to oxygen and moisture is in contrast to those of other first-row-metal complexes of this ligand. The nickel(II) complex has good thermal stability and can be purified by vacuum sublimation. The nmr spectrum (C_6D_6-tetramethylsilane) consists of a singlet at $\delta 1.77$ (12H) ppm, a singlet at $\delta 4.67$ (2H) ppm, and a singlet at $\delta 6.54$ (8H) ppm. The electronic spectrum in $CHCl_3$ contains the following absorptions (molar extinction coefficients are given in parentheses); 588 (4670), 420 (sh), (10,530), 394 (26,900), 334 (5680), 268 (21,860) nm.

B. The 5,7,12,14-Me₄-2,3: 9,10-Bzo₂[14]HEXAENEN₄ Ligand

$$Ni(C_{22}H_{22}N_4) + 4HCl \xrightarrow{C_2H_5OH} [C_{22}H_{26}H_4][NiCl_4]$$

$$[C_{22}H_{26}N_4][NiCl_4] + 2NH_4[PF_6] \xrightarrow{H_2O}$$

$$(C_{22}H_{26}N_4)[PF_6]_2 + 2NH_4Cl + NiCl_2$$

$$(C_{22}H_{26}N_4)[PF_6]_2 + 2N(C_2H_5)_3 \xrightarrow{CH_3OH} C_{22}H_{24}N_4 + 2HN(C_2H_5)_3[PF_6]$$

■ **Caution.** *Anhydrous HCl must be used in a well-ventilated hood with adequate safety precautions.*

Three steps are required to produce the free ligand in pure form. The first involves the isolation of the acid salt of the ligand as the tetrachloronickelate(II) salt. The second step involves converting this salt to a different one, not containing nickel. (Simple neutralization of the tetrachloronickelate(II) salt results in the reinsertion of nickel(II) into the macrocyclic ligand.) The final step involves the neutralization of the acid salt. Although the simple chloride salt precipitates initially as long white fibrous needles upon the addition of anhydrous HCl, it is difficult to determine when all the nickel(II) has been stripped from the macrocyclic ligand before some of the simple chloro salt is converted to the tetrachloronickelate(II) salt. Even under the best of circumstances, when the chloride salt appears pure, the neutral free ligand obtained generally has a greenish color indicative of a small amount of residual nickel(II) being coordinated by the macrocyclic ligand.

Anhydrous HCl is bubbled at a moderate rate through a constantly stirred suspension of 15 g of $Ni(C_{22}H_{22}N_4)$ in 500 mL of absolute ethanol until the solution becomes strongly acidic and a copius precipitate of bright blue-green tetrachloronickelate(II) salt of the ligand, $[C_{22}H_{26}N_4][NiCl_4]$, is observed. The following precautions are recommended for this step. Use of an HCl inlet tube at least 5 mm in diameter prevents clogging and dangerous pressure buildup. Excess HCl is vented to the hood by means of a tube with trap inserted to prevent ethanol from condensing in the vent tube. A dry trap between the HCl cylinder and the reaction vessels is needed to prevent solution backflow into the cylinder when the HCl flow is stopped. The reaction mixture becomes warm to hot during the addition of anhydrous HCl and the color of the solution turns from blue-green to green to brownish purple. The latter color is due to the formation of a small amount of a benzodiazapenium salt and is of little consequence. No attempt should be made to cool the solution. This inhibits the metal-stripping reaction. After the solution has cooled, it should be placed in a refrigerator or chilled for an hour. The product is filtered, washed with a minimum amount of ethanol, and dried (*in vacuo*, if atmosphere is humid). Yield is 90-95%.

The tetrachloronickel(II) salt of the ligand cation is readily converted to the hexafluorophosphate salt. This procedure is necessary to avoid the reinsertion of any nickel(II) into the macrocyclic ligand during the neutralization process. Ten grams of $[C_{22}H_{26}N_4][NiCl_4]$ is dissolved in a minimum amount (about 100 mL) of water, and 9.0 g (1.5 excess) of ammonium hexafluorophosphate, also dissolved in a minimum amount (about 8 mL) of water, is added slowly. A colorless precipitate of the hexafluorophosphate salt $[C_{22}H_{26}N_4][PF_6]_2$ forms immediately. The product is filtered, washed with water, and air dried. The yield is quantitative.

The free ligand, $C_{22}H_{24}N_4$, is obtained by neutralizing a suspension of 10 g of $[C_{22}H_{26}N_4][PF_6]_2$ in anhydrous methanol, by the addition of a slight excess, 5.0 mL, of triethylamine. (Any amine base, such as pyridine, may be used with equal facility.) The bright-yellow product crystallizes immediately. It is filtered on a Büchner funnel, washed with methanol, and air dried. If the product is greenish, it is because a small amount of nickel has reinserted into the macrocyclic ligand. In this case the process of acidification with anhydrous HCl, and so forth, must be repeated. Yields approach 5.0 g or about 95%. *Anal.* Calcd. for $C_{22}H_{22}N_4$: C, 76.7; H, 6.97; N, 16.3. Found: C, 76.4; H, 6.87; N, 16.4.

Properties

The free ligand is stable to the atmosphere and can be stored indefinitely. It is soluble in nonpolar solvents, for example, benzene, toluene, and halohydrocarbons, and to a lesser extent in acetonitrile. It is insoluble in primary alcohols and water. The nmr spectrum in $CDCl_3$ consists of a simple four-line spectrum: 2.12 (12H), 4.87 (2H), 6.98 (8H), 12.58 (broad, 2N—H) ppm. The compound melts sharply at 243°. The electronic spectrum contains the following absorptions: 340 (ϵ_{max} = 39,700), 269 (ϵ_{max} = 15,100), 255 nm (ϵ_{max} = 18,400).

References

1. V. L. Goedken, J. A. Molin-Case, and Y. A. Park, *Chem. Commun.*, **1973**, 337; F. A. L'Epplatenier and A. Pugin, *Helv. Chim. Acta*, **58**, 917 (1975).
2. W. H. Woodruff, R. W. Pastor, and J. C. Dabrowiak, *J. Am. Chem. Soc.*, **98**, 7998 (1976).
3. M. C. Weiss, B. Bursten, S. M. Peng, and V. L. Goedken, *J. Am. Chem. Soc.*, **98**, 8021 (1976).
4. M. C. Weiss and V. L. Goedken, *J. Am. Chem. Soc.*, **98**, 3389 (1976).
5. L. F. Warren and V. L. Goedken, *Chem. Commun.*, **1978**, 909.
6. P. L. Orioli and H. C. Lip, *Cryst. Struct. Commun.*, **3**, 3 (1974); J. L. Barltrop, C. G. Richards, D. M. Russel, and G. Ryback, *J. Chem. Soc.*, **1959**, 1132.

31. THE SYNTHESIS OF MOLYBDENUM AND TUNGSTEN DINITROGEN COMPLEXES

Submitted by JONATHAN R. DILWORTH* and RAYMOND L. RICHARDS*
Checked by GRACE J.-J. CHEN† and JOHN W. McDONALD†

The chemistry of bis(dinitrogen) complexes of molybdenum and tungsten is currently of considerable interest because one of the dinitrogen ligands may be induced to form N—H or N—C bonds under mild conditions.[1,2] Where the dinitrogen ligands have monodentate, tertiary phosphine coligands, one dinitrogen may be converted to ammonia (up to 1.98 NH_3 per W atom or 0.7 NH_3 per Mo atom) by treatment of the complexes with sulfuric acid in methanol.[2]

*A.R.C. Unit of Nitrogen Fixation, University of Sussex, Brighton, Sussex, England BN1 9QJ.
†Charles F. Kettering Research Laboratory, Yellow Springs, OH 45387.

A preparation of trans-$[Mo(N_2)_2(dppe)_2]$ (dppe = $Ph_2PCH_2CH_2PPh_2$) that uses triethylaluminum as reductant has already been reported in *Inorganic Syntheses*,[3] but it is less convenient and gives much lower yields than that described below. Other reported methods, using sodium amalgam as reductant, involve a more difficult work-up procedure.[4]

The preparations of the precursor compounds $[MoCl_4(CH_3CN)_2]$, $[MoCl_4(thf)_2]$, $[MoCl_3(thf)_3]$, $[WCl_4(PPh_3)_2]$, and $[WCl_4(dppe)]$ are also given, since they are critical to the successful synthesis of the dinitrogen complexes.

General Procedure

Conventional Schlenk-type glassware[5] is used in the procedures described below and unless otherwise stated, all manipulations are carried out under nitrogen. All solvents are anhydrous and are freshly distilled under nitrogen prior to use. Solutions are stirred with a magnetic stirrer bar unless otherwise stated.

A. BIS(ACETONITRILE)TETRACHLOROMOLYBDENUM(IV), $[MoCl_4(CH_3CN)_2]$

$$MoCl_5 + \text{excess } CH_3CN \longrightarrow [MoCl_4(CH_3CN)_2] + \text{chlorinated organic products}$$

■ **Caution.** *Acetonitrile must be regarded as a toxic material and handled in an efficient hood at all times.*

Procedure

A slight modification of the method reported by Fowles et al.[6] is used. Molybdenum pentachloride (20 g, 0.073 mole) is added slowly over 0.5 hour from a Schlenk tube through a glass connecting tube to stirred acetonitrile (100 mL, distilled from CaH_2) in a 250-mL Schlenk flask. Addition of acetonitrile to molybdenum pentachloride can produce sufficient local heating in the early stages of addition to cause the acetonitrile to froth from the flask. The reverse addition described here proceeds smoothly, an instantaneous reaction occuring with mild heat evolution and the formation of an orange-brown solid. The resulting suspension is stirred for a further 2 hours and is then allowed to stand at room temperature overnight. The bis(acetonitrile) complex is filtered under nitrogen using a 4-6 cm bore Schlenk filter, washed with acetonitrile (20 mL), and dried *in vacuo* (10^{-2} torr) at room temperature. The product is obtained as a fine orange-brown powder (18-20 g, 77-86%). *Anal.* Calcd. for $C_4H_6N_2Cl_4Mo$: C, 15.0; H, 1.9; N, 8.8. Found: C, 15.4; H, 1.8; N, 9.0.

Properties

The complex is air sensitive in the solid state and is only sparingly soluble in acetonitrile and nonpolar organic solvents. Its IR spectrum shows medium intensity bands at 2290 and 2310 cm^{-1} assignable to $\nu(C\equiv N)$. The complex is a convenient precursor for the preparation of a range of molybdenum(IV) complexes by replacement of the relatively labile acetonitrile groups with other ligands.[6]

B. TETRACHLOROBIS(TETRAHYDROFURAN)MOLYBDENUM(IV), [MoCl$_4$(thf)$_2$][7]

$$[MoCl_4(CH_3CN)_2] + THF(excess) \longrightarrow [MoCl_4(thf)_2] + 2CH_3CN$$

Procedure

Tetrahydrofuran (THF) (80 mL, freshly distilled from sodium benzophenone ketyl) is added from a 250-mL Schlenk flask to bis(acetonitrile)tetrachloromolybdenum(IV) (20 g, 0.052 mole) in a 250-mL Schlenk flask and the mixture is stirred rapidly for 2 hours to give a yellow suspension of tetrachlorobis(tetrahydrofuran)molybdenum(IV). The complex is filtered through a Schlenk filter, washed with THF (20 mL), and dried *in vacuo* (10^{-2} torr) at room temperature. The product is obtained as a microcrystalline orange-yellow powder in yields of 15-17 g, 63-71%. The product is not analytically pure. *Anal.* Calcd. for C$_8$H$_{16}$O$_2$Cl$_4$Mo: C, 25.1; H, 4.9. Found: C, 26.5; H, 4.2. However, it is sufficiently pure for subsequent reactions.

Properties

The complex is air sensitive in the solid state and almost insoluble in THF, but soluble in dichloromethane. It deteriorates even under nitrogen when stored for a period of weeks and is best used immediately for subsequent syntheses. Its IR spectrum shows an intense band at 820 cm^{-1}, characteristic of coordinated THF. The complex can be used to prepare other molybdenum(IV) complexes by displacement of the THF ligands[7] or molybdenum(VI) nitrido complexes by reaction with trimethylsilyl azide.[8]

C. TRICHLOROTRIS(TETRAHYDROFURAN)MOLYBDENUM(III), [MoCl$_3$(THF)$_3$][9]

$$[MoCl_4(thf)_2] \xrightarrow[CH_2Cl_2/THF]{Zn} [MoCl_3(thf)_3]$$

Procedure

Tetrachlorobis(tetrahydrofuran)molybdenum(IV) (18 g, 0.047 mole) (see Sec. B) and zinc shot (8-13 mesh, 30 g, 0.46 mole) in dichloromethane (90 mL, distilled from P_2O_5/anhydrous K_2CO_3) and THF (60 mL, distilled from sodium benzophenone ketyl) in a 250-mL Schlenk flask are stirred at 0° until a clear blue solution is obtained. The reaction must be stopped before the solution changes from blue to purple, otherwise the final product is pinkish-purple rather than orange and cannot be converted to bis(dinitrogen) complexes. The blue solution is filtered cold through a Schlenk filter to remove the excess zinc and evaporated to about 20 mL *in vacuo* (10^{-2} torr). The solution color may change to bluish-purple on evaporation. THF (30 mL) is added, and the brownish-orange precipitate is filtered using a Schlenk filter. The solid is washed with cold (0°) THF (2 × 20 mL) to remove any dark-colored impurities and dried *in vacuo* (10^{-2} torr) at room temperature. A yield of 12-15 g (61-76%) is obtained. *Anal.* Calcd. for $C_{12}H_{24}O_3Cl_3Mo$: C, 34.4; H, 5.8. Found: C, 32.8; H, 5.1. The complex is not analytically pure but can be used to prepare bis(dinitrogen) complexes. If required, it can be recrystallized from a THF/2-propanol mixture as orange crystals (Found: C, 34.8; H, 6.0).

Properties

The orange complex is air sensitive in the solid state. It is reasonably soluble and stable in THF but decomposes in nonpolar solvents, such as benzene, to give polymeric halogen-bridged products.[9] It is paramagnetic with a magnetic moment of 3.63 BM, and its IR spectrum shows a very intense band at about 820 cm^{-1} due to coordinated THF. Its reactions with tertiary phosphines in THF give a series of molybdenum(III) tertiary phosphine complexes,[9] and with trimethylsilyl azide Mo(V) nitrido complexes are formed.[8]

D. *trans*-BIS(DINITROGEN)BIS[ETHYLENEBIS(DIPHENYLPHOSPHINE)]-MOLYBDENUM(0), *trans*-[Mo(N$_2$)$_2$(Ph$_2$PCH$_2$CH$_2$PPh$_2$)$_2$]

$$[MoCl_3(thf)_3] + 2Ph_2PCH_2CH_2PPh_2 \xrightarrow[\text{THF}]{Mg, N_2} \textit{trans-}[Mo(N_2)_2(Ph_2PCH_2CH_2PPh_2)_2]$$

Materials

Ethylenebis(diphenylphosphine) (1,2-bis(diphenylphosphino)ethane) (dppe) is prepared by the published method[10] from triphenylphosphine or obtained commercially from Strem Chemical Co., Andover, MA. The Grignard magnesium turnings are obtained from BDH, Poole, Dorset, United Kingdom and are activated prior to use by heating *in vacuo* (10^{-2} torr) at 150° with iodine.

Procedure

Trichlorotris(tetrahydrofuran)molybdenum(III) (5 g, 0.012 mole), 1,2-bis(diphenylphosphino)ethane (12 g, 0.030 mole), Grignard magnesium turnings (6.0 g, 0.25 mole), and tetrahydrofuran (80 mL, freshly distilled from sodium benzophenone ketyl) in a 250-mL Schlenk flask are stirred as rapidly as possible under nitrogen for 16 hours. If initiation of the reduction, as shown by a darkening of the solution from orange to brown, does not occur the solution is warmed to 60° until onset of initiation; thereafter reaction proceeds at 20°. Rapid stirring, with vortexing of the solution is necessary to ensure rapid assimilation of nitrogen from the gas phase. After 16 hours the brown solution (some yellow-orange dinitrogen complex may precipitate at this stage) is evaporated to about 40 mL *in vacuo* at 10^{-2} torr and kept at 4° overnight. A mixture of dinitrogen complex and magnesium is then isolated using a Schlenk filter. As the dinitrogen complex can be rather finely divided a wide-bore (about 6 cm) Schlenk filter should be used. The mixture of magnesium and dinitrogen complex is transferred to a 500-mL Schlenk flask and sufficient warm (50-60°) THF is added to dissolve the dinitrogen complex. Filtration through a Schlenk filter, evaporation *in vacuo* at 10^{-2} torr, and addition of diethyl ether gives the complex as yellow-orange crystals (7-8 g, 62-70%), after drying *in vacuo* (10^{-2} torr) at room temperature.

Alternatively, large quantities of THF can be avoided, and the dinitrogen complex can be separated from the magnesium by extraction at reduced pressure utilizing the apparatus shown in Fig. 1. The mixture of dinitrogen complex and magnesium is filtered using the Schlenk-filter part of the apparatus with stopcock T_2 closed. The reflux condenser and the 100-mL Schlenk flask containing THF (70 mL) are then attached as shown, stopcock T_2 unopened, and the water bath is warmed to 35-40°. Stopcock T_1 is opened slowly to vacuum until the THF refluxes steadily; then it is closed and extraction is allowed to proceed until all the dinitrogen complex is transferred to the flask. Evaporation of this solution to half volume and addition of diethyl ether then gives the dinitrogen complex as before. The complex can be recrystallized from tetrahydrofuran/diethyl ether. *Anal.* Calcd. for $C_{48}H_{52}MoN_4P_4$: C, 65.8; H, 5.1; N, 5.9. Found: C, 65.8; H, 5.0; N, 5.9.

Properties

The complex is only slightly air sensitive in the solid state, but it is rapidly oxidized in solution. It is soluble with very slow decomposition (with loss of dinitrogen) in THF and toluene but reacts with halogenated solvents. Its IR spectrum (Nujol mull) shows an intense absorption at 1970 cm^{-1} and a weak absorp-

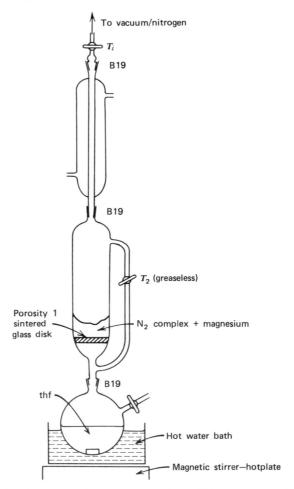

Fig. 1. *Apparatus for extraction of the dinitrogen complexes.*

tion at 2020 cm^{-1}, both assigned to $\nu(N_2)$. The complex reacts with halogen acids to give hydrazido(2-) complexes[2] and with a variety of carbon substrates to give derivatives with N—C bonds.[1]

E. TETRACHLOROBIS(TRIPHENYLPHOSPHINE)TUNGSTEN(IV), [WCl$_4$(PPh$_3$)$_2$][11]

$$WCl_6 + PPh_3 \text{(excess)} \xrightarrow[CH_2Cl_2]{Zn} [WCl_4(PPh_3)_2]$$

Procedure

This reaction is carried out in a 500-mL, single-necked flask that also has a side arm connected to the nitrogen supply by a stopcock. Dichloromethane (200 mL) is added to tungsten(VI) chloride (25 g, 0.063 mole) and dry granulated zinc (20 g). Then the flask is sealed and shaken for about 1 minute, and the pressure is periodically released through the stopcock. Triphenylphosphine (34 g, 0.13 mole) is then slowly added (2–3 min) and the vessel is sealed. It is then shaken until the tungsten(VI) chloride has reacted (about 15 min), while the pressure inside the flask is periodically released. The resulting orange solid is Schlenk filtered and dried *in vacuo* (10^{-2} torr), and any zinc is removed. Alternatively, the reaction mixture is filtered through a Schlenk filter equipped with a perforated rather than sintered-glass disc, which permits the complex to pass through but retains the large lumps of granulated zinc. The product, orange, microcrystalline [$WCl_4(PPh_3)_2$], is then washed with dichloromethane (30 mL) and dried (10^{-2} torr, 20°) (18 g, 34%). It contains 1 mole of dichloromethane of crystallization and is sufficiently pure for subsequent use. *Anal.* Calcd. for $C_{37}H_{32}Cl_6P_2W$: C, 47.4; H, 3.4. Found: C, 47.3; H, 3.5.

Properties

The orange complex is stable in the solid state in the absence of air and is very poorly soluble in organic solvents.

F. TETRACHLORO[ETHYLENEBIS(DIPHENYLPHOSPHINE)]-TUNGSTEN(IV), [$WCl_4(Ph_2PCH_2CH_2PPh_2)$][11]

$$[WCl_4(PPh_3)_2] + Ph_2PCH_2CH_2PPh_2 \xrightarrow[70°]{toluene} [WCl_4(Ph_2PCH_2CH_2PPh_2)] + 2\,PPh_3$$

Procedure

Tetrachlorobis(triphenylphosphine)tungsten(IV) (4.5 g, 5.3 mmole) and 1,2-bis-(diphenylphosphino)ethane (2.1 g, 5.3 mmole) are heated in toluene (50 mL, distilled from sodium) at 70° for 3 hours in a 100-mL flask carrying a reflux condenser. The resulting yellow-green solid is filtered using a Schlenk filter, washed with toluene (30 mL) and diethyl ether (30 mL), and dried (10^{-2} torr, 20°). Yield: 2.4 g, 89%; it is sufficiently pure for subsequent preparations. *Anal.* Calcd. for $C_{26}H_{24}Cl_4P_2W$; C, 43.1; H, 3.3. Found: C, 44.5; H, 3.6.

The color of this product is indicative of its efficiency in the subsequent preparation of the dinitrogen complex. Occasionally a pale-yellow or a brown-yellow product has been obtained; both have failed to yield the dinitrogen complex.

Properties

The yellow-green solid is only moderately air sensitive and is poorly soluble in organic solvents.

G. *trans*-BIS(DINITROGEN)BIS[ETHYLENEBIS(DIPHENYLPHOSPHINE)] TUNGSTEN(0), *trans*-[W(N$_2$)$_2$(Ph$_2$PCH$_2$CH$_2$PPh$_2$)$_2$][1,2]

$$[WCl_4(Ph_2PCH_2CH_2PPh_2)] + Ph_2PCH_2CH_2PPh_2 \xrightarrow[THF]{Mg/N_2}$$

$$trans\text{-}[W(N_2)_2(Ph_2PCH_2CH_2PPh_2)_2]$$

Procedure

Tetrachloroethylenebis(diphenylphosphine)tungsten(IV) (3.4 g, 4.7 mmole), magnesium (215 g, 0.1 mole), and 1,2-bis(diphenylphosphino)ethane (2.1 g, 5.3 mmole) are placed in a 250 mL Schlenk flask and pumped under vacuum (10^{-2} torr, 20°) for 2 hours. The flask is then filled with nitrogen, tetrahydrofuran (100 mL) is added, and the mixture is stirred for 18 hours after which it is a very dark red-brown. If any yellow-orange precipitate forms, the solution is warmed to about 60° to dissolve it, then filtered, and concentrated in a vacuum to about 40 mL. Orange crystals of *trans*-bis(dinitrogen)bis[ethylene(diphenylphosphine)]tungsten(0) deposit on cooling the solution (4°) and are Schlenk filtered, washed with diethyl ether (2 × 30 mL), and dried in a vacuum (10^{-3} torr, 20°). Yield: 3.5 g (72%). The compound may be recrystallized from tetrahydrofuran/diethyl ether. *Anal.* Calcd. for $C_{48}H_{52}N_4P_4W$; C, 60.2; H, 4.6; N, 5.4. Found: C, 60.1; H, 4.7; N, 5.4. If the preparation is carried out on a larger scale the dinitrogen complex can be isolated by extraction as described above for *trans*-[Mo(N$_2$)$_2$(dppe)$_2$].

Properties

The complex is slightly air sensitive in the solid state but is quickly oxidized in solution by air. It is soluble in tetrahydrofuran, toluene, and benzene but reacts with chlorinated solvents.[1] It is insoluble in methanol and diethyl ether. Both a strong absorption at 1946 cm^{-1} and a weak one at 2000 cm^{-1} (Nujol) in its IR spectrum are assigned to $\nu(N_2)$.

It reacts with acids to give hydride complexes with or without retention of dinitrogen.[2] The dinitrogen ligands may be replaced by such donor groups as isocyanides.[1,2] Its reactions, under appropriate conditions, with acids or various organic halides give complexes of —N_2R, N_2HR, or N—N=CR_2 ligands (R = H or organic group).[1,2]

References

1. J. Chatt, A. A. Diamantis, G. A. Heath, N. E. Hooper, and G. J. Leigh, *J. Chem. Soc., Dalton Trans.*, **1977**, 688 and references therein.
2. J. Chatt, A. J. Pearman, and R. L. Richards, *Nature*, **253**, 39 (1975); *idem., J. Chem. Soc., Dalton Trans.*, **1977**, 1852 and references therein.
3. M. Hidai, K. Tominari, Y. Uchida, and A. Misono, *Inorg. Synth.*, **15**, 25 (1974).
4. T. A. George and C. D. Siebold, *Inorg. Chem.*, **12**, 2548 (1973).
5. D. F. Shriver, *The Manipulation of Air-Sensitive Compounds*, McGraw-Hill, New York, 1969.
6. E. A. Allen, B. J. Brisdon, and G. W. A. Fowles, *J. Chem. Soc.*, **1964**, 4531.
7. E. A. Allen, K. Feenan, and G. W. A. Fowles, *J. Chem. Soc.*, **1965**, 1636.
8. J. Chatt and J. R. Dilworth, *J. Indian Chem. Soc.*, **54**, 13 (1977).
9. M. W. Anker, J. Chatt, G. J. Leigh, and A. G. Wedd, *J. Chem. Soc., Dalton Trans.*, **1975**, 2639.
10. W. Hewertson and H. R. Watson, *J. Chem. Soc.*, **1962**, 1490.
11. A. V. Butcher, J. Chatt, G. J. Leigh, and P. L. Richards, *J. Chem. Soc., Dalton Trans.*, **1972**, 1064.
12. J. Chatt, A. J. L. Pombeiro, R. L. Richards, G. H. D. Royston, K. W. Muir, and R. Walker, *Chem. Commun.*, **1975**, 708.

32. BIS[2,3-BUTANEDIONE DIOXIMATO(1-)] COBALT COMPLEXES

Submitted by J. BULKOWSKI,* A. CUTLER,† D. DOLPHIN,‡ and R. B. SILVERMAN §
Checked by L. T. TAYLOR# and W. M. COLEMAN**

Bis[2,3-butanedione dioximato(1-)]cobalt derivatives (cobaloximes(I)) appear in most respects to be good model systems for the cobalamins of the vitamin B_{12} coenzyme.[1] The central cobalt atom exhibits three "stable" oxidation states

---

*Department of Chemistry, University of Delaware, Newark, DE 19711.
†Department of Chemistry, Wesleyan University, Middletown, CT 06457.
‡Department of Chemistry, The University of British Columbia, Vancouver, B.C., Canada V6T 1W5.
§Department of Chemistry, Northwestern University, Evanston, IL 60201.
#Department of Chemistry, Virginia Polytechnic Institute and State University, Blacksburg, VA 24061.
**Naval Biosciences Laboratory, Naval Supply Center, Oakland, CA 94625.

Compounds of Biological Interest

$$\text{Structure I: } [\text{CH}_3\text{-C=N-O-H-O-N=C-CH}_3]_2 \text{ with Co(III) center, axial ligands X and B}$$

I

Co(I), Co(II), and Co(III). The monovalent species, in analogy to B_{12s}, is a powerful nucleophile that undergoes Michael additions and nucleophilic displacements to give alkyl complexes with carbon directly bonded to cobalt(III). Some cobaloximes can be prepared directly using 2,3-butanedione dioxime (dimethylglyoxime), and other reports detail alternative routes and emphasize some difficulties encountered.[3-7] We have found, however, that more reproducible procedures and more nearly pure products are obtained when a preformed cobaloxime is alkylated rather than obtaining the cobaloxime by a direct synthesis. Bromobis[2,3-butanedione dioximato(1-)](dimethyl sulfide)cobalt(III), prepared here by a modification of the procedure of Hill and Morallee,[4] serves as a very useful starting material, since both axial ligands are readily exchanged and the material can be prepared in good yield and purity. The majority of cobaloximes reported in the literature have pyridine as a sixth axial ligand, but we have found that the commercially available 4-*tert*-butylpyridine (4-*tert*-bupy) is superior, since it decreases the solubility of the cobaloxime in water (which makes isolation of products easier) and increases the solubility of cobaloxime in organic solvents. In addition, the *tert*-butyl group serves as a useful nmr marker.

In general, alkylation of a cobalt(I)-containing complex, prepared by $NaBH_4$ reduction of the cobalt(III) complex, can be carried out in methanol.[7] However, when a powerfully electrophilic alkylating agent is used, both alkylation of the periphery of the macrocycle and reaction with the solvent methanol can prevent formation of any cobalt alkylation product. In these cases dimethylformamide can be used in place of methanol for the alkylation reaction.

Neutral complexes are represented by $XCo(dh)_2B$, where X is a formally negative ligand (e.g., halide or alkyl carbanion), dh is the monoanion of dimethylglyoxime, and B is a neutral two electron-donor base.

A. BROMOBIS[2,3-BUTANEDIONE DIOXIMATO(1-)](DIMETHYL SULFIDE)COBALT(III)

$$Co(NO_3)_2 \cdot 6H_2O + 2[C(CH_3)(NOH)]_2 + NaBr + Me_2S + 0.25 O_2 \longrightarrow$$
$$BrCo(dh)_2(Me_2S) + HNO_3 + NaNO_3 + 6.5 H_2O$$

Procedure

■ **Caution.** *Dimethyl sulfide is volatile and both toxic and malodorous, and all operations involving its use, including unpacking and the opening of containers, should be conducted in an efficient hood.*

To a solution of 2,3-butanedione dioxime (23.2 g, 0.2 mole) in 500 mL of boiling 95% ethanol in a 2-L Erlenmeyer flask is added a solution of cobalt(II) nitrate hexahydrate (29.0 g, 0.10 mole) in 500 mL of hot 95% ethanol. The red-brown solution is boiled on a hot plate for 5 minutes and then a solution of sodium bromide (15.8 g, 0.15 mole) in water (20 mL) is added. The brown solution is allowed to cool to 35°. Dimethyl sulfide (11 mL, 0.15 mole) is added, and the resulting solution is stirred vigorously for 8 hours with the flask open to the atmosphere. During the stirring a brown solid precipitates from solution. The solid is collected by suction filtration on a Büchner funnel and washed successively with 200-mL portions of water, ethanol, and diethyl ether. The fine, dark-brown crystals are air dried and dissolved in dichloromethane and the system is filtered. The filtrate is evaporated to give typically 20 g of product (yield is about 50%). Its purity, as it is isolated in the above procedure, is adequate for subsequent procedures of synthesis. The product can be purified further by recrystallization from dichloromethane–hexane. *Anal.* Calcd. for $C_{10}H_{20}N_4O_4SBrCo$: C, 27.85; H, 4.68; N, 12.99; Br, 18.53. Found: C, 27.61; H, 4.59; N, 13.03; Br, 18.64.

Properties

Bromobis[2,3-butanedione dioximato(1-)](dimethyl sulfide)cobalt(III) is a brown, crystalline, air-stable solid soluble in organic solvents, such as dichloromethane, chloroform, acetone, and tetrahydrofuran. It is considerably more soluble than the chloro analog. The dimethyl sulfide is readily replaced by other ligands having nitrogen, oxygen, and phosphorus donor atoms. Characteristic infrared absorptions of the cobalt-dimethylglyoxime macrocycle in solid KBr pellets occur at: 1760–1730 cm^{-1} (vw), ν_{O-H-O}[8]; 1550 cm^{-1} (s), $\nu_{C=N}$[6,9]; 1230 and 1090 cm^{-1} (vs), ν_{N-O}[10]; and 520 and 425 cm^{-1} (m), $\nu_{Co-N(OH)}$[11]. The nmr absorptions unambiguously identify the stoichiometry of the monosubstituted bromo(dimethyl sulfide) derivative. In CDCl$_3$, two singlets occur in a ratio of 2:1 at δ_{CH_3} 2.48 ppm and $\delta_{S(CH_3)_2}$ 1.52 ppm from tetramethylsilane.

Preparation of Related Compounds

The procedure can be satisfactorily used to synthesize directly other base-substituted bromo-cobaloximes. Best results for the less volatile bases such as substituted pyridines are obtained by adding the base to the hot solution, subsequently cooling to room temperature, and bubbling air through the solution for 1 hour. The precipitate is isolated and purified as described for the bromo-(dimethyl sulfide) compound. Preparation of derivatives of bases that are sensitive to air and moisture, such as the alkyl phosphines, is best accomplished by starting with bromobis[2,3-butanedione dioximato(1-)](dimethyl sulfide)cobalt(III) and converting by the procedure described for the formation of bromobis[2,3-butanedione dioximato(1-)] (4-*tert*-butylpyridine)cobalt(III).

B. BROMOBIS[2,3-BUTANEDIONE DIOXIMATO(1-)]- (4-*tert*-BUTYLPYRIDINE)COBALT(III)

$$CoBr(dh)_2 Me_2 S + 4\text{-}t\text{-}Bu\text{-}C_5 H_5 N \longrightarrow CoBr(dh)_2 (4\text{-}tert\text{-bupy}) + Me_2 S$$

Procedure

To a magnetically stirred solution of bromobis[2,3-butanedione dioximato(1-)]-(dimethyl sulfide)cobalt(III) (8.6 g, 0.02 mole) (see above) and dry dichloromethane (100 mL) in a 500-mL round-bottomed flask is added 3.4 mL of 4-*tert*-butylpyridine (4-*tert*-bupy) (3.1 g, 0.02 mole). After the solution is stirred for 1 hour, the solvent is removed under reduced pressure, keeping the temperature below 25°. The resulting brown residue is triturated with diethyl ether, and the solid is collected by suction filtration and washed with additional diethyl ether. The brown powder is then stirred in 100 mL of water for 10 minutes, isolated by filtration, washed well with water, and air dried to give 9.2 g of bromobis-[2,3-butanedione dioximato(1-)] (4-*tert*-butylpyridine)cobalt(III) (yield is typically 90%). *Anal.* Calcd. for $C_{17}H_{27}BrN_5O_4Co$: C, 40.49; H, 5.40; N, 13.89; Br, 15.85. Found: C, 40.51; H, 5.55; N, 14.06; Br, 16.00.

Properties

Bromobis[2,3-butanedione dioximato(1-)] (4-*tert*-butylpyridine)cobalt(III) is a tan, microcrystalline solid with greatly enhanced solubility in organic solvents compared to the chloro(pyridine) analogue. It is also the compound of choice in preparing alkylcobaloximes by the subsequent procedure because of the ease of isolation of the resultant products. In addition, the bromo(4-*tert*-bupy) species react directly with electron-rich olefins, such as ethyl vinyl ether, in the presence of ethanol to yield, in this case, bis[2,3-butanedione dioximato(1-)]- (2,2-diethoxyethyl)(pyridine)cobalt(III).[12] Conversion of the dimethyl sulfide compound to the pyridine derivatives is readily detected by a characteristic infrared absorption at 1600 cm^{-1} (pyridine stretch). The ^1H nmr spectrum of bromobis[2,3-butanedione dioximato(1-)] (4-*tert*-butylpyridine)cobalt (III) has absorptions in the alkane region in the ratio of 3:4 at δ 1.25 ppm [Py—C(C$\underline{H}_3$)$_3$] and δ 2.43 ppm (dh—C$\underline{H}_3$) from tetramethylsilane.

Preparation of Related Compounds

Conversion of bromobis[2,3-butanedione dioximato(1-)](dimethyl sulfide)-cobalt(III) to other base-substituted cobaloximes is readily accomplished by substitution of an equivalent amount of the desired base for 4-*tert*-butylpyridine. If the base is air sensitive, as is the case for the trialkylphosphines, the reaction is carried out under a stream of nitrogen. Since the phosphine and phosphite derivatives tend to form the ionic derivatives, $[CoB_2(dh)_2]^+[CoX_2(dh)_2]^-$, the formation of the neutral complexes is favored by performing the reaction and evaporating the solvent at 0°. Small amounts of ionic impurities that might be formed are removed by extraction of the product with water. The 2,3-butanedione dioxime methyl absorption of the phosphine analogs are doublets with a J_{P-H} of about 2 Hz that is centered at δ 2.35 ppm from tetramethylsilane (in $CDCl_3$) for the bromobis[2,3-butanedione dioximato(1-)](tributylphosphine)-cobalt(III).

C. BIS[2,3-BUTANEDIONE DIOXIMATO(1-)] (4-*tert*-BUTYLPYRIDINE)-(2-ETHOXYETHYL)COBALT(III)

$$BrCo(dh)_2(4\text{-}tert\text{-bupy}) + EtOCH_2CH_2Br \xrightarrow[CH_3OH]{[H^-]}$$
$$(EtOCH_2CH_2)Co(dh)_2(4\text{-}tert\text{-bupy}) + 2Br^-$$

Procedure

A suspension of bromobis[2,3-butanedione dioximato(1-)](4-*tert*-butylpyridine)cobalt(III) (2.5 g, 5.0 mmole) (see above) in 20 mL of methanol in a 100 mL, three-necked flask is vigorously stirred under an N_2 purge for 10 minutes. While it is stirred and purged with N_2, solid sodium tetrahydroborate (0.5 g, 13 mmole) is gradually added in about 0.1-g portions. Upon addition of the sodium tetrahydroborate, the solution becomes warm, shows considerable effervescence due to hydrogen evolution, and becomes homogeneous. The development of the dark blue-black color is characteristic of the reduced Co(I) species. Then 2-bromoethyl ethyl ether (0.56 mL, 5.0 mmole) is added and the solution turns red-brown, indicating the presence of the alkylcobalt(III) species. The reaction mixture is then stirred for 15 minutes. To obtain maximum yields, another 0.25 g of solid sodium tetrahydroborate is added, followed by addition of an extra 0.25 mL of 2-bromoethyl ethyl ether. The reaction mixture is stirred for another 10 minutes. Further additions of sodium tetrahydroborate and 2-bromoethyl ethyl ether as described above, are continued until no noticeable darkening of the solution is observed upon addition of a small quantity of the reducing agent. The reaction mixture is then poured into 50 mL of distilled water containing 10-15 g of ice. An orange-yellow solid precipitates almost immediately. It is collected by suction filtration, washed well with distilled water, and finally air dried to give typically 1.9 g of product (yields are about 60%). The alkylcobaloxime can be purified by recrystallization from hot dichloromethane by adding hexane until the cloud point and subsequent cooling. Chromatographically pure compound can be obtained by elution of the (4-*tert*-butylpyridine)-(2-ethoxyethyl)cobaloxime on a silica gel column using 5% methanol–benzene. *Anal.* Calcd. for $C_{21}H_{36}N_5O_5Co$: C, 50.70; H, 7.29; N, 14.09. Found: C, 51.02; H, 7.27; N, 14.11.

Properties

Bis[2,3-butanedione dioximato(1-)] (4-*tert*-butylpyridine)(2-ethoxyethyl)cobalt-(III) is an air-stable, yellow-orange crystalline solid. The properties are similar to those of other alkylcobaloximes previously reported. Chromatographically pure compound can be obtained by elution of the (4-*tert*-butylpyridine)-(2-ethoxyethyl) complex on a silica gel column using 5% methanol–benzene or on an alumina column with 2% methanol–dichloromethane. The ^1H nmr spectrum in $CDCl_3$ in the alkyl region is consistent with the proposed product having the following absorptions in the appropriate intensity ratios: δ 1.09 ppm (triplet, —$CH_2C\underline{H}_3$), δ 1.28 ppm (singlet, Py—$C(C\underline{H}_3)_3$), δ 1.60 ppm (multiplet, $CoC\underline{H}_2CH_2$), δ 2.12 ppm (singlet, dh—$C\underline{H}_3$), δ 3.10 ppm (multiplet, $CoCH_2C\underline{H}_2$), δ 3.32 ppm (quartet, —O $C\underline{H}_2CH_3$) from tetramethylsilane.

D. BIS[2,3-BUTANEDIONE DIOXIMATO(1-)] (4-*tert*-BUTYLPYRIDINE)-(ETHOXYMETHYL)COBALT(III)

$$BrCo(dh)_2(4\text{-}tert\text{-bupy}) + EtOCH_2Cl \xrightarrow[DMF]{[H^-]} (EtOCH_2)Co(dh)_2(4\text{-}tert\text{-bupy}) + Br^- + Cl^-$$

Procedure

■ **Caution.** *The sodium tetrahydroborate should be added slowly to prevent foaming of the reaction mixture from the reaction vessel. Alkyl chloromethyl ethers are potent carcinogens*[13] *and should be handled and disposed of with appropriate precautions.*

A three-necked, 125-mL, round-bottomed flask fitted with a nitrogen inlet valve and a magnetic stirring bar is flamed and cooled under a gentle flow of nitrogen. A positive nitrogen atmosphere is maintained throughout the following alkylation procedure. Bromobis[2,3-butanedione dioximato(1-)] (4-*tert*-butylpyridine)cobalt(III) (1.00 g, 2.0 mmole) (see above) and dimethylformamide (35 mL, purged with nitrogen and dried over 5A molecular sieves) are placed in the flask to give a yellowish-brown solution that is cooled to $-12°$ (salt/ice bath). Sodium tetrahydroborate (0.076 g, 2.0 mmole) is added with 5 mL of DMF to give a dark-green solution (gas evolution). After 5 minutes, chloromethyl ethyl ether (0.40 mL, about 4.0 mmole) is added and the stirred mixture is warmed to room temperature. Dilution of the green solution with water (500 mL) gives a yellow-brown solution that is extracted with dichloromethane (4 × 50 mL). The combined organic layers are then washed with water (3 × 150 mL), 5% sodium hydroxide (2 × 25 mL), and saturated sodium chloride solution (2 × 50 mL) and dried over sodium sulfate. After filtration, hexane is added to the filtrate until the solution just becomes cloudy. The mixture is then reduced to half its volume, and the resulting yellow crystals (0.27 g, 29%) are collected by filtration. An analytical sample may be prepared by recrystallization from dichloromethane/hexane. *Anal.* Calcd. for $C_{20}H_{34}N_5O_5Co$: C, 49.69 H, 7.09; N, 14.49. Found: C, 49.72; H, 7.11; N, 14.56.

Properties

Bis[2,3-butanedione dioximato(1-)] (4-*tert*-butylpyridine)(ethoxymethyl)cobalt-(III) is an air-stable, yellow-orange crystalline solid with properties similar to those of the other alkylcobaloximes described above. The ^{1}H nmr spectrum in CDCl$_3$ shows δ 1.00 ppm (*t*, O—CH$_2$—CH$_3$); δ 1.28 ppm (s, PyC(CH$_3$)$_3$); δ 2.10 ppn (singlet, dh—CH$_3$); δ 3.27 ppm (quartet, O—CH$_2$—CH$_3$); δ 4.30 ppm (s, Co—CH$_2$—O).

Preparation of Related Compounds

Other alkyl(4-*tert*-butylpyridine)cobaloximes are readily synthesized using this procedure. Derivatives where the alkyl moiety is —CH$_3$, —CH$_2$CH$_3$, —CH$_2$OCH$_2$CH$_3$, —CH$_2$CH$_2$OCH$_3$, —CH$_2$CH(OCH$_3$)$_2$, —CH$_2$CH$_2$OC(O)-CH$_2$CH$_3$, and —CH$_2$Si(CH$_3$)$_3$ are isolated in about 50% yield when the alkyl bromide is used as the source of the alkyl group. If the chloroalkyl analogs are employed in the reaction, the yields, in general, are poor. Sometimes it is necessary to reflux the reaction mixture before product work-up. It should be noted also that the bromobis[2,3-butanedione dioximato(1-)](dimethyl sulfide)-cobalt(III) is not readily converted to the corresponding alkyl derivative by this

134 Compounds of Biological Interest

procedure. The phosphine-substituted cobaloximes generally react sluggishly, even with the alkyl bromides, but may be satisfactorily employed in this method if the reaction mixture is made alkaline by addition of a small quantity of aqueous NaOH or by gentle heating at reflux before carrying out the isolation step.

References

1. G. N. Schrauzer, *Acc. Chem. Res.*, **1**, 97 (1968).
2. G. N. Schrauzer, *Inorg. Synth.*, **11**, 61 (1968).
3. W. C. Trayler, R. C. Stewart, L. A. Epps, and L. G. Maryille, *Inorg. Chem.*, **13**, 1564 (1974).
4. H. A. O. Hill and K. G. Morallee, *J. Chem. Soc. (A)*, **1969**, 554.
5. G. Costa, G. Trayler, and A. Runedder, *Inorg. Chem. Acta*, **3**, 45 (1969).
6. N. Yamazaki and Y. Hokokabe, *Bull. Chem. Soc. Jap.*, **44**, 63 (1971).
7. G. N. Schrauzer and G. Kratel, *Chem. Ber.*, **102**, 2392 (1969).
8. A. Nakahara, J. Fujito, and R. Tsuchida, *Bull. Chem. Soc. Jap.*, **29**, 296 (1956).
9. G. N. Schrauzer and R. J. Windgassen, *J. Am. Chem. Soc.*, **88**, 3738 (1968).
10. R. Blenc and D. Hodge, *J. Chem. Soc.*, **1958**, 4536.
11. M. D. Benlian and G. Hernandorena, *C. R.*, **272**, 2001 (1971).
12. R. B. Silverman and D. Dolphin, *J. Am. Chem. Soc.*, **98**, 4626 (1976) and references therein.
13. *Chem. Eng. News*, Feb. 11, 1977, p. 12.

33. COBALAMINS AND COBINAMIDES

Submitted by D. DOLPHIN,* D. J. HALKO,* and R. B. SILVERMAN†
Checked by H. P. C. HOGENKAMP‡ and D. L. ANTON‡

Methylcobalamin (I, R = —CH_3) and the vitamin B_{12} coenzyme (I, R = 5′-deoxyadenosyl) are the only known naturally occurring organometallic compounds. Both are derivatives of vitamin B_{12} (cyanocobalamin, I, R = —CN) and both can be synthesized from B_{12} but are best prepared from hydroxocobalamin. These and numerous other derivatives of B_{12} containing a cobalt-carbon bond are known, and provided the cobalt is bonded to a primary carbon atom the complexes are thermally very stable, but always photochemically labile as a result of homolytic cleavage of the cobalt-carbon bond.

*Department of Chemistry, The University of British Columbia, Vancouver, B.C., Canada V6T 1W5.
†Department of Chemistry, Northwestern University, Evanston, IL 60201.
‡Department of Biochemistry, Medical School, University of Minnesota, Minneapolis, MN 55455.

Two principal methods for formation of the cobalt-carbon bond are used.[1] The first involves alkylation of B_{12s} (where the cobalt 3+ of B_{12} has been reduced to Co(1+)) with a primary alkyl halide, or tosylate, or Michael addition of B_{12s}, to an alkyne or activated alkene. While a wide variety of reducing agents will bring about the two-electron reduction of B_{12} to B_{12s}, sodium tetrahydroborate is recommended for simplicity and reproducibility. However, reduction of the commercially available cyanocobalamin may in some cases present problems when the cyanide displaces the alkyl from the cobalamin during work-up. This problem is readily overcome by using hydroxocobalamin (I, R = —OH) that is prepared from cyanocobalamin by way of methylcobalamin.

A second method for the formation of a cobalt-carbon bond directly from the Co(3+) oxidation level involves the reaction of hydroxocobalamin with an enol ether,[2] and typical examples showing both routes are given here.

For most inorganic chemists the peripheral groups of the cobalamins, which play important roles in the enzymology of the natural systems, only add to the overall complexity of the molecule, as well as impose severe restrictions on their solubility, polarity, and volatility. All these properties can be dramatically modified by converting vitamin B_{12} into dicyanocobyrinic acid heptamethyl ester (II, R = R' = —CN), which is exceedingly soluble in most organic solvents (compared to B_{12} which has a low solubility only in very polar solvent such as water, DMF, and DMSO) and is sufficiently volatile that, with care, useful mass spectra can be obtained using a direct insertion probe. Like B_{12} and the cobaloximes this dicyanocobyrinic acid heptamethyl ester can be reduced to the Co(1+) stage and alkylated with alkyl halides.*

II

A. METHYLCOBALAMIN

$$\text{cyanocobalamin (I, R = -CN)} \xrightarrow{\text{NaBH}_4} [\text{B}_{12s}] \xrightarrow{\text{CH}_3\text{I}} \text{methylcobalamin (I, R = -CH}_3)$$

Procedure

A solution of cyanocobalamin (500 mg) (available from Merck) and cobalt nitrate (2 mg) in water (40 mL) is placed in a 100-mL Erlenmeyer flask closed with a serum cap. The solution is deoxygenated with argon, and then sodium tetrahydroborate (100 mg) in water (5 mL) is slowly added to the magnetically stirred solution. The deep-red color of B_{12} immediately changes to the brown color of B_{12r} (Co(2+)) and then more slowly (about 10 min) to the blue-green† color of B_{12s}. Iodomethane (1.0 g) is added in one portion and the mixture immediately turns orange-red. Frequently, reduction to B_{12s} is not complete; nonetheless the iodomethane should be added after 10 minutes, since it intercepts B_{12s} as soon as it is formed, which drives the reaction to completion. Since alkylcobalamines are light sensitive in solution, all subsequent manipulations should be carried out in a low light intensity and all solutions should be shielded with aluminum foil.

The solution is brought to pH 7.0 with dilute HCl, to destroy excess tetrahydroborate, and methylcobalamin is desalted by extraction through phenol as

---

*The authors are grateful to Merck and Co., Inc., Rahway, NJ, for a generous gift of cyanocobalamin.
†In artificial light the color may appear red-purple.

follows. (■ **Caution.** *Phenol and solutions of phenol in dichloromethane are extremely corrosive and great care should be taken especially while working in dim light.*) The aqueous solution is extracted with 90% aqueous phenol (10 mL). The aqueous layer is extracted with additional aliquots of aqueous phenol (10 mL) until no color remains in the aqueous phase.* The combined phenolic extracts are washed with distilled water (2 × 10 mL), after which the organic layer is diluted to 10 times its original volume with dichloromethane. Methylcobalamin is then extracted from the organic layer with 15-mL portions of distilled water until most of the color is removed from the organic layer. The combined aqueous extracts are then washed with dichloromethane (5 × 100 mL). The solution of methylcobalamin is reduced in volume to about 5 mL. At this stage the methylcobalamin is pure enough for most purposes and may be crystallized (see below). However, during reduction to B_{12s} some irreversible reduction of the corrin chromophore occurs and this reduction product may be removed by chromatography on carboxymethyl cellulose (CMC).

A column (50 × 3 cm) is packed with an aqueous slurry of CMC.† The CMC is washed with 0.1 N HCl (20 mL) and then with distilled water until the washings are neutral. The concentrated solution of methylcobalamin is applied to the column, which is then eluted with distilled water. A small first fraction may run ahead of the main band of the product and it should be discarded. All the methylcobalamin is then eluted by water, leaving on the column a tightly bound orange impurity. The methylcobalamin fraction is concentrated to a volume corresponding to 15-25 mg/mL on the rotatory evaporator (bath temp <60°) and diluted with approximately 7 volumes of acetone when the mixture just becomes turbid. Upon standing overnight the mixture deposits large bright-red crystals of methylcobalamin, which are collected by filtration, washed with acetone, and air dried to yield 430 mg (85%). Combustion analyses on methylcobalamin or any other cobalamins never give satisfactory results, and the purity of the product should be checked by optical and chromatographic techniques.[1]
$\lambda_{max}^{H_2O}(\epsilon)$: 316 (12,700), 343 (13,500), 376 (11,400), 524 (8700) nm.

Properties

Methylcobalamin is a red crystalline solid that is stable in the solid state under ambient conditions. In solution, methylcobalamin is very light sensitive, undergoing homolytic cleavage of the cobalt carbon bond. In dilute acid, protonation of the axial benzimidazole group causes a dramatic change in color from red to

---

*Occasionally, emulsions may form. This can be minimized by not shaking the mixture too vigorously, by using phenol/dichloromethane for the extractions as described in Reference 1, or by breaking the emulsion by centrifugation.

†The checkers preferred to use Dowex 50 ion exchange resin in place of CMC, and elute with 0.10 M sodium acetate, pH 6.5.

yellow that can be used to assay its purity. Similarly, photolysis in the presence of cyanide gives cyano- or dicyanocobalamin, which can again be used analytically. Methylcobalamin is soluble in water and DMSO, less soluble in DMF, acetic acid, and ethanol, and insoluble in most other organic solvents. The preparation of B_{12s} and its alkylation by both S_N2 reactions and Michael addition to give alkyl-, vinyl-, and ethynyl-cobalamins have been thoroughly reviewed.[1]

B. HYDROXOCOBALAMIN

$$\text{Methylcobalamin} \xrightarrow[H_2O]{\text{light}} \text{Hydroxocobalamin}$$

A solution of methylcobalamin (850 mg) in water (250 mL) is placed in direct sunlight (or any other source of visible light such as a 200 W tungsten light bulb) until all the methylcobalamin has been photolyzed. Photolysis is complete when the optical spectrum of an aliquot, diluted in water so that its spectrum can be measured in a 1-cm cuvette, does not change upon photolysis for a further 5 minutes in the cuvette. The solution is concentrated on a rotatory evaporator (bath temp <70°) to a volume corresponding to 15-25 mg/mL, and acetone (7X) is added, at which stage the mixture just becomes turbid. After 2 days, shining burgundy crystals (650 mg) are collected by filtration, washed with acetone, and air dried. The combined filtrates are reduced in volume to about 1 mL and brought to the cloud point with acetone to give, upon standing, a second crop of 60 mg, giving a total yield of 710 mg (84%). $\lambda_{max}^{H_2O}(\epsilon)$: 357 (17,500), 513 (7200), 538 (7670) nm.

Properties

Hydroxocobalamin is a deep-red crystalline solid that is stable both in the solid state and in solution at ambient conditions. It is readily soluble in water and dimethyl sulfoxide, less soluble in dimethylformamide, ethanol, and acetic acid, and insoluble in most other organic solvents. Its purity may be checked by optical and chromatographic techniques.[1] Treatment of hydroxocobalamin with a variety of negatively charged halide and pseudohalide ligands gives the corresponding cobalamins.

C. (2,2-DIETHOXYETHYL)COBALAMIN [I, R = —CH$_2$CH(OEt)$_2$]

$$\text{Hydroxocobalamin} \xrightarrow[\text{EtOH}]{H_2C=CH(OEt)} \text{(2,2-diethoxyethyl)cobalamin}$$

Procedure

To a deaerated solution of hydroxocobalamin (25 mg, 1.8×10^{-5} mole) in absolute ethanol (4 mL) in a 10-mL Erlenmeyer flask covered with a serum cap is added triethylamine (1 drop, purfied by distillation from 1-naphthyl isocyanate) followed by ethyl vinyl ether (0.16 mL, 1.5×10^{-3} mole). The red solution is allowed to stand in the dark under nitrogen until the reaction is complete (about 3 days). The reaction is followed by observing the decrease in the optical absorption at 357 nm. Fibrous red needles of the product that crystallize from solution are collected by filtration, washed with a minimum of acetone, and air dried to give 20 mg (80%). $\lambda_{max}^{(H_2O-K_2CO_3)}(\epsilon)$: 263 (21,600), 281 (19,900), 288 (sh) (18,000), 323 (12,900), 338 (13,600), 373 (11,400), 432 (sh) (4640), 490 (sh) (6610), 525 (8420) nm.

Properties

Recrystallization from water/acetone (even at a basic pH) causes some hydrolysis of the acetal. (2,2-Diethoxyethyl)cobalamin, like all alkyl cobalamins, is light sensitive in solution; in addition, and unlike most other alkylcobalamins, it is acid sensitive, decomposing to both (formylmethyl)cobalamin and aquacobalamin.[2] The formation of other alkylcobalamins and cobaloximes by reaction with enol ethers has been described.[2]

D. DICYANOCOBYRINIC ACID HEPTAMETHYL ESTER[3] (II, R = R' = —CN)

$$\text{Vitamin B}_{12} \text{ (I)} \xrightarrow[\text{(2) 2CN}^-]{\text{(1) CH}_3\text{OH/H}_2\text{SO}_4} \text{heptamethyl dicyanocobyrinate}$$

Procedure

The reaction is carried out in a 2-L, three-necked, round-bottomed flask fitted with a nitrogen-inlet tube, condenser, magnetic stirring bar, and stopper. Boiling chips and 500 mL of absolute methanol are added to the flask, which is flushed with nitrogen for $\frac{1}{2}$ hour before cyanocobalamin (6.77 g) is added (available from Merck). To this solution 300 mL of methanolic H_2SO_4* is added with stirring. The red solution is refluxed under nitrogen for 5 days. After the reaction mixture is cooled, the volume is reduced to 150 mL on a rotary evaporator and poured into 1 L of ice-cold distilled water. After neutralization with

---

*Prepared by slowly adding 50 mL of concentrated H_2SO_4 to 250 mL of absolute methanol and outgassing with nitrogen for $\frac{1}{2}$ hour just prior to use.

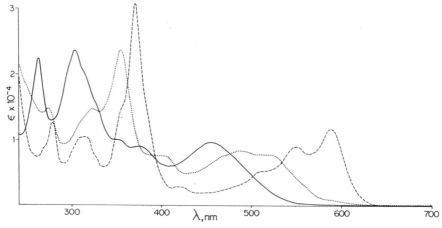

Fig. 1. Optical absorption spectra in CH_2Cl_2, of (- - -) dicyano, (· · · ·) cyanoaquo, and (———) methylaquo.

NaHCO$_3$, KCN (4.0 g) dissolved in water (15 mL) is added with stirring. Upon addition of KCN, the red solution turns purple with the formation of a purple precipitate. This mixture is first extracted with CCl$_4$ (4 × 200 mL) and then with CH$_2$Cl$_2$ (3 × 200 mL). The CH$_2$Cl$_2$ extracts are combined and taken to dryness on a rotary evaporator. This solid is saved and can be added to cyanocobalamin if the esterification reaction is repeated. After the CCl$_4$ is stripped on a rotary evaporator, the solid obtained from the combined CCl$_4$ extracts is dried under reduced pressure, dissolved in benzene (200 mL) and partitioned against 300 mL of saturated KCl/0.1% KCN. The benzene layer is filtered through 60 g of anhydrous Na$_2$SO$_4$, evaporated to dryness on a rotary evaporator, and dried overnight under vacuum. The solid is dissolved in benzene (100 mL) in a 250-mL Erlenmeyer flask and hexane is slowly added with stirring to precipitate the product. The flask is stoppered and set aside for 1 day. The dark-purple solid is filtered, washed with hexane, and dried *in vacuo*. Yield: 3.98 g (73%). *Anal.* Calcd. for C$_{54}$H$_{73}$N$_6$O$_{14}$Co: Co, 5.42, C, 59.56; H, 6.71; N, 7.72. Found: Co, 5.38; C, 59.84; H, 6.80; N, 7.50.

Properties

Dicyanocobyrinic acid heptamethyl ester is soluble in benzene, chloroform, and dichloromethane, and most other organic solvents. $\lambda_{max}^{CH_2Cl_2}$ ($\epsilon \times 10^{-4}$) 588 (1.20), 549 (0.928), 515$_{sh}$ (0.535), 420 (0.301), 371 (3.15), 353$_{sh}$ (1.44), 313 (0.990) and 277 (1.23) nm (see Fig. 1). The seven ester methyl groups appear as singlets from 3.76 to 3.63 ppm in the ^{1}H nmr spectrum (CDCl$_3$, tetramethylsilane internal standard).

E. AQUACYANOCOBYRINIC ACID HEPTAMETHYL ESTER PERCHLORATE

$$(CN)_2 Cby(OMe)_7 \xrightarrow{HClO_4/H_2O} (CN)(H_2O)Cby(OMe)_7 ClO_4$$

Procedure*

The dichloromethane (50 mL) solution of dicyanocobyrinic acid heptamethyl ester (500 mg) is placed in a 250-ml separatory funnel, and 30 mL of 30% $HClO_4$ is added. (■ *Caution. Perchlorates are explosive!*) The mixture is shaken until the solution changes from purple to orange. The organic phase is separated from the aqueous phase and washed with water (2 × 25 mL). The dichloromethane is stripped on a rotary evaporator, yielding a dark-red solid. Yield: 90%. *Anal.* Calcd. for $C_{53}H_{75}N_5O_{19}ClCo$: Co, 5.00; C, 53.94; H, 6.36; N, 5.94; Cl, 3.01. Found: Co, 4.61; C, 53.26; H, 6.11; N, 6.11; Cl, 2.89.

Properties

This perchlorate derivative is soluble in methanol, dichloromethane, and chloroform. $\lambda_{max}^{CH_2Cl_2}$ ($\epsilon \times 10^{-3}$) 520_{sh} (6.81), 487 (7.96), 400_{sh} (6.56), 354 (21.7), 322 (13.4), and 272 (12.9) nm (see Fig. 1).

F. AQUAMETHYLCOBYRINIC ACID HEPTAMETHYL ESTER PERCHLOROATE

$$(CN)(H_2O)Cby(OMe)_7 ClO_4 \xrightarrow[(2) CH_3I]{(1) NaBH_4} aq(Me)Cby(OMe)_7 ClO_4$$

Procedure

A solution of aquacyanocobyrinic acid heptamethyl ester perchlorate (515 mg) in absolute methanol (60 mL) is added to a 250-mL, three-necked, round-bottomed flask fitted with a nitrogen-inlet tube, a side arm containing $NaBH_4$ (160 mg), a serum cap, and a magnetic stirring bar. To this solution is added $Co(NO_3)_2 \cdot 6H_2O$ (5 mg) to catalyze the tetrahydroborate reduction. After nitrogen is bubbled through the solution for 45 minutes with stirring, the $NaBH_4$

---

*This procedure has been reported by Guschl and Brown; however, the product was not isolated or characterized. R. J. Guschl and T. L. Brown, *Inorg. Chem.* **12**, 2815 (1973).

is added by turning the sidearm.* After 3-4 minutes, oxygen-free iodomethane (1.5 mL)† is added through the serum cap with a hypodermic syringe. The solution immediately turns from green to red.‡ From this point, the solution should be protected from light since the cobalt-carbon bond is readily cleaved by light. After the reaction is allowed to go to completion (10 min.), the solution is poured into 75 mL of 5% aqueous $NaClO_4$/3% $HClO_4$ in a 500-mL separatory funnel. The product is then extracted with dichloromethane (60 mL). The red dichloromethane solution is separated from the colorless aqueous phase and evaporated to dryness on a rotary evaporator. Yield: 95%. *Anal.* Calcd. for $C_{53}H_{78}N_4O_{19}ClCo$: C, 54.43; H, 6.72; N, 4.79. Found: C, 55.64; H, 6.77; N. 5.29.

Properties

Heptamethyl aquamethylcobyrinate is obtained as a dark-red solid, solutions of which should be protected from light. Photolysis leads to homolytic cleavage of the cobalt-carbon bond, yielding the Co(III) complex in the presence of oxygen. $\lambda_{max}^{CH_2Cl_2}$ ($\epsilon \times 10^{-4}$) 454 (0.948), 374 (0.876), 350 (0.964), 328_{sh} (1.47), 317_{sh} (1.90), 303 (2.29), and 266 (2.15) nm (see Fig. 1). The seven ester methyl groups appear between 3.52 and 3.72 ppm and the cobalt-bound methyl appears at -0.24 ppm in the ^1H nmr spectrum ($CHCl_3$ solution, tetramethylsilane external standard).

References and Notes

1. D. Dolphin, *Methods Enzymol.*, **18C**, 34 (1971).
2. R. B. Silverman and D. Dolphin, *J. Am. Chem. Soc.*, **98**, 4626 (1976).
3. The procedure reported here was developed at Harvard (D.D.), was greatly improved upon by R. B. Woodward and his collaborators, and finally was modified to give that form presented here.

---

*Vigorous gas evolution occurs at this point, so it is necessary to have a sufficiently large nitrogen exit to avoid a pressure buildup in the reaction flask.

†Oxygen-free iodomethane is prepared in a two-necked, round-bottomed flask fitted with a nitrogen inlet-outlet tube and serum stopper. Nitrogen is bubbled through the iodomethane prior to use.

‡A green color [Co(I)] is not always obtained after reduction. A purplish-brown color sometimes results and is probably due to appreciable amounts of Co(II). These solutions still yield the desired product.

34. METALLOPORPHINES

Submitted by E. C. JOHNSON* and D. DOLPHIN*
Checked by M. A. CUSHING, JR.† and S. D. ITTEL†

Essentially all metals and metalloids can be coordinated to porphines,[1] but unlike the phthalocyanins, where a template reaction can frequently be employed to give the metallo complex, preformed macrocyclic porphines themselves must be metalated. Adler et al. have recently described a method that is general for the preparation of many metalloporphines[2]; there are, however, some systems that can neither be metalated using dimethylformamide, or other high boiling basic solvents, nor give quantitative metalation and so require special purification procedures. Complexes in this category are the nickel and vanadyl derivatives of 5,10,15,20-tetraphenyl-21H,23H-porphine (*meso*-tetraphenylporphyrin, H_2tpp)[3] and magnesium complexes, in which case other methods for metalation must be employed.

A. [5,10,15,20-TETRAPHENYL-21H,23H-PORPHINATO(2-)]-NICKEL(II)

The free base 5,10,15,20-tetraphenyl-21H,23H-porphine (H_2tpp) may be prepared from pyrrole and benzaldehyde[4] or purchased from Strem Chemicals, P.O. Box 108, Newbury Port, MA 01950 or Aldrich Chemical Co., 940 St. Paul Ave., Milwaukee, WI 53233. The material should be chlorin free, which can be effected either by dry column chromatography using alumina and chloroform or by oxidation of the chlorin to porphyrin using 2,3-dichloro-5,6-dicyano-

---

*Department of Chemistry, The University of British Columbia, Vancouver, B.C. Canada V6T 1W5.
†Central Research and Development Department, Experimental Station, E. I. duPont de Nemours and Co., Inc., Wilmington, DE 19898.

benzoquinone.[5] The purity of the tetraphenylporphine may be determined spectroscopically.[5]

Procedure

Tetraphenylporphine (1 g, 1.63 mmole) is dissolved in a minimum volume of trifluoroacetic acid (about 15 mL) and the solution is added to glacial acetic acid (200 mL) that has been heated at reflux in a 500-mL, round-bottomed flask. Nickel diacetate tetrahydrate (2 g, 8.0 mmole) is dissolved in a minimum of hot acetic acid and the resulting solution is added dropwise over a period of 5 minutes to the refluxing tetraphenylporphine solution. The mixture is allowed to heat at reflux for a further hour, and the hot solution is then filtered. The product is washed with boiling acetic acid until the filtrate is colorless and then further washed with water (3 × 25 mL) and methanol (3 × 25 mL). The purple crystalline product is dried in vacuum to give 1.05 g (96%). *Anal.* Calcd. for $C_{44}H_{28}N_4Ni$: C, 78.71; H, 4.20; N, 8.34. Found: C, 79.05; H, 4.32; N, 8.22.

Properties

The Ni(tpp) prepared by this method is of analytical purity. It is slightly soluble in $CHCl_3$, CH_2Cl_2, and ligating solvents, such as pyridine. It is insoluble in acetic acid, water, and alcohols. The material is indefinitely stable in the solid state and is stable in solution when protected from the light. The electronic absorption spectrum in dichloromethane shows maxima at 413 and 527 nm with molar absorptivities of 275,750 and 18,450, respectively.

B. OXO[5,10,15,20-TETRAPHENYL-21*H*,23*H*-PORPHINATO(2-)]-VANADIUM(IV)

Procedure

Chlorin-free H_2tpp (see above) (2.00 g, 3.26 mmole) is dissolved in refluxing dimethylformamide (DMF) (400 mL), and a solution of vanadyl sulfate (4.0 g,

2.5 mmole, dissolved in a minimum of hot water) is added dropwise, over a period of 10 minutes, to the tetraphenylporphine solution. Waters of solvation are removed by distilling off half the original solvent and then adding an additional 200 mL of DMF. The mixture is heated at reflux for an additional 7 hours or until the metal insertion is almost complete, as indicated by the appearance of the metalloporphine bands at 423 and 548 nm and the near absence of the H_2tpp bands at 417, 480, 512, 547, 588, and 646 nm. It should be noted that even upon the addition of more $VOSO_4$ and continued heating at reflux, metalation is never complete. The solution is cooled to room temperature and an equal volume of water is added. The precipitate is collected by filtration and washed with water (3 × 25 mL) and methanol (3 × 25 mL). The violet crystals are then continuously extracted with refluxing trifluoroacetic acid (TFA) through a sintered-glass frit of medium porosity to remove any unreacted H_2tpp. The remaining solid is again washed with water (3 × 25 mL) and methanol (3 × 25 mL) and dried for 24 hours in vacuum at 100° to give VO(tpp)·CH_3OH. Yield: 1.01 g (46%). *Anal.* Calcd. for $C_{45}H_{32}N_4O_2V$; C, 75.94; H, 4.53; N, 7.87. Found: C, 76.13; H, 4.43; N, 7.88.

A second crop can be recovered from the combined filtrates and washings of the TFA extraction by cooling in an ice bath for several hours and collecting the precipitate by filtration. This can increase the yield by 15%. Drying at 200° under high vacuum for 24 hours does not remove the methanol.

Properties

VO(tpp)·CH_3OH, is slightly soluble in $CHCl_3$ and CH_2Cl_2, less soluble in pyridine, and insoluble in water and alcohols. The material is stable in the solid state and stable in solution when protected from the light. The electronic spectrum contains bands at 423, 508, 547, and 581 nm with molar absorptivities of 50,100, 4900, 22,500 and 4160, respectively.

C. [2,3,7,8,12,13,17,18-OCTAETHYL-21*H*,23*H*-PORPHINATO(2-)]-MAGNESIUM(II)

Procedure

Octaethylporphine (H_2oep) can be prepared by either of two recently published procedures[6,7] or purchased from Strem Chemicals, P.O. Box 108, Newbury Port, MA 01950. Octaethylporphine (1.1 g, 1.9 mmole) is suspended in dry toluene (200 mL) under nitrogen in a predried 500-mL, round-bottomed flask fitted with a condenser and nitrogen bubbler. Bromoethylmagnesium solution (15 mL of an approximately 3 M solution) is added in one portion and the mixture is heated at reflux for 1 hour. Essentially all Grignard reagents, either commercial or prepared by various literature procedures, may be used. The checkers used bromomethylmagnesium with equal success. We have found, however, that bromophenylmagnesium and its homologs may give problems (when coupling to biphenyl occurs). After an hour, an additional 15 mL of the Grignard reagent solution is added and the mixture is refluxed for another hour. The mixture is then cooled to room temperature and saturated aqueous ammonium chloride (200 mL) is added. The remainder of the preparation can be carried out in air. After it is gently shaken, the mixture is separated, and the organic layer is washed with water (2 × 100 mL) and dried over anhydrous sodium sulfate (the use of anhydrous magnesium sulfate at this stage causes considerable demetalation). The mixture is filtered and the residue is washed with a little dry toluene. The combined filtrates are washed and taken to dryness, at room temperature, on a rotary evaporator. The residue is dissolved in dichloromethane (50 mL) (chloroform cannot be used since the traces of HCl it contains cause immediate demetalation). The mixture is boiled under efficient magnetic stirring and the volume is maintained by the addition of dry petroleum ether (bp 30-60°). When all the dichloromethane has been replaced by hydrocarbons, the mixture is allowed to stand at room temperature overnight. The product is collected by filtration, washed with petroleum ether, and dried in vacuum. Yield: 500 mg (48%). *Anal.* Calcd. for $C_{36}H_{46}N_4Mg$: C, 78.19; H, 7.24; N, 10.13. Found: C, 77.90; H, 7.40; N, 10.21.

It is our experience that attempts to recover more Mg(oep) from the filtrate is a fruitless task and that it is better to recover the octaethylporphine as the free base by the following method. The filtrate is treated with an equal volume of chloroform and washed with 5 N HCl (2 × 50 mL) and water (50 mL). The organic layer is dried over magnesium sulfate and filtered and the filtrate is taken to dryness. The residue is recrystallized from toluene to give H_2oep (500 mg).

Properties

[Octaethylporphinato(2-)] magnesium is readily demetalated by traces of acid. It is very soluble in dichloromethane and benzene but demetalates in chloroform. The solid is indefinitely stable and solutions are stable when protected from light and acid. The electronic spectrum has absorptions at 407, 542, and 579 nm with molar absorptivities of 375,350, 14,782, and 13,313, respectively (dichloromethane solution). All porphines may be metalated in the above manner unless their peripheral groups are reactive towards Grignard reagents. In such cases a new method, developed by Isenring et al.[8] for the insertion of magnesium, may be used.

References

1. J. W. Buchler, *Synthesis and Properties of Metalloporphyrins*, Vol. I, *The Prophyrins*, D. Dolphin, Ed., Academic, New York, 1978, Chapter 10.
2. A. Adler, F. R. Longo, and V. Váradi, *Inorg. Synth.*, **16**, 213 (1976).
3. F. A. Walker, E. Hui, and J. M. Walker, *J. Am. Chem. Soc.*, **97**, 2390 (1975).
4. A. D. Adler, F. R. Longo, F. Kampas, and J. Kim, *J. Inorg. Nucl. Chem.*, **32**, 2443 (1970).
5. K. Rousseau and D. Dolphin, *Tetrahedron Lett.*, **48**, 4251 (1974).
6. J. B. Paine III, W. B. Kirshner, D. W. Moskowitz, and D. Dolphin, *J. Org. Chem.*, **41**, 3857 (1976).
7. D. O. Cheng and E. Le Goff, *Tetrahedron Lett.*, **17**, 1469 (1977).
8. H.-P. Isenring, E. Zass, K. Smith, H. Falk, J.-L. Luisier, and A. Eschenmoser, *Helv. Chim. Acta*, **58**, 2357 (1975).

35. IRON PORPHINES

Submitted by C. K. CHANG,* R. K. DiNELLO,† and D. DOLPHIN†
Checked by A. B. P. LEVER‡ and B. S. RAMASWAMY‡

Iron porphines constitute the active site of hemoproteins and as such are among the most important and most widely studied series of metal complexes. The syntheses of iron protoporphyrin (heme) and its derivatives are important for the reconstitution of hemoproteins (particularly when one wishes to incorporate ^{57}Fe for Mössbauer studies), as well as for model studies. In addition to the naturally occurring porphyrins, iron complexes of both 5,10,15,20-tetraphenyl-$21H,23H$-porphine (H_2tpp) and octaethyl-$21H,23H$-porphine (H_2oep) are widely used in model studies of the natural systems.

General procedures for the metalation of porphyrins have been reviewed,[1,2] but the most commonly employed method for the insertion of iron into the por-

*Department of Chemistry, Michigan State University, East Lansing, MI 48824.
†Department of Chemistry, The University of British Columbia, Vancouver, B.C., Canada V6T 1W5.
‡Department of Chemistry, York University, Downsview, Toronto, Ontario, Canada.

phyrin macrocycle are those using iron(II) sulfate in acetic acid[3] and, more recently, refluxing dimethylformamide (DMF) as solvent.[4] While these two methods are applicable to a wide variety of porphyrins, there are other variations suitable for special applications. In addition, working with the water-soluble, naturally occurring dicarboxylic acid porphyrins presents additional considerations. It is difficult to generalize as to which method of iron incorporation should be used for a porphyrin with specific peripheral substituents: nonetheless a general rationale is outlined here.

Protoporphyrin and its derivatives, because of their labile vinyl substituents, are susceptible to acid-catalyzed hydration. For this reason aqueous solutions should be avoided in the preparation of protoporphyrin dimethyl ester from protohemin, and subsequent reinsertion of the iron. Thus the direct esterification of protohemin under basic conditions appears to be the method of choice. More stable porphyrins such as H_2oep are most conveniently prepared by the direct insertion of iron(III) ion using the iron(III) chloride method[5]; the resultant Fe(III) (oep)Cl can be separated directly from the reaction medium as shining purple crystals. However, this method fails to give successful results with protoporphyrin derivatives. In general, the iron(II) sulfate method is the mildest and should be used for all prophyrins unless one is certain that the more rapid alternatives [boiling DMF with iron(II) chloride, or preferably the more stable iron(II) perchlorate] will have no ill effect on the peripheral substituents.

A. CHLORO[DIMETHYL PROTOPORPHYRINATO(2-)]IRON(III),
Chloro[dimethyl 7,12-Diethenyl-3,8,13,17-tetramethyl-21H,23H-porphine-2,18-dipropionato-(2-)]iron(III)

Procedure 1 [Iron(II) Method]

Protoporphyrin IX dimethyl ester may be purchased (e.g., *Porphyrin Products, Sigma, Strem*) or prepared according to literature procedures.[1] A convenient

Fig. 1. Reaction vessel with a gas inlet and outlet attachment.

method for removal of iron with concurrent esterification (Grinstein) has been described in some detail.[6] The dimethyl ester should be purified by column chromatography on alumina with $CHCl_3$ as eluent. Protoporphyrin IX dimethyl ester is eluted first, followed by trace amounts of the 7(12)-(1-hydroxyethyl)-12(7)-vinyl derivatives and then hematoporphyrin dimethyl ester. Protoporphyrin IX derivatives are photolabile (to give green photoprotoporphyrins[1]). This irreversible photochemistry is especially troublesome with dilute solutions and during chromatography, but can be eliminated by working under an oxygen-free atmosphere.

The purified protoporphyrin dimethyl ester (59 mg, 0.10 mmole) is placed in a pear-shaped flask equipped with a gas-inlet tube (Fig. 1), dissolved in pyridine (1 mL), and diluted with glacial acetic acid (20 mL). A stream of argon (or nitrogen) is passed into the solution from the center tube while the mixture is placed in an oil bath preheated to 80°. A saturated aqueous solution of iron(II) sulfate or iron(II) chloride (1 mL) is syringed into the mixture, through the gas-outlet side arm. (It is not necessary to add a large excess of iron(II) salt. In this example, 0.11 mmole of iron(II) chloride is sufficient to effect quantitative metalation.) The temperature of the bath is raised to 90° and the reaction is kept at this temperature for 10 minutes. The flask is then removed from the heating bath, the argon bubbling is stopped, and the mixture is cooled to ambient temperature. A stream of air is passed into the solution briefly to allow autooxidation of the unstable [dimethyl protoporphyrinato(2-)]iron(II). The mixture is partitioned between water (60 mL) and chloroform (60 mL) in a separatory funnel. (■ *Caution. It is recommended that the manipulations using chloroform, a known carcinogen, be carried out in the hood.*) The organic phase is separated and washed first with 0.2 N HCl (60 mL) and then with water. The chloroform layer is separated and dried by filtering through a small filter paper. The chloro[dimethyl protoporphyrinato(2-)]iron(III) (48 mg, 74%) obtained in this manner is usually pure enough for most purposes, or it may be recrystallized from chloroform/methanol. The use of hard filter paper and a small funnel prevents loss of recrystallized material when it is collected by filtration.

The purity of the hemin ester can be checked by tlc (silica gel, with $CHCl_3$/methanol = 5:1), or by visible spectroscopy. A small amount of the hemin ester dissolved in pyridine can be reduced by sodium dithionite in a cuvette.[1,2] The resultant hemochrome should exhibit two absorption maxima at 557 and 526 nm. If necessary, the product can be further purified by chromatography on an alumina (grade III) column (1 × 10 cm) using a mixture of chloroform and methanol (10:1) as eluent. However, the iron prophyrin compound obtained after chromatography exists as the μ-oxo dimer, [Fe(III)-O-Fe(III)], which may be converted back to the chloride by treatment in $CHCl_3$ with dry HCl followed by crystallization from chloroform/methanol.

Procedure 2

Chloro[protoporphyrinato(2-)]iron(III) (1 g, 15.3 mmole) (Sigma) is dissolved in hexamethylphosphoramide (HMPA) (30 mL) with shaking until homogenous. (■ Caution. *The carcinogenicity of HMPA requires that all operations be carried out in the hood.*) To this solution 6 mL of aqueous KOH solution containing 0.9-g KOH pellets is added all at once. The mixture becomes very thick and precipitation soon occurs. Iodomethane (5 mL) is added and the reaction mixture is allowed to stand at room temperature for 3 hours with occasional shaking. The solution usually clears after about 30 minutes.

The solution is poured into H_2O (1 L) and filtered (Celite filtering aid may be needed). The filtrate is discarded and the residue is rinsed with chloroform until the washings become colorless. The chloroform solution is then washed with water (200 mL), separated, and taken to dryness. The crude product is chromatographed on alumina with $CHCl_3$/MeOH 10:1 as eluent. Yield: 0.80–0.88 g of the μ-oxo dimer, which can be crystallized from dichloromethane/methanol or converted to the chloride as described above.

Properties

Dimethyl protoporphyrinato(2-)iron(III), either as the chloride or the μ-oxo dimer, is fairly soluble in chloroform and dichloromethane. The iron complex is much more stable toward light than the parent protoporphyrin but is still susceptible to acid-catalyzed hydration. The ester can be hydrolyzed to yield pure protohemin by using the alcoholic-KOH procedure.[1,2] The optical spectra of these and other iron porphyrins have been reviewed recently.[7]

B. CHLORO[OCTAETHYL-21H,23H-PORPHINATO(2-)]IRON(III)

Procedure [Iron(III) Chloride Method]

Octaethyl-21H,23H-porphine (H_2oep) is now commercially available (Strem), but also may be prepared according to two recently reported syntheses.[8,9] H_2oep (54 mg, 0.10 mmole) and $FeCl_3 \cdot 6H_2O$ (33 mg, 0.12 mmole) are dissolved in glacial acetic acid (30 mL). A trace amount of sodium acetate (10 mg) is added and the solution is refluxed for 90 minutes. Fe(III)(oep)Cl crystals separate on subsequent standing at room temperature. The mixture is simply filtered (using a small filter funnel and hardened filter paper) and the crystals are washed with water, a small amount of methanol, and then diethyl ether and are then sucked to dryness. Yield: 59 mg (95%). *Anal.* Calcd. for $C_{36}H_{44}N_4FeCl$: C, 69.28; H, 7.11; N, 8.97. Found: C, 69.02; H, 7.28; N, 8.79.

Properties

Because of its symmetry and relatively high solubility in organic solvents, Fe(III)(oep)Cl has been used widely as a model compound in porphyrin chemistry.[2] It is very stable toward heat, light, acid, and base. The iron(III) ion can be removed only by concentrated sulfuric acid. The two absorption maxima in the visible region of the reduced hemochrome in pyridine are at 547 and 518 nm. Chromatography of Fe(III)(oep)Cl on alumina with $CHCl_3$ and methanol (10:1) as eluent, indicates quantitative conversion to the μ-oxo dimer.[5]

C. CHLORO[MESOPORPHYRINATO(2-)] IRON(III), (MESOHEMIN IX), CHLORO[7,12-DIETHYL-3,8,13,17-TETRAMETHYL-21H,23H-PORPHINE-2,18-DIPROPIONATO(2-)]-IRON(III)

Procedure

The preparation of metal-containing, water-soluble porphyrins presents special problems, not in the incorporation of the metal, but in the subsequent purification. We have found that these substances can be obtained in analytical purity by a final chromatographic step on polyamide, a system that is also useful for their analysis.[10]

Mesoporphyrin IX dihydrochloride may be purchased (Porphyrin Products) or prepared according to literature procedures.[11,12] Mesoporphyrin IX (600 mg, 10.6 mmole) is dissolved in a minimum volume of pyridine (about 2 mL), diluted with glacial acetic acid (25 mL), and placed in the apparatus shown in Fig. 1. The solution is heated to 90° under an inert atmosphere, and 2 mL of a freshly prepared saturated aqueous solution of iron(II) sulfate is added in one portion. When the reaction is complete (about 10 min) the solution is cooled to room temperature while air is passed through to oxidize the iron. The reaction is best monitored by following the absorption spectrum between 350 and 700 nm as the sharp Soret and four-banded visible spectrum is replaced by the more diffuse spectrum of the product. In addition, the nonmetalated starting material fluoresces strongly, while the product exhibits no fluorescence. At this stage the bulk of the iron(III) mesoporphyrin IX precipitates from solution as the chloride and is collected by filtration. The filtrate is reduced in volume to about 2 mL and water (25 mL) is added to precipitate the remaining iron(III) mesoporphyrin IX. The combined yield of iron mesoporphyrin IX is quantitative at this stage.

The crude mesohemin IX is dissolved in a minimum of pyridine, and sufficient silica gel (any grade suitable for column chromatography may be used) is added to soak up the liquid. The silica gel, with adsorbed mesohemin IX, is then dried overnight under high vacuum to remove pyridine. A mixture of benzene/ methanol/formic acid, 110:15:1 (v/v/v) is used to make a column (90 × 2.5 cm) of polyamide [100 g, Machery and Nagel polyamide (CC6 < 0.07 mm) available from Brinkman Inc., Westbury, NY]. The silica gel with adsorbed mesohemin IX is slurried in the same solvent mixture and applied, with care, to the top of the polyamide column. The column is then developed with the same solvent system at a flow rate of 6–9 mL/min. To achieve this flow rate, a positive pressure of nitrogen must be used. It is essential to use these high flow rates to prevent esterification to the methyl esters of the propionic acid side chains.

The fractions corresponding to the main (mesohemin) band are combined and taken down to dryness. The residue is dissolved in pyridine (2 mL) and filtered through a small plug of glass wool, which is then washed with pyridine/chloroform, 35:100 (2 × 1.5 mL). The combined filtrates are heated to reflux, and boiling glacial acetic acid (30 mL) is added, followed by concentrated hydrochloric acid (0.40 mL). On standing overnight, the product crystallizes and is collected by filtration. The crystals are washed with acetic acid/water (1:1), ethanol, and then diethyl ether and are thoroughly dried in vacuum to give 400 mg (67%). *Anal.* Calcd. for $C_{34}H_{36}N_4O_4FeCl$: C, 62.25; H, 5.53; N, 8.54; Cl, 8.51. Found: C, 62.47; H, 5.62; N, 8.50; Cl, 8.66.

To obtain analytically pure samples and accurate extinction coefficients, prolonged drying under high vacuum and at elevated temperatures may be necessary.[13]

Properties

Mesohemin IX prepared by this method shows a single band on polyamide tlc (benzene/methanol/formic acid, 110:15:1; and methanol/acetic acid 100:2)[10] and on silica gel tlc[14] (benzene/methanol/formic acid, 110:30:1). The electronic spectrum of the pyridine hemochrome, prepared by reducing the mesohemin (in pyridine) with an aqueous solution of sodium dithionite, contains the bands (4 N pyridine, 0.2 N KOH) λ_{max}: 407.5, 516, 546.2 nm. ϵ: 139.5, 20.9, 35.8 (mM^{-1}).

D. THE INSERTION OF ^{57}Fe INTO PORPHYRINS

$$^{57}Fe_2O_3 + 6HCl \longrightarrow 2\,^{57}FeCl_3 + 3H_2O$$
$$^{57}FeCl_3 + e^- \longrightarrow \,^{57}FeCl_2 + Cl^-$$

With the advent of Mössbauer spectroscopy, ^{57}Fe-labeled heme derivatives have become increasingly important. However, the commercially available form of

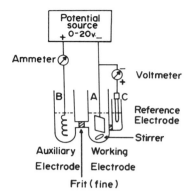

Fig. 2. Electrolysis cell for reduction of $^{57}Fe^{III}$ solutions.

this iron isotope is $^{57}Fe_2O_3$ (Oak Ridge National Laboratory), and the reduction of this oxide to atomic iron can be both time-consuming and troublesome. As far as the insertion of iron into porphyrins is concerned, the reduction of Fe_2O_3 is unnecessary. Fe_2O_3 can be dissolved in concentrated HCl to yield $FeCl_3$, which may be utilized directly to effect metalation of H_2oep, as well as other stable porphyrins, such as tetraphenyl-21H,23H-porphine and mesoporphyrin IX esters, by the aforementioned iron(III) chloride method. Alternatively, $FeCl_3$ can be reduced electrochemically to the iron(II) form before the more conventional iron(II) salt methods are applied. It should be noted that both the iron(II) and the iron(III) salt methods require only stoichiometric amounts of iron, as exemplified by the above procedures.

The electrochemical reduction is best carried out using a controlled potentiostat and cell, as shown in Fig. 2. $^{57}Fe_2O_3$ (80 mg, 0.05 mmole) is dissolved in concentrated HCl (10 mL); the resultant yellow solution is then evaporated to near dryness on a rotovap with a bath temperature of about 60°. The residue is diluted to 15 mL with water and transferred to compartment A of the electrolysis cell. Compartments B and C are immediately filled with 0.1 M $LiClO_4$. The solutions are purged with argon (or nitrogen) for 10 minutes before electrolysis starts.

If a potentiostat is used for reduction, the potential between the working and the SCE reference electrode should be set at −0.1 V. Alternatively, a dc power supply can be used as the potential source. In such cases the potential has to be constantly adjusted manually such that the voltage across the reference and the working electrodes is kept around 0.1 V (Fig. 2). The power supply should at least have a 0–20 V, 1-A rating. Initially, the reduction may draw a large current, but as the reaction proceeds toward completion the resistance of the solution increases and the current decreases. The electrolysis can be stopped when the solution no longer gives a positive test with KSCN.[15] The $FeCl_2$ solution (nearly colorless) is then transferred to a small flask and concentrated to about 1–2 mL with a vacuum pump. This solution is immediately used for iron insertion as described above.

References

1. K. M. Smith, *Porphyrins and Metalloporphyrins*, Elsevier, Amsterdam, 1975, p. 757.
2. J. E. Falk, *Porphyrins and Metalloporphyrins*, Elsevier, Amsterdam, 1964, p. 129.
3. D. B. Morell, J. Barrett, and P. S. Clezy, *Biochem. J.*, **78**, 793 (1961).
4. A. D. Alder, F. R. Longo, and V. Varadi, *Inorg. Synth.*, **16**, 213 (1976).
   A. D. Alder, F. R. Longo, F. Kampas, and J. Kim, *J. Inorg. Nucl. Chem.*, **32**, 2443 (1970).
5. J.-H. Fuhrhop, K. M. Kadish, and D. G. Davis, *J. Am. Chem. Soc.*, **95**, 5140 (1973).
6. W. S. Caughey, J. O. Alben, W. Y. Fujimoto, and J. L. York, *J. Org. Chem.*, **31**, 2631 (1966).
7. F. Adar, *The Porphyrins*, D. Dolphin, Ed., Chapter 2, *Electronic Absorption Spectra of Hemes and Hemeproteins*, Vol. III, Academic, 1978.
8. J. B. Paine III, W. B. Kirshner, D. W. Moskowitz, and D. Dolphin, *J. Org. Chem.*, **41**, 3857 (1976).
9. D. O. Cheng and E. LeGoff, *Tetrahedron Lett.*, **17**, 1469 (1977).
10. D. W. Lampson, A. F. W. Coulson, and T. Yonetani, *Anal. Chem.*, **45**, 2273 (1975).
11. W. S. Caughey, J. O. Albern, W. V. Fujimoto, and J. L. York, *J. Org. Chem.*, **31**, 2631 (1966).
12. A. H. Corwin and J. G. Erdman, *J. Am. Chem. Soc.*, **68**, 2473 (1946).
13. A. D. Alder and J. L. Harris, *Anal. Biochem.*, **14**, 472 (1966).
14. R. K. DiNello and D. Dolphin, *Anal. Biochem.*, **64**, 449 (1975).
15. A. I. Vogel, *A Text Book of Macro and Semimicro Qualitative Inorganic Analysis*, 4th ed., Longmans, 1954, p. 266.

36. METALLOPHTHALOCYANINS AND BENZOPORPHINES

Submitted by D. DOLPHIN,* J. R. SAMS,* and T. B. TSIN*
Checked by S. D. ITTEL† and M. A. CUSHING, JR.†

The benzoporphines (I) and phthalocyanins (II) together constitute an unusual class of organic ligands of exceptionally high thermal and chemical stability. As with porphines, a large number of metals and nonmetals can replace the weakly acidic hydrogen atoms at the center of the macrocyclic systems to form compounds that frequently contain unusual structural features and unexpected stability. As an example, copper phthalocyanine can be heated in air to 550°

---

*Department of Chemistry, The University of British Columbia, Vancouver, B.C. Canada V6T 1W5.
†Central Research and Development Department, Experimental Station, E. I. duPont de Nemours and Co., Wilmington, DE 19898.

156   Compounds of Biological Interest

for hours without decomposition, and it is unaffected by cold concentrated hydrochloric acid or molten sodium hydroxide.

The enormous amount of work on phthalocyanins[1] reflects their prime importance to the dye and pigment industry and is attested to by the numerous patents in the literature. It is our experience, however, that these materials are rarely pure and that their purification, from unknown impurities, is invariably more troublesome than performing the complete synthesis. Phthalocyanins are stronger σ-donors than porphines, and the former stabilize higher oxidation states of coordinated metals,[2] while the tetrabenzoporphyrins have properties more characteristic of porphines.

Phthalocyanine itself is best prepared[3] by self-condensation of phthalimidine, which is available from the reaction of phthalonitrile with ammonia. However, in many cases, direct metalation of the macrocycle cannot be achieved. Instead, metalation by means of dilithium phthalocyanine or a template reaction, whereby the macrocycle is formed around the metal using phthalonitrile (or one of its derivatives), must be employed for the synthesis of metallophthalocyanins.

The tetrabenzoporphines can, like the phthalocyanins, be prepared by a template reaction, or can, like the porphyrins, be directly metalated.

A. [1,4,8,11,15,18,22,25-OCTAMETHYL-29H,31H-TETRABENZO[b,g,l,q] PORPHINATO(2-)]COBALT(II)

Procedure

The following reaction should be carried out under an atmosphere of nitrogen to avoid slow air oxidation of the isoindole. Ammonium sulfate (350 g, 2.7

mole) and 2,5-hexanedione (180 g, 1.6 mole) are added to deoxygenated distilled water (3 L) in a 5-L round-bottomed flask and the solution is heated at reflux under nitrogen for 24 hours. The red-brown solution is cooled to room temperature and treated with 10% sodium hydroxide (about 300 mL) until it is basic to litmus. At this stage, an off-white precipitate appears, and the mixture is left to stand for several hours or overnight to complete precipitation of the isoindole. The precipitate is collected by filtration, washed thoroughly with water, and dried under vacuum for 24 hours to give 75 g of crude 1,3,4,7-tetramethylisoindole (III).

The crude isoindole is dissolved in diethyl ether (2 L) under nitrogen. The solution is filtered and reduced in volume to 300 mL. It is then left under nitrogen for several hours until crystallization is complete. The fluffy white precipitate is collected by filtration, washed with a little diethyl ether, and dried under vacuum to give pure 1,3,4,7-tetramethylisoindole (mp,[2] 144° *in vacuo*). Yield: 60 g (33%). This product is used for the preparation of metalloporphines. (Pink coloration due to isoindole oxidation does not reduce yields in subsequent steps).

■ **Caution.** *The Carius tube will explode if the reaction is scaled up or the tube size is reduced at this point. Face shield and heavy gloves should be worn when handling the tube after reaction. The tube should be cooled to $-198°$ before opening.*

1,3,4,7-Tetramethylisoindole (6 g, 34.6 mmole) and cobalt powder (30 g, 0.51 mole) are mixed and loaded into a Carius tube (4 × 35 cm; volume about 450 mL). After it is evacuated and sealed, the tube is placed in a shielded oven preheated to 390° and heated at this temperature for 4 hours. After the tube is cooled, the dark-blue (or black) powder is placed in a Soxhlet extractor and extracted with pentane (300 mL) until a red-brown impurity is removed (about 24 hours). The pentane is replaced by toluene (300 mL), and extraction is continued until the extract (initially green) is colorless (about 8 hr). Finally, pyridine (300 mL) is used to extract (Soxhlet extractor) the porphine, which requires about 10 hours. The pyridine solution is concentrated to 20 mL, and 100 mL of hexane is added. The precipitate is collected, washed with hexane, and dried overnight in vacuum at 60°, to give [octamethyltetrabenzoporphinato-(2-)]cobalt(II). Yield: 2-3 g (35-55%). *Anal.* Calcd. for $C_{44}H_{36}N_4Co$: C, 77.79; H, 5.30; N, 8.24. Found: C, 77.87; H, 5.40; N, 8.48.

Properties

[Octamethyltetrabenzoporphinato(2-)]cobalt(II) is a green-blue powder that is stable towards atmospheric oxidation in the solid state. The compound is readily soluble in pyridine bases and tetrahydrofuran, and to a lesser extent in diethyl ether, benzene, ethanol, and acetone. The electronic spectrum in pyridine shows bands at 320 (ϵ = 13,200), 455 (ϵ = 71,500), and 645 nm (ϵ = 36,700).

B. [1,4,8,11,15,18,22,25-OCTAMETHYL-29H,31H-TETRABENZO[b,g,l,q]PORPHINE][4]

Procedure

1,3,4,7-Tetramethylisoindole (III) (6 g, 34.6 mmole) (see above) and magnesium powder (8 g, 0.33 mole) are allowed to react, in the manner described above for the cobalt complex, to give [octamethyltetrabenzoporphinato(2-)]bis(pyridine)-magnesium. Yield: 4.6 g (67%). The magnesium complex (0.3 g, 0.37 mmole) is dissolved in trifluoroacetic acid (25 mL), and the solution is stirred for 2 hours. The green solution is filtered and water is slowly added until a dark-blue precipitate is formed. This precipitate is collected by filtration, washed with water, and dried in high vacuum for 6 hours. The dark-blue material, which may still contain some unreacted magnesium octamethyltetrabenzoporphine, is suspended in boiling pyridine (100 mL) for 5 minutes. After it is cooled to room temperature, the mixture is filtered and the filtrate is discarded. The residue is again suspended in hot pyridine and the process is repeated until the filtrate is colorless. The solid is then dissolved in boiling quinoline (100 mL) and is filtered hot; the filtrate is set aside to crystallize overnight. The product is collected by filtration and dried in vacuum at 150°. Yield: 0.25 g (83%). *Anal.* Calcd. for $C_{44}H_{38}N_4$: C, 84.85; H, 6.15; N, 9.00. Found: C, 84.55; H, 6.18; N, 8.73.

Properties

Octamethyltetrabenzoporphine is a dark-blue powder that, although much more soluble than tetrabenzoporphine, has a low solubility in most common organic solvents. It is, however, very soluble in trifluoroacetic acid, giving a green solution of the diprotonated macrocycle. This molecule is a square-planar tetradentate ligand. It reacts with a variety of metal salts to form porphine–metal complexes. In general such metalations are best carried out in quinoline. The electronic spectrum contains the following bands: 415 (ϵ = 75,000), 440 (ϵ = 280,000), 530 (ϵ = 10,000), 630 (ϵ = 7,500), 690 nm (ϵ = 80,000).

C. DILITHIUM PHTHALOCYANINE[5]

$$4 \text{ } \underset{\text{CN}}{\underset{\text{CN}}{\bigcirc}} \xrightarrow{\text{Li}} \text{[phthalocyanine structure with 2 Li]}$$

Procedure

Clean lithium metal (0.2 g, 29 mmole) is placed in reagent grade pentyl alcohol (30 mL) in a 50-mL, round-bottomed flask fitted with a reflux condenser and Drierite tube (the reactants and products are sensitive to moisture in *all* subsequent operations), and the mixture is warmed to dissolve the lithium. Phthalonitrile (3 g, 18.3 mmole) is then added to the mixture, which initially turns yellow, then rapidly dark blue. The mixture is heated at reflux for 0.5 hour, taking care for the first 5 minutes against too vigorous a reaction and foaming. The solvent is removed on a rotary evaporator using a vacuum pump rather than a water aspirator. The oily residue is heated at 250° under vacuum to produce a green powder, which is placed in a Soxhlet extractor and extracted with freshly distilled acetone (200 mL) (reagent grade acetone is heated at reflux for 13 hr over $CaSO_4$ before distillation). When all the blue pigment has been extracted (usually 8 hr), the acetone is reduced in volume to 20 mL. Hexane (100 mL) is added and the mixture is left to stand overnight. The solvated product is collected by filtration and heated under vacuum at 250° for 3 hours to give the blue dilithium phthalocyanine 2.3 g (75%).

A portion of the product (0.5 g, 0.95 mmole) is dissolved in freshly distilled dry acetone (100 mL) and the solution is filtered. The filtrate is reduced in volume to about 10 mL, dry toluene/hexane, 49:1 (50 mL) is added, and the mixture is allowed to stand overnight to give crystals of the solvated product, which, after filtration and washing with a little toluene/hexane (49:1), are heated under vacuum at 250° for 2 hours to give pure dilithium phthalocyanine, 0.45 g (90%). *Anal.* Calcd. for $C_{32}H_{16}N_4Li_2$: C, 73.0; H, 3.04; N, 21.30. Found: C, 72.8; H, 3.08; N, 21.10.

Properties

Dilithium phthalocyanine is obtained as dark-blue crystals. The compound has high thermal stability, as is typical of many phthalocyanines. It is soluble in acetone, giving a deep-blue solution that deposits phthalocyanine when in contact with even trace amounts of water. The material is also soluble in ethanol and tetrahydrofuran, but it is insoluble in diethyl ether, hexane, or chloroform. Solutions of the dilithium complex in ethanol react rapidly and quantitatively with a variety of metal salts to give the metallophthalocyanines, which precipitate, in very pure form, from solution. The electronic spectrum contains the bands (acetone solution): 370 (ϵ = 24,800), 596 (ϵ = 17,300), 630 (ϵ = 16,100), 655 nm (ϵ = 11,100).

D. [PHTHALOCYANINATO(2-)]IRON(II)[6]

$$4 \; \text{o-C}_6\text{H}_4(\text{CN})_2 \xrightarrow{\text{Fe}} \text{Fe(II)(pc)}$$

Procedure

Phthalonitrile (2.9 g, 17.7 mmole) and iron powder 5 g, 0.09 mole) are intimately mixed and placed in a 25-mL conical flask. The mixture is heated in a Wood's metal or sand bath at 260° for 5 hours. Phthalonitrile that condenses on the cooler upper portion of the flask should be scraped down to the bottom periodically. The blue product is finely ground and extracted in a Soxhlet extractor with ethanol (200 mL) for 5 hours. The ethanol is replaced by pyridine (200 mL), and the extraction is continued for a further 5 hours. The pyridine extract is reduced in volume to 20 mL, and hexane is added to precipitate the blue [phthalocyaninato(2-)] bis(pyridine)iron(II), which is collected by filtration, washed with ethanol, and dried in vacuum to give 0.8 g of Fe(pc)py$_2$ (pc = phthalocyanine, py = pyridine). A portion (0.1 g) of this material can be further purified by dissolving in pyridine (30 mL) and filtering the solution by suction through alumina (3 × 10 cm Woelm alumina grade I). Hexane (100 mL) is added to the filtrate and the blue precipitate is collected by filtration, washed with petroleum ether, and sublimed in vacuum at 300° to give the dark-blue [phthalocyaninato(2-)]iron(II). Yield: 0.05 g (50%). *Anal.* Calcd. for $C_{32}H_{16}N_8Fe$: C, 67.5; H, 2.8; N, 19.7. Found: C, 67.1; H, 2.8; N, 19.7.

Properties

[Phthalocyaninato(2-)] iron(II) is a dark blue, thermally stable solid that can be sublimed *in vacuo* at 300°. It is very soluble in pyridine, giving deep blue solutions of the bis(pyridine) adducts. It also forms an unstable purple hexaaniline adduct when dissolved in aniline. It is soluble in concentrated sulfuric acid and dimethyl sulfoxide (slightly) but is insoluble in most other organic solvents. The iron(II) complex, unlike the corresponding iron(II) porphines, is relatively stable toward oxidation to the iron(III) state. The electronic spectrum shows the following absorption bands: (1-chloronaphthalene solution) 595 (ϵ = 16,000), 630 (ϵ = 17,000), 658 (ϵ = 63,000); (pyridine solution) 333 (ϵ = 45,000), 415 (ϵ = 15,000), 395 (ϵ = 2000), 658 nm (ϵ = 8000).

References

1. A. H. Jackson, *Azaporphyrins*, Vol. I of The Porphyrins, D. Dolphin, Ed., Academic, New York, 1978.
2. C. O. Bender and R. Bonnett, *J. Chem. Soc. (C)*, 1968, 3036.
3. P. J. Brach, S. J. Grammatica, O. A. Ossanna, and L. Weinberger, *J. Heterocyclic Chem.*, 7, 1403 (1970).
4. C. O. Bender, R. Bonnett, and R. G. Smith, *J. Chem. Soc. (C)*, 1970, 1251.
5. P. A. Barrett, D. A. Frye, and R. P. Linstead, *J. Chem. Soc.*, 1938, 1157.
6. R. P. Linstead, *J. Chem. Soc.*, 1934, 1016.

37. (DIOXYGEN)(N-METHYLIMIDAZOLE)[(all-cis)-5,10,15,20-TETRAKIS[2-(2,2-DIMETHYLPROPIONAMIDO)-PHENYL] PORPHYRINATO(2-)] IRON(II)

Submitted by THOMAS N. SORRELL*
Checked by CHARLES M. BUMP[†] and JOSEPH JORDAN[†]

(Dioxygen) (*N*-methylimidazole) [5,10,15,20-tetrakis [2-(2,2-dimethylpropionamido)phenyl] porphyrinato(2-)] iron(II) has been of considerable interest in recent years because it represents one of the few isolated, well-characterized (dioxygen)(porphyrinato)iron complexes that can be regarded as low-molecular-weight analogs of the oxygen-binding hemoproteins, myoglobin and hemoglobin.[1,2] Moreover, the macrocyclic ligand, which has been known more simply as the "picket-fence porphyrin," has been shown to be useful for the preparation of other complexes, notable examples being the cobalt analogue[3] and the tetrahydrofuran[1,4]- and tetrahydrothiophene[5]-coordinated iron complexes.

*Department of Chemistry, University of North Carolina, Chapel Hill, NC 27514.
†Department of Chemistry, The Pennsylvania State University, University Park, PA 16802.

These molecules, which also reversibly bind the dioxygen ligand, may help in unraveling the mysteries of oxygen-binding and activation.[2,5]

The procedures detailed below include a considerable amount of organic synthesis that is indispensable if one wishes to study the coordination chemistry of the ultimate product, and hence it is included. All the steps may be scaled down but only the last two steps may be scaled up efficiently. Many of the compounds in the sequence are used in subsequent steps without further purification. Analytical data are included for those materials that must be pure if the following steps are to succeed.

A. 5,10,15,20-TETRAKIS(2-NITROPHENYL)PORPHYRIN, H$_2$TNPP

$$4 \text{ (2-NO}_2\text{-C}_6\text{H}_4\text{)-CHO} + 4 \text{ pyrrole} \longrightarrow \text{H}_2\text{TNPP}$$

Procedure

One hundred grams of 2-nitrobenzaldehyde (0.6 mole) is dissolved in 1700 mL of glacial acetic acid in a 3-L, three-necked, round-bottomed flask fitted with an efficient condenser and a dropping funnel. The solution is then heated just to its boiling point while being vigorously stirred with a magnetic stirring bar. Forty-six milliliters of pyrrole (47.6 g, 0.71 mole) is added dropwise to the solution at such a rate that the reaction does not become uncontrollable. (■ **Caution.** *This highly exothermic reaction sometimes becomes so violent that the solvent sprays out of the top of the condenser.*) (The yield is often lower, however, if the reaction is not vigorous enough.) The resulting black mixture is allowed to reflux for 30 minutes before it is cooled in an ice bath to 35°. Tars are formed if the solution is cooled below this temperature. During the cooling process, 250 mL of chloroform is slowly added to prevent the formation of tars. (■ **Caution.** *The chloroform will boil if the solution is too hot.*) The purple crystalline product is filtered by suction, washed with five 100-mL portions of CHCl$_3$, and dried at 100° overnight. Yield: 17.3–22.0 g (13.2–16.8%).

Properties

H$_2$TNPP is a bright-purple crystalline solid that is insoluble in most organic solvents with the exception of dimethylformamide (DMF), dimethyl sulfoxide (DMSO), and nitrobenzene. λ_{max}(DMF): 409, 518, 551, 594, and 652 nm.

B. 5,10,15,20-TETRAKIS(2-AMINOPHENYL)PORPHYRIN, H_2TAPP

$$H_2\text{TNPP} \xrightarrow[\text{HCl}_{(aq)}]{\text{SnCl}_2} H_2\text{TAPP}$$

Procedure

A 3-L beaker is charged with 12.0 g (0.015 mole) of H_2TNPP and 600 mL of concentrated hydrochloric acid (sp gr 1.18). To this is added a solution of 50 g (0.22 mole) of $SnCl_2 \cdot H_2O$* dissolved in 50 mL of concentrated hydrochloric acid. The solution is stirred for 90 minutes at room temperature. The beaker is then placed in a hot water bath atop a hot plate-magnetic stirrer, making sure that the bottom of the beaker is not in contact with the bottom of the heating bath. A thermometer is suspended such that it does not touch the bottom or sides of the beaker. The temperature is raised to 65° in a 10-minute period and is held between 65 and 70° for 25 minutes. Good stirring is maintained during this time. (Heating this solution above 75° results in a low yield of impure product.) After the specified time, the beaker is placed in an ice bath and swirled to bring the contents to room temperature. The solution is then neutralized by the slow addition (20-30 min) of about 600 mL of concentrated ammonium hydroxide. (■ *Caution. This is a highly exothermic and vigorous reaction.*) After it is cooled to room temperature again, the highly basic solution (pH > 10) is stirred for at least 12 hours with 1 L of chloroform.

The organic layer is separated and the aqueous phase is transferred to a 4-L separatory funnel. Fifteen-hundred milliliters of water is added and the solution is extracted with three 150-mL portions of $CHCl_3$. The combined extracts are returned to the separatory funnel and washed with 1 L of dilute NH_4OH. This in turn is extracted with two 50-mL portions of chloroform. The combined organic portion is evaporated to 600 mL on a rotary evaporator and then filtered by suction. The filtrate and washings are concentrated to 250 mL, 150 mL of 95% ethanol containing 10 mL of conc. aqueous NH_3 is added, and the solvent slowly evaporated until the remaining volume is about 200 mL. The sides of the flask are washed down with chloroform and 100 mL of ethanol is added. The volume is then reduced to 75 mL, and the resulting crystals are filtered, washed with five 10-mL portions of 95% ethanol, and dried in an oven at 100° for several hours. Yield: 6.6-8.1 g (65-80%).

---

*A reagent grade tin(II) chloride must be used. A good test for the quality of this reagent is to mix 1 g of the $SnCl_2 \cdot 2H_2O$ in 1.5 mL of concentrated hydrochloric acid with 0.25 g of 3-nitrobenzaldehyde. The mixture should become warm and turn red-orange within about 10 minutes.

Properties

H_2TAPP is a purple-black microcrystalline compound that is a mixture of four atropisomers (see below). Tlc on silica (1:1 benzene/diethyl ether) shows four spots; R_f values (relative intensity): 0.77 (1), 0.64 (2), 0.43 (4), 0.04 (1). It is quite soluble in organic solvents with the exception of alcohols and saturated hydrocarbons. λ_{max}(CHCl$_3$): 418, 517, 550, 589, 645 nm.

C. SEPARATION OF *all-cis*-H_2 TAPP

Procedure

As obtained above, H_2TAPP is a mixture of four atropisomers in statistical abundance. The most polar one, the *all-cis* isomer, is easily separated from the others by column chromatography. All solvents should be reagent quality, although technical grade chloroform may be used. Twenty-four grams of H_2TAPP is dissolved in 2 L of CHCl$_3$ and the solution is poured onto a 10 × 25 cm column of silica (W. R. Grace & Co., Davidson Chemical Division, 60–200 mesh, grade 62) prepared as a slurry in CHCl$_3$. The solvent is passed through the column as rapidly as possible.* After all the material is loaded, an additional 2 L of CHCl$_3$ is passed through the column. At this point, 1:1 CHCl$_3$/diethyl ether (about 2.5 L) is used as the eluent, and this process is continued until the eluate becomes very pale. The desired *all-cis*-isomer (4:0) is then removed by elution with 1:1 acetone/diethyl ether, and the solvent is removed at a temperature lower than 30° using a rotary evaporator.† The checkers recommend following the chromatographic separation by tlc on silica using 1:1 benzene-diethyl ether. In this manner, less of the desired *all-cis* isomer is lost.

In the meantime, the previous eluates are concentrated to 2.5 L by boiling off the solvent in a 4-L filter flask under N_2. Toluene is then added gradually to raise the temperature of the solution. By judiciously adding either chloroform or toluene, a volume of 2 L and a temperature of 95–100° are maintained over at least 6 hours to ensure complete equilibration of the atropisomers. After that amount of time, the volume is reduced to about 200 mL and the residue is cooled to ambient temperature. It may then be dissolved in 2 L of chloroform and rechromatographed by the above procedure to give more of the 4:0 isomer. A convenient procedure is to begin with 24 g of mixed isomers and, after re-

---

*In other words, this is more of a filtration through silica rather than a normal "chromatographic separation." If the time elapsed from loading the column to removing the 4:0 isomer is greater than 2 hours, then the column is running too slowly. Check to make sure the silica is not too fine.

†This product is about 90% pure and must be chromatographed on a second column (see below).

moving the 4:0 isomer (theoretical yield, 3 g), to reequilibrate the other isomers and then add an additional 3 g of mixed isomers. By this method, the "stockpot" always contains 24 g of H_2TAPP and each column gives about 3 g of the 4:0 isomer.

Five grams of the crude 4:0 isomer obtained as described above is dissolved in 500 mL of chloroform and loaded onto an 8 × 30 cm column of 60–200 mesh silica prepared as a slurry in chloroform. After it is loaded, the column is eluted with 1 L of chloroform and then with 1:1 benzene/diethyl ether (about 2.5 L) until the eluates are essentially colorless. The pure 4:0 isomer is then eluted using acetone/diethyl ether (1:1) and the solution is subsequently taken to dryness at ambient temperature using a rotary evaporator. The residue is dissolved in chloroform, and after the addition of methanol, the solution is concentrated slowly to 75 mL at room temperature using a rotary evaporator. The resulting purple crystals are filtered, washed with methanol, and dried in air at room temperature. Yield: 4.0–4.5 g. Again, the separation may be followed by tlc analysis of the eluate. The checkers found that the 4:0 isomer needed to be chromatographed a third time to purify it completely.

D. (*all-cis*)-5,10,15,20-TETRAKIS[2-(2,2-DIMETHYL-PROPIONAMIDO)PHENYL]PORPHYRIN, H_2TpivPP

$$\textit{all-cis-}H_2\text{TAPP} + 4(CH_3)_3CC\underset{Cl}{\overset{O}{\|}} \xrightarrow[\text{py}]{CH_2Cl_2} \textit{all-cis-}H_2\text{TpivPP}$$

Procedure

In a 1-L, round-bottomed flask, 5.0 g (7.4 mmole) of *all-cis*-H_2TAPP is dissolved with stirring in 500 mL of CH_2Cl_2 and 5 mL of pyridine. Five milliliters (5.1 g, 42 mmole) of 2,2-dimethylpropionyl chloride (pivaloyl chloride)* is added, and the solution is allowed to stir for 2 hours. Two hundred milliliters of 10% aq NH_4OH is then added and the solution is stirred for an additional hour. The organic layer is separated and subsequently washed with two 100-mL portions of water. The dichloromethane is evaporated using a rotary evaporator, and the residue is dissolved in a minimum amount of chloroform (about 50 mL). The solution is applied to a 4 × 60 cm column of 60–200 mesh silica and the product is eluted from the column with 4:1 chloroform–diethyl ether. The solu-

---

*Trimethylacetyl chloride was purchased from the Aldrich Chemical Co. The purity, determined by nmr, was at least 99%. If it is less pure than this, the material should be distilled under nitrogen.

tion is evaporated to dryness, and the residue is dissolved in chloroform. Crystallization is then induced by adding ethanol and heptane, and the solvent is slowly evaporated using a rotary evaporator. Yield: 6.1-6.3 g (81-84%) *Anal.* Calcd. for $C_{64}H_{66}N_8O_4$: C, 76.2; H, 6.60; N, 11.2. Found: C, 75.7; H, 6.75; N, 11.2.

Properties

all-cis-H_2 TpivPP forms red-violet crystals (large crystals are deep purple) that are soluble in most organic solvents. The nmr spectrum shows a peak at $\delta = 0.05$ ppm that is attributed to the four equivalent *tert*-butyl groups of the pivaloyl moiety. If other peaks are observed at $\delta \simeq 0.1$ ppm, then some of the other atropisomers are present.

E. BROMO {(*all-cis*)-5,10,15,20-TETRAKIS-[2-(2,2-DIMETHYLPROPIONAMIDO)PHENYL] PORPHYRINATO (2-)}-IRON (III); [Fe(*all-cis*)-TpivPP] Br

$$\textit{all-cis-}H_2 \text{TpivPP} + \text{FeBr}_2 \xrightarrow[\text{(CH}_2\text{OCH}_3)_2]{\text{py}} \text{Fe}(\textit{all-cis-}\text{TpivPP})\text{Br}$$

Procedure

In a 500-mL round-bottomed flask equipped with a side arm and connected to a nitrogen source, 3.0 g (3.0 mmole) of *all-cis*-H_2TpivPP, 3.0 g (13.9 mmole) of anhydrous $FeBr_2$,[6] and 1 mL of pyridine are dissolved in 200 mL of 1,2-dimethoxyethane or tetrahydrofuran* (distilled from CaH_2 under N_2) under *rigorously* oxygen-free conditions, and the resulting solution is allowed to reflux for about 1-4 hours under N_2. The progress of the reaction is followed by observing the absorption spectrum of small aliquots of the solution that have been exposed to air and then treated with a drop of mineral acid. Any free-base porphyrin shows an absorption under these conditions at about 450 nm due to $H_4\text{TpivPP}^{2+}$. The solution is brought to dryness (this and subsequent manipulations are carried out in air) and the residue is dissolved in 100 mL of chloroform. The resulting solution (some insoluble material is also found) is chromatographed on a 3 × 30 cm column of basic alumina (Woelm, activity I) using chloroform as the eluent. All porphyrinic material is removed (a single orange band of iron bromide complexes should be left at the top of the column) and the solution is evaporated to dryness. The residue is suspended in a solution of

---

*The checkers obtained no product using tetrahydrofuran as the solvent. In accordance with a suggestion by J. C. Marchon,[8] they substituted 1,2-dimethoxyethane as solvent and achieved iron insertion in four (4) hours with a 50% yield.

1 mL of concentrated hydrobromic acid in 100 mL of methanol. The mixture is heated on a steam bath, and dichloromethane is added slowly until all the solid dissolves. The resulting solution is cooled in a refrigerator overnight to give crystals that are collected by filtration, washed with methanol, and dried at 70°. Yield: 3.3–4.3 g (58–75%). A second crop may often be obtained.

Properties

Fe(*all-cis*-TpivPP)Br forms almost black crystals that are soluble in many organic solvents. The material is high-spin iron(III) [$\mu = 5.9$ BM]. λ_{max} (CHCl$_3$): 418, 508, 575, 648, 675 nm. The complex may be stored at room temperature for extended periods of time.

F. BIS(N-METHYLIMIDAZOLE)[(*all-cis*)-5,10,15,20-TETRAKIS-[2-(2,2-DIMETHYLPROPIONAMIDO)PHENYL]PORPHYRINATO(2-)]-IRON(II), Fe(*all-cis*-TpivPP)(N-MeC$_3$H$_3$N$_2$)$_2$

This complex should be prepared in an inert-atmosphere box or in any system capable of providing a rigorously oxygen- and water-free atmosphere. Solvents should be correspondingly dried and deoxygenated.

Fe(*all-cis*-TpivPP)Br + Cr(acac)$_2$ + 2N-MeC$_3$H$_3$N$_2$ $\xrightarrow{THF}$

Fe(*all-cis*-TpivPP)(N-MeC$_3$H$_3$N$_2$)$_2$

Procedure

One gram (0.87 mmole) of Fe-(*all-cis*-TpivPP)Br and 0.35 g (0.7 mmole) of [Cr(acac)$_2$]$_2$[7] (acac = 2,4-pentanedionato) are heated in 35 mL of boiling tetrahydrofuran containing 1 mL (0.97 g, 11.8 mmole) of N-methylimidazole (distilled from solid potassium hydroxide under vacuum) for 5 minutes. The resulting red-orange solution is filtered hot and allowed to cool to ambient temperature while 100 mL of heptane is slowly added. The crystals are filtered, washed with heptane, and dried in a stream of nitrogen. Yield: 0.8–0.85 g (76–80%). The complex may be recrystallized from tetrahydrofuran–heptane–N-methylimidazole. *Anal.* Calcd. for C$_{72}$H$_{76}$N$_{12}$O$_4$Fe: C, 70.2; H, 6.35; N, 13.6; Fe, 4.55. Found: C, 70.0; H, 6.60; N, 13.7; Fe, 4.30.

Properties

Fe-(*all-cis*-TpivPP)(N-MeC$_3$H$_3$N$_2$)$_2$ forms large red-violet crystals that often appear orange or red when powdered. The complex is stable to air in the solid state and the crystals are diamagnetic. Other substituted imidazole complexes may be prepared by this same method. All such complexes are quite soluble in organic solvents. λ_{max}(benzene): 432, 537, 562 nm.

G. (DIOXYGEN)(N-METHYLIMIDAZOLE)[(*all-cis*)-5,10,15,20-TETRAKIS[2-(2,2-DIMETHYLPROPIONAMIDO)-PHENYL]PORPHYRINATO(2-)]IRON(II), Fe(*all-cis*-TpivPP)(*N*-MeC$_3$H$_3$N$_2$)(O$_2$)

$$\text{Fe}(\textit{all-cis}\text{-TpivPP})(N\text{-MeC}_3\text{H}_3\text{N}_2)_2 + \text{O}_2 \xrightarrow{\text{toluene}}$$
$$\text{Fe}(\textit{all-cis}\text{-TpivPP})(N\text{-MeC}_3\text{H}_3\text{N}_2)(\text{O}_2) \cdot \text{C}_7\text{H}_8$$

Procedure

Four hundred milligrams (0.32 mmole) of Fe(*all-cis*-TpivPP)(*N*-MeC$_3$H$_3$N$_2$)$_2$ is dissolved in a warm solution of 50 mL of toluene and 0.1 mL (0.097 g, 1.18 mmole of *N*-methylimidazole under N$_2$. The deep-red solution is cooled to ambient temperature, purged with oxygen for 15 seconds, and treated with 5 mL of heptane. An additional 30 mL of heptane is added slowly over a period of 15 minutes and the resulting microcrystals are filtered, washed with heptane, and dried in air. Yield: 0.31-0.36 g (80-94%). *Anal.* Calcd. for C$_{68}$H$_{70}$N$_{10}$O$_6$Fe · 0.5 C$_7$H$_8$: C, 70.1; H, 6.05; N, 11.5; Fe, 4.6. Found: C, 70.1; H, 6.20; N, 11.5; Fe, 4.4. C$_7$H$_8$(by glc): 0.45 ± 0.1 equiv.

Properties

Fe(*all-cis*-TpivPP)(*N*-MeC$_3$H$_3$N$_2$)(O$_2$) forms blood-red(!) crystals and powders that are soluble with some decomposition in most organic solvents. The complex is diamagnetic. It appears to be stable indefinitely in the solid state at ambient temperature. The dioxygen can be removed under vacuum to give the five coordinate Fe(*all-cis*-TpivPP)(*N*-MeC$_3$H$_3$N$_2$) or displaced by carbon monoxide to give the carbonyl. Substituting CO for O$_2$ in the above procedure gives the carbonyl adduct directly.

References

1. J. P. Collman, R. R. Gagné, C. A. Reed, T. R. Halbert, G. Lang, and W. T. Robinson, *J. Am. Chem. Soc.*, **97**, 1427 (1975).
2. J. P. Collman, *Acc. Chem. Res.*, **10**, 265 (1977).
3. J. P. Collman, J. I. Brauman, K. M. Doxsee, T. R. Halbert, S. E. Hayes, and K. S. Suslick, *J. Am. Chem. Soc.*, **100**, 2761 (1978).

4. J. P. Collman, R. R. Gagné, and C. A. Reed, *J. Am. Chem. Soc.*, **96**, 2629 (1974).
5. J. P. Collman and T. N. Sorrell in *Drug Metabolism Concepts*, D. M. Jerina, Ed., American Chemical Society Symposium Series, No. 44, American Chemical Society, Washington, DC, 1977, Chap. 2.
6. G. Winter, *Inorg. Synth.*, **14**, 101 (1973).
7. L. R. Ocone and B. P. Block, *Inorg. Synth.*, **8**, 125 (1966).
8. J. C. Marchon, private communication to J. Jordan, Grenoble, 1976.

Chapter Six
ORGANOMETALLIC COMPOUNDS

38. SILYL AND GERMYL SELENIDES AND TELLURIDES

Submitted by JOHN E. DRAKE,* BORIS M. GLAVINČEVSKI,*
RAYMOND T. HEMMINGS,† and H. ERNEST HENDERSON*
Checked by K. M. MACKAY‡

Studies of silyl and germyl groups attached to selenium and tellurium are notably fewer than those of the well-known oxides and sulfides.[1] In part this may be attributed to the obnoxious nature of these compounds, as well as the scarcity of synthetic procedures. Many of the objections to the study of such materials are removed when high vacuum techniques are employed as discussed herein.

Disproportionation of initially formed $(CH_3)_3Si(E)C_6H_5$ (E = Se, Te) is a likely rationale for the unexpected formation of $[(CH_3)_3Si]_2E$ in the reaction of $(CH_3)_3SiCl$ with $C_6H_5(E)MgBr$.[2] The compounds $[(CH_3)_3Si]_2E$ have been obtained more directly by way of Li_2E species.[3,4] Digermyl selenide, $(GeH_3)_2Se$, has been reported from the cleavage of germylarsine[5] or digermylcarbodiimide[6,7] with H_2Se or, alternatively, by exchange of germyl bromide with lithium selenide[8] or disilyl selenide.[6] The latter method has also afforded digermyl telluride.[6] Routes leading to comparable derivatives of organo selenols are described in the literature.[9]

Our synthetic routes are based on two assumptions: first, that exchange reac-

---

*Department of Chemistry, University of Windsor, Windsor, Ontario, Canada N9B 3P4.
†Department of Chemistry, University of the West Indies, Mona, St. Andrew, Jamaica, West Indies.
‡School of Science, University of Waikato, Private Bag, Hamilton, New Zealand.

172 *Organometallic Compounds*

tions of germyl halides with silyl derivatives are generally more facile than alternative procedures for preparing germyl derivatives, and second, that $(CH_3)_3Si$ species are more stable thermally than SiH_3 species and hence are easier to prepare, handle, and store (the avoidance of potentially hazardous SiH_4 sources being an added consideration).

We therefore describe the syntheses of bis(trimethylsilyl) selenide, $[(CH_3)_3Si]_2Se$, and bis(trimethylsilyl) telluride, $[(CH_3)_3Si]_2Te$, from the metathesis of $(CH_3)_3SiCl$ with the corresponding dilithium salts and their subsequent exchange with GeH_3F to give digermyl selenide and digermyl telluride. This route represents a considerable saving in laboratory time over the original methods, which require extended reaction times at low temperature. Preparation of the $(CH_3)_3Si$ compound starting from Li and Se takes approximately 90 hours, whereas the formation of the GeH_3 compound is complete within 30 minutes, including purification.

Starting Materials

The manipulation of volatile compounds is carried out on a Pyrex-glass vacuum system using conventional techniques.[10] The authors' system consisted of two manifolds, interconnected by four U-traps, and a central manifold leading to two liquid nitrogen backing-traps and mercury diffusion and rotary oil pumps. The vacuum in the system was monitored by a Pirani-type gauge fitted to the central manifold. Pressure readings in excess of 1 torr were registered by mercury manometers. Any comparable system is suitable provided working pressures of $<1 \times 10^{-2}$ torr can be achieved.

Freshly prepared lithium wire,* and selenium* and tellurium* powders were used as supplied. Germyl fluoride may be prepared from germyl iodide[11] and lead(II) fluoride.[12] Chlorotrimethylsilane* is distilled through traps at $-45°$ (chlorobenzene–liquid N_2 slush), $-78°$ (methanol–dry ice slush), and $-196°$ (liquid N_2); the fraction retained in the trap at $-78°$ is the desired purified chlorotrimethylsilane. Peroxide-free diethyl ether was distilled in a vacuum system from fresh $Li[AlH_4]$ and stored *in vacuo*. ■ **Caution.** *The use of Li[AlH₄] for drying ethers can lead to explosions.*[13] Commercial "dry" NH_3 (10-15 mL portion) was further dried by condensation onto Na pellets (about 1 g) in an ampul (about 50 mL) held at $-78°$ on the vacuum line. After 30 minutes the dried NH_3 was pumped from the ampul and degassed in a trap held at $-196°$ before use.

■ **Caution.** *The selenium and tellurium compounds in these preparations are toxic and vile smelling. Rapid decomposition, particularly for the silanes, will occur on exposure to air and/or moisture. Manipulations should be carried out*

---

*Available from Alfa Products, Ventron Corp., P.O. Box 299, Danvers, MA 01923.

in a vacuum system of sound construction in a well-ventilated area or fume hood. The nauseating smell of the selenides and tellurides usually results from their dissolution in stopcock lubricants. To minimize this, greaseless high-vacuum Teflon-in-glass valves and a minimum of silicone-type lubricant for detachable glass joints are highly recommended. Personal contamination with Se compounds will lead to weeks of foul odors!

A. HEXAMETHYLDISILASELENANE AND HEXAMETHYLDISILATELLURANE, [BIS(TRIMETHYLSILYL) SELENIDE AND BIS(TRIMETHYLSILYL) TELLURIDE]

$$2Li + Se \longrightarrow Li_2Se*$$

$$\{2Li + Te \longrightarrow Li_2Te*\}$$

$$2(CH_3)_3SiCl + Li_2Se \longrightarrow [(CH_3)_3Si]_2Se + 2LiCl$$

$$\{2(CH_3)_3SiCl + Li_2Te \longrightarrow [(CH_3)_3Si]_2Te + 2LiCl\}$$

Procedure

The reactor is a bulb (volume about 70 mL) the neck of which is extended into a tube 10 cm long (about 10 mm od), terminating in a high-vacuum Teflon-in-glass valve and glass joint for attachment to the vacuum line. The reactor is evacuated and purged with dry N_2. The stopcock plug is removed in a dry box, and clean lithium wire (0.0694 g, 10.00 mmole), powdered selenium (0.3960 g, 5.02 mmole), and a small stirring bar are added to the reactor. The plug is replaced and the vessel is then thoroughly evacuated on the vacuum line. Dry NH_3 (about 10-15 mL) is then distilled into the vessel held at $-196°$. The stopcock is closed and the reactor is maintained at $-78°$ with occasional shaking and stirring until the blue color of dissolved lithium is discharged (about 10-12 hr). This will be slower if lithium has an oxide coating. With cautious warming and shaking, the reaction may be accelerated to completion in about 1 hour. (■ *Caution. It is advisable to incorporate a bubbler manometer into the vacuum line to offset any difficulties arising from sudden expansion of the NH_3*). The NH_3 is then distilled out of the reactor under vacuum and the residual solid is heated (130-160°) *in vacuo* for 4-6 hours. The dilithium selenide thus prepared is light grey and should show no signs of residual Li or Se. Polyselenides impart a characteristic red color to the material. Dilithium telluride is prepared in the same way from Li (0.0694 g, 10.00 mmole) and powdered tellurium (0.6392 g,

---

*By writing Li_2Se and Li_2Te for lithium selenide and lithium telluride, we do not wish to claim that the exact nature of these molecular formulae is known.

5.01 mmole). It is characteristically a darker grey than the selenide but should appear homogenous. Inhomogenous Li_2Se or Li_2Te gives poor yields. Both materials may be stored indefinitely under nitrogen at room temperature.

Diethyl ether (about 10-15 mL) and $(CH_3)_3SiCl$ (1.1393 g, 10.50 mmole) are distilled *in vacuo* into the reactor held at $-196°$ and containing dilithium selenide (about 5 mmole) prepared as above. The mixture is brought to room temperature and stirred for 72 hours, after which time the supernatant liquid is pale yellow, and a whitish-gray precipitate is present. The volatile material is then distilled under vacuum out of the reactor and fractionated through traps at -45 and $-196°$. Diethyl ether and unreacted $(CH_3)_3SiCl$ may be identified in the $-196°$ trap. The contents of the trap that was maintained at $-45°$ are redistilled through traps held at -23 (CCl_4-liquid N_2 slush), -45, and $-196°$. (■ *Caution. Carbon tetrachloride is toxic and a suspected carcinogen. It should be handled in an efficient hood.*) The former retains pure bis(trimethylsilyl) selenide $\{[(CH_3)_3Si]_2Se,$ 4.0 mmole, 80%$\}$. Traces of hexamethyldisiloxane may be found in the trap held at $-45°$ (identified by the characteristic strong band at 1055 cm^{-1} in the IR spectrum[14] of the gas). High yields of bis(trimethylsilyl) telluride, $[(CH_3)_3Si]_2Te$, in the region 70-80%, may be obtained by an analogous procedure using Li_2Te (about 5 mmole), diethyl ether (about 10-15 mL), and $(CH_3)_3SiCl$ (10.50 mmole).

■ *Caution. The Se and Te residues in the reactor should be handled in an efficient fume hood. Smell contamination may be considerably reduced by treatment of the residues with a strong bleach solution followed by an acid wash. These solutions should be disposed of in the normal way prescribed for toxic wastes.*

Properties

Bis(trimethylsilyl) selenide (mp $-7°$, bp $58-59°/11$ torr)[14] and bis(trimethylsilyl) telluride (mp $13.5°$, bp $74°/11$ torr)[14] are liquids of low volatility with revolting and persistent odors resembling garlic and rotten cabbage. They exhibit first-order ^{1}H nmr spectra:[2,3] $[(CH_3)_3Si]_2Se$; $\delta(CH_3)$ 0.43 ppm, $J(H-^{13}C)$ 120.5 Hz, $J(H-C-^{29}Si)$ 7.2 Hz; $[(CH_3)_3Si]_2Te$; $\delta(CH_3)$ 0.60 ppm, $J(H-^{13}C)$ 121.5 Hz (shifts downfield from tetramethylsilane in carbon tetrachloride solution). The infrared spectra show strong lines above 300 cm^{-1} : $[(CH_3)_3Si]_2Se$[14]; 2960 (s), 2900 (m), 1406 (m), 1257 (vs), 843 (vs), 823 (vs), 747 (s), 686 (s), 623 (s), 369 (vs) cm^{-1}; $[(CH_3)_3Si]_2Te$[14]; 2959 (s), 2900 (m), 1404 (w), 1247 (vs), 843 (vs), 823 (vs), 758 (m), 695 (m), 620 (s), 323 (vs) cm^{-1}. Further identification comes from the mass spectra, which give peaks at m/e 220-230 $(C_6H_{18}Si_2Se)^+$ and m/e 268-278 $(C_6H_{18}Si_2Te)^+$, respectively. Bis(trimethylsilyl) selenide and telluride may be stored indefinitely at room temperature in break-seal glass ampules.

B. DIGERMASELENANE AND DIGERMATELLURANE, (DIGERMYL SELENIDE AND DIGERMYL TELLURIDE)

$$2GeH_3F + [(CH_3)_3Si]_2Se \longrightarrow (GeH_3)_2Se + 2(CH_3)_3SiF$$

$$2GeH_3F + [(CH_3)_3Si]_2Te \longrightarrow (GeH_3)_2Te + 2(CH_3)_3SiF$$

Procedure

The reactor is a glass ampul (about 10 mL) fitted with a high-vacuum Teflon-in-glass valve for attachment to the vacuum line. The ampul is evacuated and cooled to $-196°$, and GeH_3F (0.50 mmole) and a slight deficit of $[(CH_3)_3Si]_2Se$ (0.236 mmole) are distilled in. The mixture is allowed to warm to room temperature for 10 minutes. The volatile material is separated *in vacuo* using traps held at -45 and $-196°$. Unreacted GeH_3F (about 0.02 mmole), $(CH_3)_3SiF$ (0.47 mmole), and traces of GeH_4 are identified by ^1H nmr spectra of the contents of the trap held at $-196°$. The trap held at $-45°$ retains pure digermyl selenide $[(GeH_3)_2Se, 0.235$ mmole, 99%]. In some reactions the trap held at $-45°$ may also contain slight traces of bis(trimethylsilyl) selenide or germyl trimethylsilyl selenide.[15]

In a typical reaction by an identical procedure, GeH_3F (0.75 mmole) and $[(CH_3)_3Si]_2Te$ (0.356 mmole) react to give $(GeH_3)_2Te$ (0.353 mmole, 99%), which is retained in a trap at $-23°$.

Properties

Digermyl selenide (mp $-41 \pm 1°$; vp 2-3 torr at $25°$)[6] and digermyl telluride (mp $-75° \pm 1°$)[6] are sparingly volatile liquids [$(GeH_3)_2Te$ is only just volatile in the vacuum system] with disgusting garlic-cabbage-like odors. They are colorless when pure but become yellow if allowed to stand for long periods at room temperature. They are best characterized by their ^1H nmr spectra,[6] which are singlet resonances with characteristic satellites arising from interaction with ^{77}Se ($I = \frac{1}{2}$, 7.58%) and ^{125}Te ($I = \frac{1}{2}$, 6.99%): $(GeH_3)_2Se$; $\delta(GeH)$ 4.23 ppm, $J(HGeSe)$ 12.1 Hz; $(GeH_3)_2Te$; $\delta(GeH)$ 3.59 ppm, $J(HGeTe)$ 19.4 Hz (measured downfield from tetramethylsilane 5% by volume in cyclopentane). The infrared spectra[6] show the following principal features: $(GeH_3)_2Se$; 2105 (s), 2086 (s), 874 (m, br), 828 (s), 801 (s), 285 (w) cm^{-1}; $(GeH_3)_2Te$; 2083 (s), 866 (mw), 816 (m), 781 (vs) 516 (vw) cm^{-1}. The molecular ions for digermyl selenide [m/e 216-238, $(H_nGe_2Se)^+$] and telluride [m/e 264-285, $(H_nGe_2Te)^+$] are observed in their respective mass spectra. Both materials are best stored at $-196°$ in break-seal glass ampules.

References

1. J. E. Drake and C. Riddle, *Q. Rev.*, **24**, 263 (1970).
2. K. A. Hooton and A. L. Allred, *Inorg. Chem.*, **4**, 671 (1965).
3. M. Schmidt and H. Ruf, *Angew. Chem.*, **73**, 64 (1961); *Z. anorg. allg. Chem.*, **321**, 270 (1963).
4. H. Bürger and U. Goetze, *Inorg. Nucl. Chem. Lett.*, **3**, 549 (1967).
5. J. E. Drake and C. Riddle, *J. Chem. Soc. (A)*, **1969**, 1573.
6. S. Cradock, E. A. V. Ebsworth, and D. W. H. Rankin, *J. Chem. Soc. (A)*, **1969**, 1628.
7. J. E. Drake, R. T. Hemmings, and H. E. Henderson, *J. Chem. Soc., Dalton Trans.*, **1976**, 366.
8. J. D. Murdoch, D. W. H. Rankin, and C. Glidewell, *J. Mol. Struct.*, **9**, 17 (1971).
9. J. E. Drake and R. T. Hemmings, *J. Chem. Soc., Dalton Trans.*, **1976**, 1730.
10. D. F. Shriver, *The Manipulation of Air-Sensitive Compounds*, McGraw-Hill, New York, 1969.
11. S. Cradock, *Inorg. Synth.*, **15**, 161 (1974); J. E. Drake and R. T. Hemmings, *Synth. Org. Metal-Org. Chem.*, **3**, 125 (1975).
12. S. Cradock, *Inorg. Synth.*, **15**, 164 (1974).
13. *Inorg. Synth.*, **12**, 317 (1970).
14. H. Burger, U. Goetze, and W. Sawodny, *Spectrochim. Acta*, **24A**, 2003 (1968).
15. J. E. Drake, B. M. Glavinčevski, R. T. Hemmings, and H. E. Henderson, *Can. J. Chem.*, **56**, 465 (1978).

39. DIGERMOXANE AND 1,3-DIMETHYL-, 1,1,3,3-TETRAMETHYL-, AND HEXAMETHYLDIGERMOXANE

Submitted by JOHN E. DRAKE,* BORIS M. GLAVINČEVSKI,*
RAYMOND T. HEMMINGS,† and H. ERNEST HENDERSON*
Checked by CHARLES H. VAN DYKE‡ and N. VISWANATHAN §

Studies of Ge—O containing species are notably less abundant than those of the related Si—O species. Procedures useful for the synthesis of siloxanes are generally unsuccessful for the preparation of germoxanes. Thus $(GeH_3)_2O$ is not obtained by hydrolysis of GeH_3Cl[1] or GeH_3CN[2] nor by the interaction of GeH_3Br with Ag_2O or Ag_2CO_3.[3] Reactions of HgO with $(GeH_3)_2S$ or GeH_3I are similarly unsatisfactory, producing either low yields or inseparable mixtures.[4] Recently, high yields of germoxanes have been obtained by hydrolysis of digermylcarbodiimides,[5,6] chloromethylgermane, and chlorodimethylgermane in the presence of ammonia[7] and by the interaction of bromogermanes with lead(II)

---

*Department of Chemistry, University of Windsor, Windsor, Ontario, Canada N9B 3P4.
†Department of Chemistry, Scarborough College, University of Toronto, Toronto, Ontario, Canada.
‡Department of Chemistry, Carnegie-Mellon University, Pittsburgh, PA 15213.
§ Pennsylvania State University, Uniontown, PA 15401.

oxide.[8,9] The carbodiimide route,[5,6] although essentially quantitative, is suited only to small-scale preparations and appears to yield a product under conditions that render it less stable than that produced by the lead(II) oxide route.

We describe in detail the syntheses of digermoxane and its methyl analogues from the reaction of lead(II) oxide with bromogermanes. The procedures are convenient and well suited to large-scale preparations in a relatively short time (about 1 hr), using a minimum of special apparatus. The yields are in the range 87–98%.

Apparatus

The manipulation of all volatile material is carried out on a Pyrex-glass vacuum system of conventional design.[10] The system may consist of two manifolds interconnected by four U-traps and a central manifold leading to two liquid nitrogen backing-traps and mercury diffusion and rotary oil pumps. The vacuum in the system is monitored by a Pirani-type gauge fitted to the central manifold. Pressure readings in excess of 1 torr are registered by mercury manometers. High-vacuum Teflon-in-glass valves and a silicone-type grease for ground-glass joints are preferred because of the marked solubility of the materials in hydrocarbon grease.*

Starting Materials

Commercial sources for some of the starting materials include germane (Matheson Gas Products, East Rutherford, NJ), dichlorodimethylgermane, bromotrimethylgermane (Alfa Inorganics Inc., Beverly, MA), and lead(II) oxide (Fisher Scientific Co., Fair Lawn, NJ). The preparations of bromogermane†[11] and methylgermane‡[12] are well documented. Dimethylgermane‡ may be recovered in yields >95% from the reduction of dichlorodimethylgermane with Li[AlH$_4$] in dry dibutyl ether[13] and its purity can be confirmed by ^{1}H nmr[14] and ir[15] measurements. Methylgermanes are conveniently converted to chloromethylgermanes with BCl$_3$ (Matheson) and subsequently to bromogermanes with HBr[13,14,16] (Matheson).

■ **Caution.** *The digermoxanes are toxic and their exposure to air and/or moisture is likely to promote rapid decomposition. Manipulations should be*

---

*The checkers used a conventional vacuum line equipped with high-vacuum glass stopcocks that were lubricated with Dow Corning high-vacuum silicone lubricant.

†The checkers recommend the preparation of GeH$_3$Br as described in *Inorg. Synth.*, **15**, 157 (1974).

‡The checkers recommend the preparation of MeGeH$_3$ and Me$_2$GeH$_2$ by the sodium tetrahydroborate(1-) reduction of the corresponding chloromethylgermane in aqueous acidic solution as described in *Inorg. Synth.*, **2**, 375 (1963).

178 Organometallic Compounds

performed under vacuum conditions in a well-ventilated area. All preparations of the methyl digermoxanes may be scaled up to use 10 mmole of starting material.

A. DIGERMOXANE

$$2GeH_3Br + PbO \longrightarrow (GeH_3)_2O + PbBr_2$$

Procedure

An excess of anhydrous lead(II) oxide (about 15 g, 67 mmole) is placed in a reaction vessel (volume about 45 mL equipped with a Teflon-in-glass valve and a 18/9 ball joint) attached to the vacuum line. Glass wool should be packed above the vessel, to prevent the lead(II) oxide from entering the vacuum line, and the vessel should be degassed. Upon evacuation, bromogermane (GeH_3Br, 0.1633 g, 1.05 mmole, measured in the gas phase using a mercury manometer) is condensed into the vessel at $-196°$. The valve is closed and the reactants are allowed to warm to room temperature. After the reaction has proceeded for 15 minutes* with shaking, the valve is opened and the volatile materials are allowed to distill through U-traps held at -126 (methylcyclohexane/N_2) and $-196°$ (liquid N_2), with continuous pumping to remove noncondensable gas (H_2). The trap at $-126°$ contains pure digermoxane* [$(GeH_3)_2O$, 0.0769 g, 0.46 mmole] identified by its ^1H nmr[17] and vibrational[4,18] parameters. Traces of germane (GeH_4, identified spectroscopically[19]) are retained in the trap at $-196°$. The yield† of digermoxane based on the bromogermane consumed is 88%.

Properties

Digermoxane (vp[4] about 66 torr at $0°$) is a clear colorless liquid that is best stored at liquid nitrogen temperature in break-seal glass ampules. The ^1H nmr spectrum,[17] measured in cyclohexane, consists of a singlet [$\delta(GeH)$] at 5.28 ppm, while the infrared spectrum[4,18] shows prominent absorptions at 2112 (s), 928, 882, 798 (vs), 784 (vs), 674, and 452 cm^{-1}. The mass spectrum confirms the presence of the molecular ion at m/e 156-172 $[H_nGe_2O]^+$.

---

*The checkers found that shorter reaction times can result in incomplete conversion. GeH_3Br is extremely difficult to separate from $(GeH_3)_2O$. When unreacted GeH_3Br is detected, the sample should be distilled back onto a fresh sample of PbO for additional reaction.
†Low yields may be obtained as a result of the partial decomposition of bromogermane in contact with the grease.

B. 1,3-DIMETHYL-, 1,1,3,3-TETRAMETHYL-, AND HEXAMETHYL DIGERMOXANE

$$2(CH_3)_n GeH_{3-n} Br + PbO \longrightarrow [(CH_3)_n GeH_{3-n}]_2 O + PbBr_2$$

$$n = 1, 2, 3$$

Procedure

The reaction vessel and conditions are identical to those described in Sec. A. The bromogermane [CH_3GeH_2Br, 0.4949 g, 2.92 mmole; $(CH_3)_2GeHBr$, 0.6780 g, 3.70 mmole; $(CH_3)_3GeBr$, 0.7624 g, 3.86 mmole] is condensed onto an excess of PbO (about 15 g, 67 mmole) and allowed to react for 20–35 minutes with some shaking. The volatile products are fractionated using cold traps at −78 (methanol/solid CO_2 slush) and −196°. The first trap retains the corresponding methyl digermoxane,* namely, $(CH_3GeH_2)_2O$, 89%; $[(CH_3)_3 \cdot GeH]_2O$, 95%; $[(CH_3)_3Ge]_2O$, 97%. Any traces of methylgermane are collected in the trap at −196°.

Properties

1,3-Dimethyl-, 1,1,3,3-tetramethyl-, and hexamethyldigermoxane are all clear, colorless liquids and the latter two are stable at room temperature in break-seal glass ampuls. Hexamethyldigermoxane (analyzed sample) gives[20]: mp −61°, bp (extrapolated) 129°, vapor pressure relationship: $\log_{10}P(mm) = 8.580 − (2290/T)$ over the range 18.2–72.4°, ΔH_{vap} 10,540 cal/mole, Trouton's constant 26.0. The methyl digermoxanes exhibit first-order ^1H nmr spectra.[6,7,9] (CH_3-$GeH_2)_2O$: $\delta(CH_3)$ 0.59, $\delta(GeH_2)$ 5.28 ppm, J_{HH}^{vic} 2.91, J_{13CH} 129.1 Hz. $[(CH_3)_2GeH]_2O$: $\delta(CH_3)$ 0.40, $\delta(GeH)$ 5.40 ppm, J_{HH}^{vic} 2.43, J_{13CH} 128.2 Hz. $[(CH_3)_3Ge]_2O$: $\delta(CH_3)$ 0.29 ppm, J_{13CH} 125.9 Hz. (Shifts are downfield from tetramethylsilane in carbon tetrachloride solution.) The purity of the methyl digermoxanes may be further checked by their infrared spectra. $(CH_3GeH_2)_2O$[9]: 2997 (m), 2923 (w), 2077 (sh), 2057 (vs), 1418 (vw), 1259 (m), 901 (sh), 876 (s), 840 (s), 800 (vs), 725 (s), 610 (ms), 486 (m), 454 cm^{-1}. $[(CH_3)_2GeH]_2O$[6,9]: 2997 (s), 2926 (m), 2040 (vs), 1422 (m), 1255 (s), 882 (s), 849 (s), 805 (vs), 762 (m), 702 (s), 664 (w), 613 (s), 591 (s), 480 cm^{-1}. $[(CH_3)_3Ge]_2O$[21]: 2983 (s), 2914 (m), 1408 (w), 1236 (s), 881 (vs), 823 (vs), 794 (vs), 753 (m), 607 (s),

---

*If spectral evidence indicates the presence of unreacted bromogermane in the germoxane after the suggested purification, the checkers suggest that the sample be distilled onto a fresh sample of PbO for additional reaction.

566 (m), 467 cm^{-1}. Likely impurities include unreacted bromogermane,* which is in fast exchange (on the ^{1}H nmr time scale)^9 and has a prominent ν(GeBr) absorption at about 272 cm^{-1} in the Raman spectrum.

References

1. L. M. Dennis and R. W. Work, *J. Am. Chem. Soc.*, **55**, 4486 (1933).
2. S. Sujishi and T. D. Goldfarb, Abstracts, 140th Meeting of the American Chemical Society, Inorganic Chemistry Division, 1958, p. 35N.
3. T. N. Srivastava, J. E. Griffiths, and M. Onyszchuk, *Can. J. Chem.*, **40**, 739 (1962).
4. T. D. Goldfarb and S. Sujishi, *J. Am. Chem. Soc.*, **86**, 1679 (1964).
5. S. Cradock and E. A. V. Ebsworth, *J. Chem. Soc. (A)*, **1968**, 1423.
6. J. E. Drake, R. T. Hemmings, and H. E. Henderson, *J. Chem. Soc. (A)*, **1976**, 366.
7. V. F. Mironov, L. N. Kalinina, E. M. Berliner, and T. K. Gar, *Zh. Obshch. Khim.*, **40**, 2597 (1970).
8. P. C. Angus and S. R. Stobart, *J. Chem. Soc. (A)*, **1975**, 2342.
9. J. E. Drake, B. M. Glavinčevski, R. T. Hemmings, and H. E. Henderson, *Can. J. Chem.*, **56**, 465 (1978).
10. D. F. Shriver, *The Manipulation of Air-Sensitive Compounds*, McGraw-Hill, New York, 1969.
11. J. W. Anderson, G. K. Barker, J. E. Drake, and R. T. Hemmings, *Can. J. Chem.*, **50**, 1607 (1972).
12. D. S. Rustad, T. Birchall, and W. L. Jolly, *Inorg. Synth.*, **11**, 128 (1968).
13. J. E. Drake, B. M. Glavinčevski, R. T. Hemmings, and H. E. Henderson, *Inorg. Synth.*, **18**, 154 (1978).
14. G. K. Barker, J. E. Drake, and R. T. Hemmings, *Can. J. Chem.*, **52**, 2622 (1974).
15. J. E. Griffiths, *J. Chem. Phys.*, **38**, 2879 (1963).
16. J. E. Drake, R. T. Hemmings, and C. Riddle, *J. Chem. Soc. (A)*, **1970**, 3359.
17. S. Cradock, E. A. V. Ebsworth, and D. W. H. Rankin, *J. Chem. Soc. (A)*, **1969**, 1628.
18. S. Cradock, *J. Chem. Soc. (A)*, **1968**, 1426.
19. C. D. McKean and A. A. Chalmers, *Spectrochim. Acta*, **23A**, 777 (1967).
20. J. E. Griffiths and M. Onyszchuk, *Can. J. Chem.*, **39**, 339 (1961).
21. A. Marchand, M. T. Forel, M. Lebedeff, and J. Volade, *J. Organomet. Chem.*, **26**, 69 (1971).

*The checkers suggest that infrared spectroscopy might also be used to detect unreacted methyl bromogermanes in the methyl germoxanes. This is most easily achieved by comparison of their representative infrared spectra.[9,13,14,16]

40. DI-μ-CHLORO-DICHLOROBIS(ETHYLENE)-DIPLATINUM(II), DI-μ-CHLORO-DICHLORO-BIS(STYRENE)DIPLATINUM(II), AND DI-μ-CHLORO-DICHLOROBIS(1-DODECENE)-DIPLATINUM(II)

Submitted by PAUL J. BUSSE,* BENJAMIN GREENE,* and MILTON ORCHIN*
Checked by ROBERT ZAHRAY† and JACK DOYLE†

Di-μ-chloro-dichlorobis(ethylene)diplatinum(II) is customarily prepared by evaporating an aqueous solution of $H[PtCl_3(C_2H_4)]$ to dryness and then recrystallizing the dimer.[1] However, this procedure has not been applied as a general synthesis of other related olefin complexes. The dimer has also been prepared, but with less success, by the reaction of Na_2PtCl_6 with boiling ethanol.[2] The direct reaction of ethylene with platinum(IV) chloride also provides the dimer,[3] but yield data are not available and presumably the method is unsatisfactory.

Dimers of the general structure $[PtCl_2(olefin)]_2$ may be obtained by displacing the ethylene in $[PtCl_2(C_2H_4)]_2$ with an excess of the desired olefin. Styrene[4] and 1-dodecene[5] complexes have been prepared in this manner. However, $[PtCl_2(C_2H_4)]_2$ is relatively sensitive to moisture and air and accordingly is not conveniently stored.

A convenient direct synthesis of $[PtCl_2(olefin)]_2$ is reported below. The reaction consists of treating a diethyl ether solution of the stable, readily available trans-$[PtCl_2(C_2H_4)(pyridine)]$ with a sulfonic acid polymer resin (®—SO_3H) that removes the amine and generates the dimer.

A. trans-DICHLORO(ETHYLENE)(PYRIDINE)PLATINUM(II)

Procedure

$$K[PtCl_3(C_2H_4)] + C_5H_5N \longrightarrow trans\text{-}[PtCl_2(C_2H_4)(C_5H_5N)] + KCl$$

Potassium trichloro(ethylene)platinum(II),[6] Zeise's salt (2.00 g, 5.43 mmole), is placed in a 100-mL Erlenmeyer flask containing a magnetic stirrer, and 50 mL of water is added. The mixture is stirred until complete dissolution of the salt, whereupon 0.44 mL (5.46 mmole) of pyridine is added dropwise (syringe) with

---

*Department of Chemistry, University of Cincinnati, Cincinnati, OH 45221.
We thank Engelhard Industries for kindly supplying us with K_2PtCl_4.
†Department of Chemistry, The University of Iowa, Iowa City, IA 52242.

stirring, over a period of 1-2 minutes, to the yellow solution. A yellow precipitate forms immediately and the reaction mixture is allowed to stir for 30 minutes. The precipitate is collected by suction filtration and washed three times with 10 mL portions of water. The yellow powder is dried *in vacuo* overnight. Yield: 1.82 g (90% based on platinum); mp 125-127°. The compound is stable to air and moisture and can be used without further purification. *Anal.* Calcd. for [PtCl$_2$(C$_2$H$_4$)(C$_5$H$_5$N)]: C, 22.52; H, 2.41. Found: C, 22.66; H, 2.42.

B. Di-μ-CHLORO-DICHLOROBIS(ETHYLENE)DIPLATINUM(II)

$$2\textit{trans-}[PtCl_2(C_2H_4)(C_5H_5N)] + 2\ \textcircled{R} -SO_3H \longrightarrow$$
$$[PtCl_2(C_2H_4)]_2 + 2\ \textcircled{R} -SO_3^-C_5H_5NH^+$$

Procedure

A solution of 400 mg (1.07 mmole) of *trans*-dichloro(ethylene)(pyridine)platinum(II) in 25 mL of diethyl ether is added to 5.0 g (12 mequiv) of Dowex 50W-X8 ion exchange resin in a 50-mL Erlenmeyer flask. The reaction mixture is stirred vigorously for 1 hour with a magnetic stirrer and filtered through a sintered-glass frit. While on the frit, the resin is stirred twice with 5 mL of diethyl ether and filtered. The combined filtrates are treated with 5.0 g of fresh resin and stirred for 1 hour. The mixture is filtered, and the insoluble polymer is washed twice with 5 mL of diethyl ether as above. The combined filtrates are treated a third time in the same manner. The yellow filtrate is dried overnight over sodium sulfate at 0°. The sodium sulfate is removed by filtration and washed three times with 5-mL portions of anhydrous benzene. The combined filtrates are evaporated to dryness under reduced pressure, providing an orange solid. The solid is added to approximately 10 mL of benzene, 1-2 mg of decolorizing charcoal is added, and the mixture is quickly brought to boiling and immediately filtered. On cooling, orange crystals separate and are collected, mp 180-185° (dec.). Yield: 130 mg (40% based on platinum). Extended heating of solutions of [PtCl$_2$·(C$_2$H$_4$)]$_2$ must be avoided. *Anal.* Calcd. for [PtCl$_2$(C$_2$H$_4$)]$_2$: C, 8.16; H, 1.36. Found: C, 8.75; H, 0.99.

C. DI-μ-CHLORO-DICHLOROBIS(STYRENE)DIPLATINUM(II)

$$\textit{trans-}[PtCl_2(C_2H_4)(C_5H_5N)] + C_6H_5CH=CH_2 \longrightarrow$$
$$\textit{trans-}[PtCl_2(C_6H_5CH=CH_2)(C_5H_5N)] + C_2H_4$$
$$2\textit{trans-}[PtCl_2(C_6H_5CH=CH_2)(C_5H_5N)] + 2\ \textcircled{R} -SO_3H \longrightarrow$$
$$[PtCl_2(C_6H_5CH=CH_2)]_2 + 2\ \textcircled{R} -SO_3^-C_5H_5NH^+$$

Procedure

A solution of 400 mg (1.07 mmole) of *trans*-dichloro(ethylene)(pyridine)-platinum(II) in 25 mL of anhydrous diethyl ether is heated at reflux for 30 minutes with 0.13 mL (1.13 mmole) of styrene in a 40-mL, round-bottomed flask protected by a calcium sulfate drying tube. The yellow solution is allowed to cool to room temperature. Three treatments with Dowex 50W-X8 and drying with sodium sulfate are performed as described in Sec. B. The sodium sulfate is separated by suction filtration and washed three times with 5-mL portions of anhydrous benzene. Solvent is removed from the combined filtrates under reduced pressure, providing a bright-orange solid. The solid is dissolved in approximately 10 mL of refluxing benzene containing 1-2 mg of decolorizing charcoal. The mixture is filtered hot and the filtrate is cooled to 10°, providing orange crystals, mp 201-202° (dec.). Yield: 220 mg (55%). *Anal.* Calcd. for $[PtCl_2(C_6H_5CH=CH_2)]_2$: C, 25.96; H, 2.18; Cl, 19.15. Found: C, 25.72; H, 2.16; Cl, 19.23.

D. Di-µ-CHLORO-DICHLOROBIS(1-DODECENE)DIPLATINUM(II)

$$\textit{trans-}[PtCl_2(C_2H_4)(C_5H_5N)] + C_{10}H_{21}CH=CH_2 \longrightarrow$$
$$\textit{trans-}[PtCl_2(C_{10}H_{21}CH=CH_2)(C_5H_5N)] + C_2H_4$$
$$2\textit{trans-}[PtCl_2(C_{10}H_{21}CH=CH_2)(C_5H_5N)] + 2\,®\!-\!SO_3H \longrightarrow$$
$$[PtCl_2(C_{10}H_{21}CH=CH_2)]_2 + 2\,®\!-\!SO_3^-C_5H_5NH^+$$

Procedure

A solution of 400 mg (1.07 mmole) of *trans*-dichloro(ethylene)(pyridine)-platinum(II) in 25 mL of anhydrous diethyl ether is heated at reflux for 30 minutes with 0.25 mL (1.13 mmole) of 1-dodecene in a 50-mL, round-bottomed flask protected by a calcium sulfate drying tube. The yellow solution is allowed to cool to room temperature. Three treatments with Dowex 50W-X8 and drying with sodium sulfate are conducted as described in Sec. B. The sodium sulfate is separated by suction filtration and washed three times with 5-mL portions of anhydrous benzene. Solvent is removed from the combined filtrates under reduced pressure, providing an orange solid. Twenty milliliters of hexane is added to the solid. After being heated to reflux, the orange solution is filtered hot through a sintered-glass frit, and the filtrate is cooled to 0°. A waxy, orange solid precipitates and is separated by suction filtration and washed twice with 10 mL of cold hexane. Vacuum drying overnight provides an orange solid, mp 73°. Yield: 304 mg (65%).

Properties

The dimers are moderately stable in air and can be handled without special precautions. However, extended exposure to air or moisture results in decomposition, and the dimers are best stored refrigerated under an inert gas. All three dimeric complexes are soluble in acetone, diethyl ether, tetrahydrofuran, and alcohols. They are less soluble in halogenated or aromatic solvents, with the ethylene complex being the least soluble.

trans-[PtCl$_2$(C$_2$H$_4$)(C$_5$H$_5$N)] is stable in air when dry and can be stored indefinitely at 0°. It is soluble in most organic solvents other than aliphatic hydrocarbons. The extent of contamination by unreacted *trans*-[PtCl$_2$(olefin)(C$_5$H$_5$N)] in the dimeric complexes may be evaluated by the nmr spectrum of a chloroform-*d* slurry (ethylene) or solution (styrene and dodecene) of the appropriate dimer. The nmr spectrum of *trans*-[PtCl$_2$(C$_2$H$_4$)(C$_5$H$_5$N)] in DCCl$_3$ (TMS internal standard) exhibits a singlet for the ethylene protons with two satellite peaks due to ^{195}Pt-^{1}H coupling, J_{PtH} = 61.5 Hz, centered at δ 4.93 ppm; a triplet with splitting J_{HH} = 6 Hz centered at δ 7.53 ppm (*meta*-pyridine hydrogen); a triplet with splitting J_{HH} = 7 Hz centered at δ 7.99 ppm (*para*-hydrogen); and a doublet J_{HH} = 5 Hz centered at δ 8.88 ppm (*ortho*-hydrogen). In the absence of excess pyridine,[7] the doublet at δ 8.88 exhibits ^{195}Pt-^{1}H coupling with satellite doublets having J_{PtH} = 35 Hz. The nmr spectrum of [PtCl$_2$(C$_2$H$_4$)]$_2$ in acetone-d_6 (TMS internal standard) consists of a broad singlet centered at δ 4.53 ppm. The nmr spectrum of [PtCl$_2$(C$_6$H$_5$CH=CH$_2$)]$_2$ in DCCl$_3$ (TMS internal standard) exhibits broad peaks centered at δ 4.8 ppm and at δ 6.5 ppm, and a multiplet centered at δ 7.44 ppm. The nmr spectrum of [PtCl$_2$(*n*-C$_{10}$H$_{21}$CH=CH$_2$)]$_2$ in DCCl$_3$ (TMS internal standard) consists of a triplet with splitting J_{HH} = 5 Hz centered at δ 0.90 ppm, a singlet centered at δ 1.31 ppm, and three broad peaks centered at δ 1.8, δ 4.7, and δ 5.5 ppm.

The choice of *trans*-[PtCl$_2$(olefin)(C$_5$H$_5$N)] for the preparation of dimers is a matter of convenience. Actually almost any complex having the structure *trans*-[PtCl$_2$(olefin)(L)] where *L* is an amine may be used. Because pyridine is readily available and convenient to use, it is the amine of choice.

An alternate route to the desired dimers consists of a two-step sequence of reactions:

$$K[PtCl_3(C_2H_4)] + \text{olefin} \longrightarrow K[PtCl_3(\text{olefin})] + C_2H_4$$

$$K[PtCl_3(\text{olefin})] + M^+ \longrightarrow [PtCl_2(\text{olefin})]_2 + K^+ + MCl(s)$$

As metal salts to remove chloride and provide the dimer, we have investigated AgPF$_6$, AgBF$_4$, and Pb(ClO$_4$)$_2$. Although the dimer can be obtained by this sequence, yields, product purity, and convenience are inferior to those of the method recommended above.

References

1. J. Chatt and M. L. Searle, *Inorg. Synth.*, 5, 210 (1957).
2. J. S. Anderson, *J. Chem. Soc.*, 1934, 971.
3. M. S. Kharasch and T. A. Ashford, *J. Am. Chem. Soc.*, 58, 1733 (1936).
4. S. S. Hupp and G. Dahlgren, *Inorg. Chem.*, 15, 2349 (1976).
5. J. R. Joy and M. Orchin, *J. Am. Chem. Soc.*, 81, 305 (1959).
6. P. B. Chock, J. Halpern, and F. E. Paulik, *Inorg. Synth.*, 14, 90 (1973); C. Y. Hsu, B. T. Leshner, and M. Orchin, *Inorg. Synth.*, 19, 114 (1978).
7. F. Pesa and M. Orchin, *J. Organomet. Chem.*, 78, C26 (1974).

41. DIBROMODIMETHYLPLATINUM(IV) AND DIHYDROXODIMETHYLPLATINUM(IV) SESQUIHYDRATE

Submitted by J. R. HALL,* D. A. HIRONS,* and G. A. SWILE*
Checked by W. H. PAN† and J. P. FACKLER, JR.†

Although iodotrimethylplatinum(IV)[1] is a readily prepared starting material for the preparation of other trimethylplatinum(IV) compounds, a convenient starting material for synthesis of dimethylplatinum(IV) compounds remains to be established. One synthetic route is the oxidative addition of either CH_3X to monomethylplatinum(II) compounds[2,3] or of halogens to dimethylplatinum(II) compounds.[3,4] An alternative is to start with $PtBr_2(CH_3)_2$.[5] The procedure described below yields $PtBr_2(CH_3)_2$ in 90% yield from the readily obtainable $PtI(CH_3)_3$. The dibromo compound is suitable for use in nonaqueous solvents. Preparation of derivatives may be accomplished in an aqueous medium by starting with a sulfuric, nitric, or perchloric acid solution of $Pt(OH)_2(CH_3)_2 \cdot 1.5H_2O$, which is obtainable in quantitative yield from the dibromo compound.

A. DIBROMODIMETHYLPLATINUM(IV)

$$PtI(CH_3)_3 + Br_2 \longrightarrow PtBr_2(CH_3)_2 + CH_3I$$

Procedure

Yellow iodotrimethylplatinum(IV) (5.0 g, 0.014 mole), bromine (25 mL, 0.45 mole), and hydrobromic acid (0.8 mL of 46–49% by weight) are placed in a 100-mL, round-bottomed flask fitted with a water condenser, and the solution

---

*Chemistry Department, University of Queensland, Brisbane, Australia 4067.
†Chemistry Department, Case Western Reserve University, Cleveland, OH 44106.

186 Organometallic Compounds

is heated at reflux for 8 hours. All liquid is then evaporated by heating the flask on a water bath, leaving a yellow residue contaminated by a small amount of reddish-brown material. The solid is washed with water until the washings are colorless and then with acetone (50 mL), after which it is air dried. The yield of crude $PtBr_2(CH_3)_2$ is 5.0 g, 0.013 mole (95%). The crude product is purified by way of its bis(pyridine) derivative. The solid is dissolved in hot pyridine (15 mL), the pyridine is evaporated on the water bath in a fume hood, and the yellow crystalline residue is extracted with hot chloroform (50 mL). The chloroform solution is filtered and concentrated to a small volume (10 mL) on the water bath, and small portions of ethanol (5 mL) are added to maintain this volume until yellow crystals of the product are formed. The solution is cooled in ice and then filtered. The crystalline solid is washed with cold ethanol (5 mL) and air dried. The yield is 6.7 g (85-95%). *Anal.* Calcd. for $C_{12}H_{16}N_2Br_2Pt$: Pt, 35.9; C, 26.5; H, 3.0; N, 5.2; Br, 29.4. Found: Pt, 35.4; C, 26.3; H, 3.1; N, 5.2; Br, 29.6.

■ **Caution.** *Perchloric acid should always be treated as a hazardous substance.*[6]

$[PtBr_2(CH_3)_2(py)_2]$ (5.0 g) is added to water (50 mL) and 60% perchloric acid (12.5 mL) in a round-bottomed flask fitted with a reflux condenser. The mixture is heated at reflux for 2.5 hours, at which stage a yellow solid has precipitated and the supernatant liquid is almost colorless (there may be a slight yellow color or a trace of iodine may crystallize on the walls of the condenser). The solution is cooled and the yellow solid is filtered, washed with water and then acetone (20 mL), and air dried. The yield is 3.35 g (95%). *Anal.* Calcd. for $C_2H_6Br_2Pt$: Pt, 50.7; C, 6.2; H, 1.6; Br, 41.5. Found: Pt, 50.3; C, 6.3; H, 1.6; Br, 41.4.

Properties

Dibromodimethylplatinum(IV) is a yellow crystalline solid that decomposes at 180-190°. It is insoluble in water and sparingly soluble in organic solvents. Although no X-ray crystal structure data are available, the complex is expected to be polymeric if the usual six-coordination of platinum(IV) is maintained. In methanol solution it is monomeric. Its reactivity with a variety of ligands has been described.[7] Sharp infrared absorption bands at 1220 (s), 1222 (w, sh), 1245 (w), and 1252 (m) cm^{-1} in the CH_3 deformation region are characteristic of the compound.

B. DIHYDROXODIMETHYLPLATINUM(IV) SESQUIHYDRATE

$$PtBr_2(CH_3)_2 + 4NaOH \longrightarrow Pt(OH)_4(CH_3)_2{}^{2-} + 4Na^+ + 2Br^- \xrightarrow{H_3O^+}$$
$$Pt(OH)_2(CH_3)_2 \cdot 1.5H_2O$$

Procedure

■ **Caution.** *Perchloric acid should always be treated as a hazardous substance.*[6]

Recrystallized $[PtBr_2(CH_3)_2]_n$* (1 g, 0.0026 mole) is suspended in 30 mL of an aqueous solution of sodium hydroxide (2.5 g, 0.063 mole) in a 100-mL beaker. On boiling the suspension, the solid dissolves completely in 15–20 minutes to give a yellow solution. When heated for a further 45–60 minutes, the solution becomes almost colorless (the solution sometimes remains pale yellow, possibly because of the formation of a little $Pt(OH)_6^{2-}$). The solution is then cooled in an ice bath. While it is monitored with a pH meter, 60% perchloric acid solution is added in a dropwise manner until the pH reaches 10. At this point, 10% perchloric acid is used to reduce the pH to 7, at which stage a fine, white precipitate forms. The solid is allowed to settle and is washed by decantation several times and then either centrifuged or filtered slowly through a low-porosity sintered-glass filter funnel. The solid is washed with water and dried over P_4O_{10} in a vacuum desiccator. The yield is quantitative. The solid has a glassy appearance but is a white powder when crushed. It is sometimes contaminated with metallic platinum or oxide, which may be removed by the following recrystallization procedure. Crude $Pt(OH)_2(CH_3)_2 \cdot 1.5H_2O$ (1 g, 0.0035 mole) is suspended in water (15 mL) and 60% $HClO_4$ (1 mL) and the mixture is heated for 5 minutes. After it is cooled, the solution is filtered through a fine sintered-glass filter funnel, leaving a residue that is probably $PtO_2 \cdot nH_2O$. The filtrate is neutralized with dilute NaOH (1 *M*), using a pH meter. The resultant white precipitate is filtered off, washed with water, and dried in a vacuum desiccator. The freshly prepared substance analyzes for between 1 and 2 moles of water per mole of Pt. The calculated percentages correspond to $Pt(OH)_2(CH_3)_2 \cdot 1.5H_2O$. *Anal.* Calcd. for $C_4H_{22}O_7Pt_2$: Pt, 68.2; C, 8.4; H, 3.9. Found: Pt, 68.8; C, 8.3; H, 3.7. The analytical results for an aged sample indicate diminished water content.

Properties

Dihydroxodimethylplatinum(IV) sesquihydrate is a white solid that turns black at 126–130° and *explodes* at about 170°. It loses weight in an evacuated drying pistol at 110°, but darkening of the solid begins at about 80°. The substance is insoluble in organic solvents. It behaves as a typical amphoteric hydroxide, dissolving in warm caustic solution to yield a single species, $[Pt(OH)_4(CH_3)_2]^{2-}$, and readily in sulfuric acid, for example, to yield the cation $[Pt(CH_3)_2]^{2+}$ (aq.).[8] Excess hydrobromic and hydrochloric acids produce the complex ions $[PtX_4(CH_3)_2]^{2-}$ (X = Br and Cl), but hydriodic acid yields the insoluble $PtI_2(CH_3)_2$. Some other reactions have been reported.[8]

*If crude $[PtBr_2(CH_3)_2]_n$ is used, the reaction rapidly produces platinum metal.

References

1. D. E. Clegg and J. R. Hall, *Inorg. Synth.*, **10**, 71 (1967).
2. J. R. Hall and G. A. Swile, *J. Organomet. Chem.*, **76**, 257 (1974).
3. J. D. Ruddick and B. L. Shaw, *J. Chem. Soc. (A)*, **1969**, 2801, *ibid.*, **1969**, 2964.
4. H. C. Clark and L. E. Manzer, *J. Organomet. Chem.*, **59**, 411 (1973).
5. J. R. Hall and G. A. Swile, *Aust. J. Chem.*, **24**, 423 (1971).
6. See L. A. Muse, *J. Chem. Educ.*, **49**, A 463 (1972).
7. J. R. Hall and G. A. Swile, *J. Organomet. Chem.*, **56**, 419 (1973); *ibid.*, **67**, 455 (1974).
8. J. R. Hall and G. A. Swile, *J. Organomet. Chem.*, **122**, C19 (1976).

42. HALOCYCLOPENTADIENYL COMPLEXES OF MANGANESE AND RHODIUM

$$MnX(CO)_5 + C_5H_4N_2 \longrightarrow Mn(\eta^5\text{-}C_5H_4X)(CO)_3 + 2CO\uparrow + N_2\uparrow$$

$$X = Cl, Br, I$$

$$MnX(CO)_5 + C_5Cl_4N_2 \longrightarrow$$
$$Mn(\eta^1\text{-}C_5Cl_4X)(CO)_5 + Mn(\eta^5\text{-}C_5Cl_4X)(CO)_3 + CO\uparrow + N_2\uparrow$$

$$X = Cl, Br$$

$$[RhXL_2]_2 + 2C_5R_4N_2 \longrightarrow 2Rh(\eta^5\text{-}C_5R_4X)(L_2) + 2N_2\uparrow$$

$$R = C_6H_5, L_2 = 1,5\text{-}C_8H_{12}, X = Cl \text{ or } Br$$
$$R = C_6H_5, L_2 = (CO)_2, X = Cl$$
$$R = H \text{ or } Cl, L_2 = 1,5\text{-}C_8H_{12}, X = Cl$$

Submitted by KENNETH J. REIMER* and ALAN SHAVER†
Checked by MICHAEL H. QUICK‡ and ROBERT J. ANGELICI‡

Halocyclopentadienyl complexes are very useful starting materials for the synthesis of other substituted compounds. Until recently, the compounds $Mn(\eta^5\text{-}C_5H_4X)(CO)_3$ (X = Cl, Br, I) were obtained in low yield by means of multistep synthetic procedures.[1] As a result, they were not suitable as synthetic intermediates. The method described here supplies these compounds easily and in high yield by means of the insertion of 5-diazo-1,3-cyclopentadiene[2] into the manganese–halogen bond of $MnX(CO)_5$ (X = Cl, Br, I). The preparations of some halocyclopentadienyl rhodium complexes are also described.

*Department of Chemistry, Royal Roads Military College, Victoria, B.C., Canada.
†Department of Chemistry, McGill University, Montreal, Quebec, Canada.
‡Department of Chemistry, Iowa State University, Ames, IA 50011.

The complexes containing the pentachlorocyclopentadienyl ligand are of great importance. Not only are they of spectroscopic and theoretical interest, but the few known examples of this class of compounds often exhibit interesting chemical properties, such as resistance to oxidation.[3] Tetrachloro-5-diazo-1,3-cyclopentadiene can be inserted into the metal–chlorine bond of certain manganese and rhodium complexes to give the compounds outlined in the scheme. These have further indicated the unusual nature of the $C_5Cl_5^-$ ligand. For example, $Rh(\eta^5\text{-}C_5Cl_5)$ cod (cod = 1,5-cyclooctadiene) has been shown to have unique structural features[4] and $Mn(\eta^1\text{-}C_5Cl_5)(CO)_5$ is[5] the only known σ-bonded transition metal complex of $C_5Cl_5^-$. There is no $C_5H_5^-$ analog for this compound. The chemical reactivity of these types of compounds remains to be investigated.

■ **Caution.** *Diazo compounds are potentially explosive. Although neat 5-diazo-1,3-cyclopentadiene is known to be highly explosive,[6] it may be handled conveniently and safely in a pentane solution.[2,7] Tetrachloro-5-diazo-1,3-cyclopentadiene is reported to be stable,[8] but due caution should be exercised in the manipulation of this compound. Carbon monoxide, volatile metal carbonyls, and some diazo compounds are highly toxic, and reactions with these species should be conducted in a well-ventilated fume hood. Avoid inhalation or contact with skin.*

General Procedure

Inert-atmosphere techniques[9] are used in the synthesis and purification of all the metal complexes described here. Nitrogen gas (Liquid Carbonic, "Hi pure") can be used without further purification. All solvents, although of "spectroquality," should be dried over molecular sieves and vacuum degassed or purged with nitrogen before use. Yields are quantitative if not specified. $RhCl_3 \cdot 3H_2O$* and hexachloro-1,3-cyclopentadiene† were used without further purification. $[RhX(cod)]_2$[10] (X = Cl, Br), $[RhCl(CO)_2]_2$[11], $C_5Ph_4N_2$,[12] and $MnBr(CO)_5$ (X = Cl, Br)[13] were prepared by published procedures.

A. TETRACHLORO-5-DIAZO-1,3-CYCLOPENTADIENE

The synthesis of $C_5Cl_4N_2$ follows the series of reactions:

$$C_5Cl_6 \longrightarrow C_5Cl_4(NNH_2) \longrightarrow C_5Cl_4N_2$$

The preparation of tetrachloro-2,4-cyclopentadien-1-one hydrazone has been reported in the patent literature[14] and elsewhere.[15,16] The synthesis presented

---
*Available from Alfa Products, Ventron Corp., P.O. Box 299, Danvers, MA 01923.
†Available from Aldrich Chemical Co., 940 W. St. Paul Ave., Milwaukee, WI 53233.

here represents a considerable modification of previous methods and provides a convenient route to reasonable yields of this compound in a high degree of purity. A variety of reagents, such as mercury(II) oxide,[15] sodium hypochlorite,[16] and silver(I) oxide,[8] have been used for the oxidation of the hydrazone, but only silver oxide is used in this work. This preparation of $C_5Cl_4N_2$ is used with some minor alterations.

1. Procedures for Tetrachloro-2,4-cyclopentadien-1-one Hydrazone, $C_5Cl_4NNH_2$

■ **Caution.** *Hydrazine hydrate and hexachlorocyclopentadiene are hazardous; avoid inhalation or skin contact.*

Hydrazine hydrate (29.4 mL, 30.32 g, 0.61 mole) is added dropwise from a pressure-equalizing addition funnel to a cooled (0°) solution of hexachloro-1,3-cyclopentadiene (16 mL, 27.3 g, 0.10 mole) in 200 mL of methanol in a 500-mL Erlenmeyer flask. The solution turns dark red immediately. After the addition, the flask is securely stoppered (with a glass stopper) and the solution is stirred in the dark for 5 days at 5°, during which time a red solid collects on the walls of the flask. The mixture is poured into 5 L of water and filtered. The solid is air dried, dissolved in diethyl ether, and filtered. The filtrate is evaporated to dryness and the residue is extracted with cyclohexane in a Soxhlet extractor. The solution is concentrated and cooled and the red-brown crystals are filtered off. Additional product can be obtained by washing the material remaining in the thimble with water and extracting again with cyclohexane. Total yield: 15 g (65%).

2. Procedures for Tetrachloro-5-diazo-1,3-cyclopentadiene, $C_5Cl_4N_2$

Silver oxide (28.3 g, 0.122 mole), $C_5Cl_4(NNH_2)$ (11.32 g, 0.049 mole), and anhydrous magnesium sulfate (2-3 g) are placed in a 500-mL Erlenmeyer flask, and 125 mL of diethyl ether is added. The original paper[8] incorrectly reported 0.125 mole of Ag_2O as being 15.5 g. The correct amount, used above, is required for the reaction to be completed in a reasonable length of time. A glass stopper is secured with an elastic band and the flask is wrapped with a wet towel to dissipate the heat evolved in the early stages of the reaction. The mixture is then agitated on a mechanical shaker for 20 hours (stirring with a magnetic stir bar is inadequate; the checkers used a motor-driven paddle stirrer). The mixture is filtered and the filtrate is evaporated to dryness. The residue is recrystallized from boiling methanol to give 8.8 g (78%) of orange needles. Recrystallization from methanol rather than the reported[8] chromatography was found to be a more convenient purification step. If kept in the dark and under nitrogen, $C_5Cl_4N_2$ may be stored at room temperature unchanged for several months.

■ **Caution.** *Tetrachloro-5-diazo-1,3-cyclopentadiene, $C_5Cl_4N_2$, is said to be very stable[8] and we exerienced no difficulty. However, it is advisable to use care when subjecting the compound to extreme reaction conditions.*

B. 5-DIAZO-1,3-CYCLOPENTADIENE, $C_5H_4N_2$

The ligand is prepared according to the general method of Weil and Cais.[17] A mixture of 1,3-cyclopentadiene (7 mL, 5.6 g, 0.085 mole), diethylamine (4.7 mL, 3.34 g, 0.046 mole), and *p*-toluenesulfonyl azide[18] (10.4 g, 0.053 mole) contained in a 50-mL Erlenmeyer flask is allowed to stand at 5° for 5 days. Water (25 mL) is added and the product is extracted with pentane (several times, with filtering) from the brown heterogeneous reaction mixture. The extracts are combined and washed with water until a test with litmus paper indicates the absence of amine. After the solution is dried over $MgSO_4$, the volume is concentrated to slightly less than 10 mL. (■ *Caution. The pure compound is known to be highly explosive and should always be kept in solution.*[6,7])

This solution is transferred to a 10-mL volumetric flask and the volume is brought to the calibration mark by the addition of pentane. The concentration of the $C_5H_4N_2$ is determined by adding a known amount of this solution to an ethereal solution containing excess triphenylphosphine. The concentration can be conveniently calculated from the amount of the adduct, $C_5H_4N_2 \cdot PPh_3$,[17] that crystallizes from solution. A typical reaction yields 10 mL of 2.5 *M* solution (2.3 g, $C_5H_4N_2$, 47% yield based on *p*-$CH_3C_6H_4SO_2N_3$ used*).

C. (CHLOROTETRAPHENYL-η^5-CYCLOPENTADIENYL)(η^4-1,5-CYCLOOCTADIENE)RHODIUM, $Rh(\eta^5-C_5Ph_4Cl)(cod)$

Benzene (10 mL) is added to a mixture of $[RhCl(cod)]_2$ (0.20 g, 0.40 mmole) and $C_5Ph_4N_2$ (0.32 g, 0.81 mmole) in a nitrogen-filled 60-mL Schlenk tube connected to a nitrogen line equipped with a mercury bubbler.[9] Bubbles of gas are evolved and on stirring the red-orange solution lightens in color. The reaction is allowed to proceed overnight and then the solvent is removed *in vacuo*. The product is recrystallized by dissolving it in a minimum of pentane and

---

*The checkers obtained 30–35% yields and found their sample contained dicyclopentadiene. To overcome this they used a small excess of C_5H_5 and a large excess of 2-aminoethanol (ethanolamine). The reaction mixture was stirred 1–2 hours at 0° and then stored for 24 hours at 0°. Extraction with pentane/hexane 9:1 (omitting the aqueous washing) gave, after concentration, a final solution of $C_5H_4N_2$ (30% yield) largely in hexane, which facilitates syringe manipulations. The compound decomposes over a period of a week at room temperature, but samples stored at −78° are stable for several months.

cooling to $-20°$. Mp 208-210°. *Anal.* Calcd. for $Rh(\eta^5-C_5Ph_4Cl)(cod)$: C, 72.29; H, 5.21. Found: C, 72.08; H, 5.28.

D. (BROMOTETRAPHENYL-η^5-CYCLOPENTADIENYL)-(η^4-1,5-CYCLOOCTADIENE)RHODIUM, $Rh(\eta-C_5Ph_4Br)(cod)$ AND DICARBONYL(CHLOROTETRAPHENYL-η^5-CYCLOPENTADIENYL)RHODIUM, $Rh(\eta^5-C_5Ph_4Cl)(CO)_2$

Both compounds are prepared as above, using $[RhBr(cod)]_2$ and $[RhCl(CO)_2]_2$, respectively, and the appropriate stoichiometric quantities of C_5Ph_4-N_2. The latter red complex is more soluble than the cod complex, and it is recrystallized from hexane as above. Mp 202-204° (dec.) and 78-81°, respectively. *Anal.* Calcd. for $Rh(\eta^5-C_5Ph_4Br)(cod)$: C, 67.41; H, 4.85. Found: C, 67.56; H, 5.02. Calcd. for $Rh(\eta^5-C_5Ph_4Cl)(CO)_2$: C, 66.17; H, 3.55. Found: C, 66.39; H, 3.73.

E. (CHLORO-η^5-CYCLOPENTADIENYL)(η^4-1,5-CYCLOOCTADIENE)-RHODIUM, $Rh(\eta^5-C_5H_4Cl)(cod)$

To a solution of $[RhCl(cod)]_2$ (0.20 g, 0.40 mmole) in benzene (5 mL) in a Schlenk tube equipped as in Sec. C is added a portion of 2.5 M pentane solution of 5-diazo-1,3-cyclopentadiene (0.32 mL, 0.80 mmole). The solution is stirred overnight and then the solvent is removed *in vacuo*. The product is recrystallized from hexane at $-78°$. Mp 63-64°. *Anal.* Calcd. for $Rh(\eta^5-C_5H_4Cl)$-(cod): C, 50.29; H, 5.15. Found: C, 50.14; H, 5.12.

F. TRICARBONYL(CHLORO-η^5-CYCLOPENTADIENYL)-MANGANESE, $Mn(\eta^5-C_5H_4Cl)(CO)_3$

To $MnCl(CO)_5$ (0.2 g, 0.87 mmole) suspended in pentane (10 mL) in a Schlenk tube as in Sec. C is added 0.35 mL (0.875 mmole) of a 2.5 M pentane solution of 5-diazo-1,3-cyclopentadiene and the mixture is stirred overnight. It is then filtered under nitrogen[9] into a 100-mL, round-bottomed flask and the volume is reduced to about 5 mL by a nitrogen stream. Cooling to $-78°$ causes the product to crystallize. The remainder of the solvent is removed with a syringe and the product is dried by brief evacuation. The flask is then fitted with a water-cooled sublimation probe and evacuated. Then the stopcock connecting the flask to the vacuum pump is closed. The product is sublimed under static vacuum at room temperature. The purified product is transferred quickly in air (because of its low melting point) to a nitrogen-filled Schlenk tube where it is stored at 5°. Mp 24-25°. *Anal.* Calcd. for $Mn(\eta^5-C_5H_4Cl)(CO)_3$: C, 40.30; H, 1.68. Found: C, 40.44; H, 1.57.

G. (BROMO-η^5-CYCLOPENTADIENYL)TRICARBONYL-MANGANESE, Mn(η^5-C_5H_4Br)(CO)$_3$

5-Diazo-1,3-cyclopentadiene (0.22 mL of a 2.5 M pentane solution, 0.55 mmole) is added to a suspension of MnBr(CO)$_5$ (0.150 g, 0.55 mmole) in pentane (20 mL) contained in a Schlenk tube as in Sec. C. The mixture is stirred overnight, filtered, and concentrated to about 5 mL in a nitrogen stream. Cooling to $-78°$ induces crystallization. The mother liquor is removed and the product is dried by brief evacuation. Mp 43-45°. *Anal.* Calcd. for Mn(η^5-C_5H_4Br)(CO)$_3$: C, 33.96; H, 1.41. Found: C, 33.90; H, 1.53.

H. TRICARBONYL(IODO-η^5-CYCLOPENTADIENYL)-MANGANESE, Mn(η^5-C_5H_4I)(CO)$_3$

This compound is prepared as above using MnI(CO)$_5$, but from 48 to 72 hours are required for complete reaction. Mp 29-30°. *Anal.* Calcd. for Mn(η^5-C_5H_4I)-(CO)$_3$: C, 29.12; H, 1.22. Found: C, 29.13; H, 1.28.

I. PENTACARBONYL(PENTACHLORO-η^1-CYCLOPENTADIENYL)-MANGANESE, Mn(η^1-C_5Cl_5)(CO)$_5$

Pentane (80 mL) is added to a mixture of MnCl(CO)$_5$ (1.0 g, 4.35 mmole) and $C_5Cl_4N_2$ (1.0 g, 4.35 mmole) in a nitrogen-filled, 100-mL, two-necked flask connected to a nitrogen line as in Sec. C. The flask is closed, using stoppers secured with elastic bands to prevent loss of CO, which leads to reduced yields, and the mixture is stirred for 24 hours at room temperature. The mixture is then filtered and cooled to $-78°$ overnight. During this time large yellow cubic crystals are formed. The mother liquors are removed by means of a syringe and the crystals (0.99 g) are dried, first in a nitrogen stream and then briefly *in vacuo*. Infrared analysis usually indicates that this crop contains mainly Mn-(η^1-C_5Cl_5)(CO)$_5$, but it may be contaminated with a small amount of Mn(η^5-C_5Cl_5)(CO)$_3$.* Recrystallization from pentane by slow cooling of a satu-

---

*Mn(η^1-C_5Cl_5)(CO)$_5$ slowly forms Mn(η^5-C_5Cl_5)(CO)$_3$ in solution. Therefore, the relative amounts of the two compounds depend greatly on the reaction time. To optimize the yield of the η^1 compound, the reaction should be monitored by observing the infrared spectrum [ν(NN) for $C_5Cl_4N_2$ is 2110 cm^{-1}] and worked up as soon as the reaction is completed. Depending on the amount of pentahapto compound present in the reaction mixture, the purity of the initial crop of crystals is highly variable. However, the less soluble *monohapto* product is easily isolated by fractional recrystallization from pentane at $-20°$. These mother liquors may be stripped and combined with the second crop of crystals [which are always mainly Mn(η^5-C_5Cl_5)(CO)$_3$] and sublimed to obtain the pentahapto product. Any remaining traces of η^1 complex are converted to the pentahapto compound during sublimation.

rated solution to $-20°$ gives 0.91 g (48% yield) of pure $Mn(\eta^1\text{-}C_5Cl_5)(CO)_5$. Mp 110° (with effervescence, to give the pentahapto analog). *Anal.* Calcd. for $Mn(\eta^1\text{-}C_5Cl_5)(CO)_5$: C, 27.79; Cl, 41.01. Found: C, 27.94; Cl, 41.23.

The mother liquors from the initial preparation are concentrated to about 50% of their volume and cooled to $-78°$ to give an additional 0.44 g of gummy yellow needles. This second crop (IR usually indicates only the pentahapto compound) is purified by sublimation (40°, 0.1 torr), yielding 0.43 g (27%) of $Mn(\eta^5\text{-}C_5Cl_5)\text{-}(CO)_3$. Mp 83–85°. The overall yield of the two products is 75%.

J. (BROMOTETRACHLORO-η^1-CYCLOPENTADIENYL)-PENTACARBONYLMANGANESE, $Mn(\eta^1\text{-}C_5Cl_4Br)(CO)_5$

This compound is prepared by a procedure that is identical to that for $Mn(\eta^1\text{-}C_5Cl_5)(CO)_5$. A reaction employing $MnBr(CO)_5$ (0.4 g, 1.46 mmole) and $C_5Cl_4N_2$ (0.34 g, 1.46 mmole) in 20 mL of pentane gives, after 2 days,* 0.22 g (35%) of $Mn(\eta^1\text{-}C_5Cl_4Br)(CO)_5$ and 0.25 g (41%) of $Mn(\eta^5\text{-}C_5Cl_4Br)\text{-}(CO)_3$. The overall yield is 76%. Mp 103° (with effervescence, to give the pentahapto analog). *Anal.* Calcd. for $Mn(\eta^1\text{-}C_5Cl_4Br)(CO)_5$: C, 25.20; Cl, 29.75; Br, 16.76. Found: C, 25.16; Cl, 29.78; Br, 16.75.

K. (η^4-1,5-CYCLOOCTADIENE)(PENTACHLORO-η^5-CYCLOPENTADIENYL)RHODIUM, $Rh(\eta^5\text{-}C_5Cl_5)(cod)$

A benzene solution (50 mL) containing $[RhCl(cod)]_2$ (1.06 g, 2.15 mmole) and $C_5Cl_4N_2$ (0.99 g, 4.30 mmole) is stirred for 10 hours at room temperature in a Schlenk tube as in Sec. C. The orange solution is stripped of solvent and the product is recrystallized from hexane to give 1.80 g (94%) of large orange crystals. The complex decomposes at 92–94°. Although this complex slowly decomposes at room temperature, it may be stored at $-15°$ unchanged for several months. *Anal.* Calcd. for $Rh(\eta^5\text{-}C_5Cl_5)(cod)$: C, 38.43; H, 2.68. Found: C, 38.43; H, 3.03.

L. TRICARBONYL(PENTACHLORO-η^5-CYCLOPENTADIENYL)-MANGANESE, $Mn(\eta^5\text{-}C_5Cl_5)(CO)_3$

The following procedure is used when only the pentahapto compound is desired. Octane (60 mL) (the checkers used hexane) is added to 1.62 g (7.04 mmole) each of $MnCl(CO)_5$ and $C_5Cl_4N_2$ in a 100-mL, round-bottom flask under nitrogen. The flask is stoppered and the mixture is stirred vigorously for 36 hours. The solution is filtered, the receiving flask is fitted with a reflux condenser, and the orange solution is heated to 80° for about 15 minutes. (Some darkening of the solution occurs, but this represents an insignificant amount of

---

*See note for preparation of $Mn(\eta^1\text{-}C_5Cl_5)(CO)_5$.

decomposition.) The solution is cooled to room temperature and the octane is removed *in vacuo*. During the solvent removal the solution is kept at room temperature to prevent thermal decomposition of the product. The flask is then fitted with a sublimation probe and the product is sublimed onto the water-cooled finger (40°, 1.0 torr). Because of the presence of residual octane the sublimate is often oily. This is easily corrected by recrystallization from pentane to give 2.20 g (83%) of yellow crystals. *Anal.* Calcd. for $Mn(\eta^5\text{-}C_5Cl_5)(CO)_3$: C, 25.54; Cl, 47.14. Found: C, 25.49; Cl, 47.12.

M. (BROMOTETRACHLORO-η^5-CYCLOPENTADIENYL)-TRICARBONYLMANGANESE, $Mn(\eta^5\text{-}C_5Cl_4Br)(CO)_3$

As above, $MnBr(CO)_5$ (0.30 g, 1.1 mmole) and $C_5Cl_4N_2$ (0.252 g, 1.1 mmole) in 35 mL of octane give 0.39 g (85%) of product. Mp 81–82°. *Anal.* Calcd. for $Mn(\eta^5\text{-}C_5Cl_4Br)(CO)_3$: C, 22.84; Cl, 33.71; Br, 19.00. Found: C, 22.87; Cl, 33.66; Br, 18.88.

Properties

Although these reactions are often conducted overnight, the progress of the reaction may be conveniently monitored by observing the decrease in intensity of the band due to $\nu(NN)$ of the diazocyclopentadiene. This band occurs at about 2080, 2100, and 2110 cm^{-1} for $C_5Ph_4N_2$, $C_5H_4N_2$, and $C_5Cl_4N_2$, respectively.

The complexes are generally intensely yellow and, with the exception of $Mn(\eta^5\text{-}C_5H_4Cl)(CO)_3$, are highly crystalline. The carbonyl stretching frequencies of the halogen-substituted cyclopentadienyl compounds are useful in their characterization and are given in Table I. The presence of an electron-withdraw-

TABLE I Infrared Carbonyl Stretching Frequencies for Cyclopentadienyl Manganese and Rhodium Complexes[a]

Complex	$\nu(C\equiv O)$ (cm^{-1})			
$Mn(\eta^5\text{-}C_5H_4Me)(CO)_3$	2025 (s)	1943 (vs)		
$Mn(\eta^5\text{-}C_5H_5)(CO)_3$	2030 (s)	1945 (vs)		
$Mn(\eta^5\text{-}C_5H_4Cl)(CO)_3$	2034 (s)	1952 (vs)		
$Mn(\eta^5\text{-}C_5H_4Br)(CO)_3$	2034 (s)	1952 (vs)		
$Mn(\eta^5\text{-}C_5H_4I)(CO)_3$	2034 (s)	1952 (vs)		
$Mn(\eta^5\text{-}C_5Cl_5)(CO)_3$	2048 (s)	1982 (vs)		
$Mn(\eta^5\text{-}C_5Cl_4Br)(CO)_3$	2047 (s)	1980 (vs)		
$Mn(\eta^1\text{-}C_5Cl_5)(CO)_5$	2126 (m)	2076 (w)	2043 (vs)	2012 (vs)
$Mn(\eta^1\text{-}C_5Cl_4Br)(CO)_5$	2127 (m)	2076 (w)	2043 (vs)	2012 (vs)
$Rh(\eta^5\text{-}C_5H_5)(CO)_2$	2051 (s)	1987 (s)		
$Rh(\eta^5\text{-}C_5Ph_4Cl)(CO)_2$	2048 (s)	1987 (s)		

[a] Cyclohexane solution.

ing halogen substituent causes a shift of $\nu(CO)$ to higher frequencies for the compounds $Mn(\eta^5\text{-}C_5H_4X)(CO)_3$ X = Cl, Br, I, when compared to the unsubstituted parent compound. More detailed descriptions of properties are presented in the literature.[2,5]

References

1. M. Cais and N. Narkis, *J. Organomet. Chem.*, **3**, 269 (1965).
2. K. J. Reimer and A. Shaver, *J. Organomet. Chem.*, **93**, 239 (1975).
3. F. L. Hedberg and H. Rosenberg, *J. Am. Chem. Soc.*, **95**, 870 (1973).
4. V. W. Day, K. J. Reimer, and A. Shaver, *Chem. Commun.*, 403 (1975).
5. K. J. Reimer and A. Shaver, *Inorg. Chem.*, **14**, 2707 (1975).
6. W. B. DeMore, H. D. Pritchard, and N. Davidson, *J. Am. Chem. Soc.*, **81**, 5874 (1959); F. Ramirez and S. Levy, *J. Org. Chem.*, **23**, 2036 (1958).
7. R. A. Moss, *J. Org. Chem.*, **31**, 3296 (1966).
8. E. T. McBee, J. A. Bosoms, and C. J. Morton, *J. Org. Chem.*, **31**, 768 (1966).
9. D. F. Shriver, *The Manipulation of Air-Sensitive Compounds*, McGraw-Hill, New York, 1969, chapter 7.
10. J. Chatt and L. M. Venanzi, *J. Chem. Soc.*, **1957**, 4735 and G. Giordano, and R. H. Crabtree, *Inorg. Synth.*, **19**, 218 (1978).
11. J. A. McCleverty and G. Wilkinson, *Inorg. Synth.*, **8**, 211 (1966) and R. Cramer, *Inorg. Synth.*, **15**, 14 (1974).
12. D. Lloyd and M. I. C. Singer, *J. Chem. Soc. (C)*, 2939 (1971).
13. K. J. Reimer and A. Shaver, *Inorg. Synth.*, **19**, 160, 162 (1978).
14. E. T. McBee, U.S. Patent, 3,141,043 (July 14, 1964, Appl. March 24, 1961), 4 pp; *Chem. Abstr.*, **61**, 8205f (1964).
15. F. Klages and K. Bott, *Chem. Ber.*, **97**, 735 (1964).
16. H. Disselnkotter, *Angew. Chem. Int. Ed.*, **3**, 379 (1964).
17. T. Weil and M. Cais, *J. Org. Chem.*, **28**, 2472 (1963).
18. W. Von E. Doering and C. H. DePuy, *J. Am. Chem. Soc.*, **75**, 5955 (1953).

43. (η^6-BENZENE)(η^5-CYCLOPENTADIENYL)MOLYBDENUM AND (η^6-BENZENE)(η^5-CYCLOPENTADIENYL)- MOLYBDENUM DERIVATIVES

Submitted by JOHN A. SEGAL*
Checked by WILLIAM E. SILVERTHORN†

(η^6-Benzene)(η^5-cyclopentadienyl)molybdenum was first reported by Fischer and his co-workers. They describe two procedures for its preparation.[1-3] One starts with the system composed of $(C_5H_5)Mo(CO)_3Cl$, $AlCl_3$, and benzene, and

---

*ICI Plastics Division, Welwyn Garden City, Hertfordshire AL7 1HD, England.
†Department of Chemistry, Oregon State University, Corvallis, OR 97331.

the other starts with $MoCl_5$, $(i\text{-}C_3H_7)MgBr$, cyclopentadiene, and 1,3-cyclohexadiene. However, by these methods, the compound was obtained only in small quantities and in the low yields of 5.4 and 1.2%, respectively. The present method, which with modification also leads to the halides $(C_5H_5)Mo(C_6H_6)X$ (X = Cl, Br, I), gives much improved yields and so provides a new route to the extensive chemistry of the system monocyclopentadienylmolybdenum.[4]

Cyclopentene is used as a source of the η-cyclopentadienyl ligand; the hydrogen transfer properties[5] of the system $[(C_6H_6)Mo(C_3H_5)Cl]_2/(EtAlCl_2)_2$ are employed to remove three methylene hydrogen atoms that are transferred to the η-allyl group and lead to its liberation as propane. A cationic intermediate, $[(C_5H_5)Mo(C_6H_6)]^+$, is thus formed, and this is then either reduced to the neutral $(C_5H_5)Mo(C_6H_6)$ or treated with halide ion to give $(C_5H_5)Mo(C_6H_6)X$. The iodide complex $(C_5H_5)Mo(C_6H_6)I$ is best obtained by oxidation of $(C_5H_5)Mo(C_6H_6)$ with iodine.

All procedures are carried out under oxygen-free nitrogen and all solvents are dried and deaerated before use. It is convenient to use standard Schlenk-tube techniques in conjunction with a small vacuum/nitrogen manifold. Liquids are transferred through rubber serum caps by means of hypodermic syringes or stainless steel hypodermic tubing. The filtration of mixtures and all transfers are best conducted using low positive pressures of nitrogen. These methods are amply described by Shriver.[6] Dichloroethylaluminum dimer may be purchased from Alpha Ventron Corporation and is diluted with toluene to give the solution strength used.

A. (η^6-BENZENE)(η^5-CYCLOPENTADIENYL)MOLYBDENUM

$$[(C_6H_6)Mo(C_3H_5)Cl]_2 + [EtAlCl_2]_2 + 2C_5H_8 \longrightarrow$$
$$2[(C_6H_6)Mo(C_5H_5)]^+ [EtAlCl_3]^- + 2C_3H_8$$

$$[(C_5H_5)Mo(C_6H_6)]^+ \xrightarrow{Na_2[S_2O_4]/aq\ KOH} (C_5H_5)Mo(C_6H_6)$$

Procedure

The pure allyl dimer[7] $[(C_6H_6)Mo(C_3H_5)Cl]_2$ (5.0 g, 9.98 mmole) is treated with $[EtAlCl_2]_2$ in toluene solution (22.0 mL at 230 mg/mL, 19.9 mmole) at 0°. The mixture is shaken (at 0°) for 5 minutes, giving a very intense blue-violet solution, which is cooled to −50°. Freshly distilled cyclopentene (8.0 mL, 90.7 mmole) is then added dropwise. The solution is allowed to warm gradually to room temperature with stirring. At about 0°, a color change commences; slowly the violet color is dissipated, giving way to an oily, red-brown lower layer and an almost colorless upper layer. The mixture is allowed to stand for 3 hours at 20°. Then the upper layer is discarded, and the lower one is washed with light petroleum ether (bp 30-40°), separated, and pumped down to a tarry solid. After having been pumped for about 20 minutes, this residue is cooled to −50°, and a similarly cooled 30% aqueous KOH solution (300 mL), containing $Na_2[S_2O_4]$ (20 g, 115 mmole), is added. The mixture is vigorously stirred and allowed to warm gradually to 20°, giving a deep-red solid, which is filtered, washed with water, and pumped dry. This material is then extracted with benzene (100 mL), filtered, and treated with light petroleum ether (bp 100-120°). Concentration of the resulting solution gives the crude product as a red crystalline solid, which is obtained in pure form by sublimation at 0.01 torr and 85°. Yield: 80% (3.8 g). *Anal.* Calcd. for $C_{11}H_{11}Mo$: C, 55.2; H, 4.6. Found: C, 55.4; H, 4.7.

B. (η^6-BENZENE)CHLORO(η^5-CYCLOPENTADIENYL)MOLYBDENUM

$$[(C_5H_5)Mo(C_6H_6)]^+ + Cl^- \longrightarrow (C_5H_5)Mo(C_6H_6)Cl$$

Procedure

Initially the procedure described in Sec. A is followed. The tarry residue obtained on evacuation, following the reaction with $[EtAlCl_2]_2$ and cyclopentene, is again cooled to −50°. At this stage, however, a solution of anhydrous lithium chloride (10 g) in analytical reagent grade methanol (50 mL), also cooled to −50°, is added, and the resulting mixture is shaken vigorously while it warms to room temperature. At about −10°, vigorous effervescence occurs, and it may be necessary to control this by cooling the reaction vessel in liquid nitrogen. A deep red-black precipitate forms. After 15 minutes, the solvent is evaporated, and the residue is immediately extracted with a mixture of benzene (500 mL) and water (50 mL). Extraction should follow the evaporation step directly, as storage of the residue can lead to a low final yield. The benzene layer is collected and combined with a second benzene extract (200 mL), and the solvent is then evaporated. The residue is pumped dry and taken up in dry benzene (about 300 mL). The red solution is filtered through diatomaceous earth (Kieselguhr), concentrated to about 50 mL, and treated with light petroleum ether (bp 100-120°, 100 mL). The solution is evaporated until most of the product has precipitated from solution. The material is collected and pumped dry. Yield: 70% (3.8 g). Further purification may be effected by slow recrystallization from benzene/light petroleum ether (bp 100-120°) to give red-black crystals. *Anal.* Calcd. for $C_{11}H_{11}ClMo$: C, 48.1; H, 4.0; Cl, 12.9. Found: C, 48.4; H, 4.1; Cl, 13.0.

This synthesis has also been carried out on a larger scale using 10 g of allyl chloride dimer. Here, it is found to be more convenient to filter the crude product from the excess methanolic LiCl rather than to evaporate to dryness and extract. The crude product is washed with water (30 mL), dried, and then purified as above.

C. (η^6-BENZENE)BROMO(η^5-CYCLOPENTADIENYL)MOLYBDENUM

Procedure

On a proportionally smaller scale, the procedure in Sec. B is followed, using 2.0 g (4.0 mmole) of the allyl chloride dimer. However, the hydrolysis is carried out (at $-50°$), using a solution of anhydrous LiBr (6 g) in analytical reagent grade methanol (30 mL). The yield of $(C_5H_5)Mo(C_6H_6)Br$ is 60% (1.5 g). *Anal.* Calcd. for $C_{11}H_{11}BrMo$: C, 41.4; H, 3.5. Found: C, 41.0; H, 3.7.

D. (η^6-BENZENE)(η^5-CYCLOPENTADIENYL)IODOMOLYBDENUM

$$2[(C_5H_5)Mo(C_6H_6)] + I_2 \longrightarrow 2[(C_5H_5)Mo(C_6H_6)I]$$

Procedure

A pure sample of (η^6-benzene)(η^5-cyclopentadienyl)molybdenum (3.0 g, 12.5 mmole), prepared as in Sec. A above, is dissolved in benzene (60 mL). The resulting deep-yellow solution is stirred and treated with iodine (1.7 g, 6.70 mmole) in light petroleum ether (bp 100-120°, 150 mL). Immediate reaction produces a dark precipitate and a purple solution. The solution is concentrated to about 100 mL, and the precipitate is collected and recrystallized from acetone/ethanol. Yield: 50% (2.3 g). *Anal.* Calcd. for $C_{11}H_{11}IMo$: C, 36.1; H, 3.0. Found: C, 35.8; H, 3.0.

Properties

(η^6-Benzene)(η^5-cyclopentadienyl)molybdenum is a volatile, paramagnetic, and highly air-sensitive compound that is readily oxidized by halogens to the slightly more air-stable, 18-electron (η^6-benzene)(η^5-cyclopentadienyl)molybdenum halides. These species, all dark red-black in color, are soluble in polar solvents and are sparingly soluble in benzene. They react with oxygen in solution and in the solid state, losing benzene to give cyclopentadienyloxomolybdenum(V) and (VI) complexes.[8,9] Each shows a molecular ion in the mass spectrum and exhibits two ^{1}H nmr singlets in a ratio of 5:6 in the region about $\tau = 5$ (see Reference 10). The compounds $(C_5H_5)Mo(C_6H_6)X$ have proven to be more useful precursors for monocyclopentadienylmolybdenum chemistry than the paramagnetic $(C_5H_5)Mo(C_6H_6)$. Thus with Grignard reagents, the species $(C_5H_5)Mo(C_6H_6)R$ are obtained, and in the presence of Tl$^+$ the reaction with small molecules, such as carbon monoxide, triphenylphosphine, and ethylene, leads to the cations [(C$_5$H$_5$)Mo(C$_6$H$_6$)L]$^+$. With chelating ligands, such as diphosphines, the arene may be readily displaced, giving, for example, [(C$_5$H$_5$)Mo(dppe)$_2$]$^+$, [dppe = 1,2-bis(diphenylphosphino)ethane].

References

1. E. O. Fischer and F. J. Kohl, *Z. Naturforsch.*, **18b**, 504 (1963).
2. E. O. Fischer and F. J. Kohl, *Chem. Ber.*, **98**, 2134 (1965).
3. H. W. Wehner, E. O. Fischer, and J. Muller, *Chem. Ber.*, **103**, 2258 (1970).
4. See, for example, J. A. Segal, M. L. H. Green, J.-C. Daran, and K. Prout, *Chem. Commun.*, **1976**, 766, and M. L. H. Green, J. Knight, and J. A. Segal, *J. Chem. Soc. Chem. Commun.*, **1975**, 283.
5. M. L. H. Green and J. Knight, *J. Chem. Soc., Dalton Trans.*, **1974**, 311.
6. D. F. Shriver, *The Manipulation of Air-Sensitive Compounds*, McGraw-Hill, 1969, pp. 145–158; see also J. P. Collman, N. W. Hoffman, and J. W. Hosking, *Inorg. Synth.*, **12**, 10 (1970).
7. W. E. Silverthorn, *Inorg. Synth.*, **17**, 54 (1977).
8. M. Cousins and M. L. H. Green, *J. Chem. Soc.*, **1964**, 1567.
9. M. Cousins and M. L. H. Green, *J. Chem. Soc. (A)*, **1969**, 16.
10. M. L. H. Green, J. Knight, and J. A. Segal, *J. Chem. Soc., Dalton Trans.*, **1977**, 2189.

44. cis-[ACETYLTETRACARBONYL(1-HYDROXY-ETHYLIDENE)RHENIUM] (cis-Diacetyltetracarbonylrhenium Hydrogen)

Submitted by K. P. DARST,* C. M. LUKEHART,* L. T. WARFIELD,* and JANE V. ZEILE*
Checked by J. A. GLADYSZ† and J. C. SELOVER†

cis-[Acetyltetracarbonyl(1-hydroxyethylidene)rhenium] is prepared readily by treating acetylpentacarbonylrhenium with 1 molar equivalent of methyllithium at 0° followed by the addition of 1 molar equiv of anhydrous hydrogen chloride as a diethyl ether solution at −78°. This complex has been characterized structurally as a metalla analogue of the enol tautomer of acetylacetone (2,4-pentanedione) where the methine group is replaced formally by the cis-tetracarbonylrhenium moiety.

This structural similarity has been confirmed by X-ray crystallography.[1] The "metalla-acetylacetone" complex possesses idealized C_{2v} molecular symmetry. The atoms comprising the backbone of the metalla ligand system are essentially coplanar, having a maximum atomic deviation from planarity of 0.08 Å. The

*Department of Chemistry, Vanderbilt University, Nashville, TN 37235.
†Department of Chemistry, University of California, Los Angeles, CA 90024.

two acyl oxygen atoms are located in a relative position of closest approach, affording an O—O bite distance of 2.398(15) Å. This very short contact distance strongly supports the presence of a symmetrical hydrogen bond as is observed in the structure of the dienol tautomer of sym-tetraacetylethane.[2] The enolic hydrogen atom appears in the ^{1}H nmr spectrum as a single resonance at δ 21.79 ppm versus TMS in a carbon disulfide solution. The chemical reactivity of this metalla-acetylacetone complex shows many similarities to that of acetylacetone also.[3-5]

This procedure is a general preparative method for metalla-β-diketone molecules where the metalla moiety may be either the cis-tetracarbonylrhenium or the carbonyl-η^5-cyclopentadienyliron fragment.[6] In each case the diacylmetallate monanion is prepared from an acyl complex, and then this anion is protonated at low temperature. The synthesis of the metalla-acetylacetone complex presented here utilizes acetylpentacarbonylrhenium as the acyl complex. The preparation of this acetyl complex from acetyl chloride and sodium pentacarbonylrhenium is provided, also.

A. ACETYLPENTACARBONYLRHENIUM

$$Re_2(CO)_{10} + 2Na/Hg \longrightarrow 2Na[Re(CO)_5]$$

$$Na[Re(CO)_5] + CH_3C(O)Cl \longrightarrow [[(CH_3C(O)]Re(CO)_5] + NaCl$$

Procedure

A 100-mL, two-necked flask is removed from a 130° drying oven and is flushed well with prepurified nitrogen, while a gas inlet is attached to the side neck. This flask is charged with 4 mL of mercury and a stirring bar, and then 0.35 g (15.2 mmole) of sodium metal is added to the stirred mercury puddle in small pieces, one at a time, under a continuous nitrogen flush. (■ **Caution.** *This operation should be performed in a hood since the dissolution of the sodium metal is a highly exothermic reaction.*) After the sodium amalgam has cooled to room temperature, 50 mL of freshly distilled tetrahydrofuran[7] is introduced into the flask followed by the addition of 4.0 g (6.1 mmole) of dirhenium decacarbonyl.* (■ **Caution.** *Metal carbonyl compounds are toxic chemicals and they should be handled in the hood.*) The red-orange solution is stirred at 25° under nitrogen for 1 hour.

After this time, the solution of Na[Re(CO)$_5$] is transferred under a nitrogen flush, by means of a syringe into another 100-mL, two-necked flask containing a stirring bar and fitted with a gas inlet. The solution of the anion is cooled to -78° (Dry Ice-acetone bath), and 0.86 mL (12.1 mmole) of acetyl chloride is

---

*Dirhenium decacarbonyl was purchased from Strem Chemicals Inc., Danvers, MA 01923.

added dropwise by means of a syringe. The reaction solution is stirred at −78° for 5 minutes, at which time the bath is removed and stirring is continued for 15 minutes more at −20°. The solvent is removed at reduced pressure (5 torr) using a 0° ice-water bath, after which the solid residue is stirred with 75 mL of hexane for 20 minutes at 30°. The hexane solution is filtered through a Schlenk frit,[8] and the filtrate is cooled at −20° for 16 hours. The residue is extracted with 40 mL more of hexane at 30° and filtered through a Schlenk frit. The two crops of crystallized solid are collected by syringing the supernatant solutions into a third flask, followed by drying the solids at reduced pressure (5 torr, 25°) for 5 minutes. The volume of the combined supernatants is reduced to 20 mL at reduced pressure, and this solution is cooled to −20° for 16 hours. The third crop of product is collected by the above procedure. The total yield of product is 1.98 g (44%). *Anal.* Calcd. for $C_7H_3O_6Re$: C, 22.76; H, 0.82. Found: C, 22.60; H, 0.70.

Properties

Acetylpentacarbonylrhenium is a slightly volatile, pale-yellow solid, mp 79.5–80.5°. It is air-stable over at least a 2-day period, and it has excellent solubility in most organic solvents. The infrared spectrum in cyclohexane shows $\nu(C\equiv O)$ bands at 2130 (w), 2020 (vs), 2005 (s) cm^{-1} and a ν(acyl) band at 1622 (m) cm^{-1}. The ^{1}H nmr spectrum (CS$_2$, versus TMS) shows a singlet for the methyl resonance at τ 7.56.

B. *cis*-[ACETYLTETRACARBONYL(1-HYDROXYETHYLIDENE)-RHENIUM] (*cis*-Diacetyltetracarbonylrhenium Hydrogen)

$[[CH_3C(O)] Re(CO)_5] + CH_3Li + HCl/Et_2O \longrightarrow$

$+ LiCl$

Procedure

A 100-mL, two-necked flask is removed from a 130° drying oven and is flushed well with prepurified nitrogen while a gas inlet is attached to the side neck. This flask is charged with 1.0 g (2.7 mmole) of acetylpentacarbonylrhenium (see Sec. A), 30 mL of distilled diethyl ether, and a stirring bar. The pale-yellow solution mixture is cooled to 0°, and 1.48 mL of a 1.84 *M* solution of methyl-lithium (2.7 mmole) in diethyl ether is added dropwise by means of a syringe over a 10-minute period. During this time the reaction solution changes to a deep-yellow color, which is characteristic of the diacetylmetallate anion, and any undissolved acetyl complex goes into solution. The solution is stirred at 0° for an additional 30 minutes.

A stock solution of hydrogen chloride is prepared by bubbling hydrogen chloride gas for 10 minutes into a weighed flask containing a known volume of distilled diethyl ether at 0°. Prolonged bubbling should be avoided to minimize any loss of diethyl ether. The weight of the flask and solution after this addition permits the calculation of the molar concentration of hydrogen chloride.

The reaction solution is cooled to $-78°$ (Dry Ice–acetone bath), and 0.10 g (2.8 mmole) of HCl (as the diethyl ether solution) is added dropwise by means of a syringe over a 5-minute period. During this addition a white precipitate forms. After the reaction mixture is stirred at $-78°$ for an additional 10 minutes, the Dry Ice–acetone bath is replaced by an ice-water bath and stirring is continued for 1 hour. The solvent is removed at reduced pressure (5 torr, 10°), and the solid residue is extracted with 20 mL of dry dichloromethane for 15 minutes at 25°. The mixture is filtered through a Schlenk frit,[8] and the solvent is removed from the filtrate at reduced pressure (5 torr, 10°), affording 0.90 g (86%) of the product as a yellow solid. This solid is sufficiently pure for most purposes. Recrystallization from 10 mL of distilled diethyl ether at $-20°$ affords nearly colorless single crystals. *Anal.* Calcd. for $C_8H_7O_6Re$: C, 24.94; H, 1.83; Re, 48.32. Found: C, 24.85; H, 1.59; Re, 48.62.

Properties

cis-[Acetyltetracarbonyl(1-hydroxyethylidene)rhenium] is an off-white solid that crystallizes from diethyl ether as monoclinic needles. It melts at 73-74° and sublimes readily under vacuum (1 torr) at 25°. The complex is air stable for at least 1 day, and it has good to excellent solubility in most organic solvents. The infrared spectrum in cyclohexane shows $\nu(C\equiv O)$ bands at 2095 (m), 2005 (s, sh), 1990 (vs), 1965 (s) cm^{-1} and a $\nu(C\mathrel{\vcenter{\hbox{..}}}O)$ band at 1520 (m) cm^{-1}. The ^{1}H nmr spectrum (CS$_2$, versus TMS) shows a singlet for the resonance of the two equivalent methyl groups at τ 7.22 and a broad singlet for the enolic proton resonance at τ -11.79. This resonance disappears upon the addition of methanol-d_4 to the sample solution. A parent peak at m/e 386 is observed in the mass spectrum, followed by a fragmentation pattern showing the loss of methyl and acetyl radicals and methane to give Re(CO)$_5^+$. The base peak is observed at m/e 43 and is assigned to the CH$_3$CO$^+$ ion.

References

1. C. M. Lukehart and J. V. Zeile, *J. Am. Chem. Soc.*, **98**, 2365 (1976).
2. J. P. Schaefer and P. J. Wheatley, *J. Chem. Soc. (A)*, **1966**, 528.

3. C. M. Lukehart and J. V. Zeile, *J. Organomet. Chem.*, **140**, 309 (1977).
4. C. M. Lukehart and L. T. Warfield, *Inorg. Chem.*, **17**, 201 (1978).
5. C. M. Lukehart and J. V. Zeile, *J. Am. Chem. Soc.*, **100**, 2774 (1978).
6. C. M. Lukehart and J. V. Zeile, *J. Am. Chem. Soc.*, **99**, 4368 (1977).
7. Appendix, *Inorg. Synth.*, **12**, 317 (1970).
8. R. B. King, in *Organometallic Syntheses*, Vol. 1, J. J. Eisch and R. B. King, Eds. Academic, New York, 1965, p. 56.

45. *cis*-[ACETYL(1-AMINOETHYLIDENE)-TETRACARBONYLRHENIUM] [*cis*-(Acetimidoyl)(Acetyl)-Tetracarbonylrhenium Hydrogen]

Submitted by D. T. HOBBS,* C. M. LUKEHART,* G. PAUL TORRENCE* and JANE V. ZEILE*
Checked by J. A. GLADYSZ† and J. C. SELOVER†

cis-[Acetyl(1-aminoethylidene)tetracarbonylrhenium] is prepared readily by treating *cis*-[acetyltetracarbonyl(1-hydroxyethylidene)rhenium][1] with an excess of ammonia at 0°. This complex is structurally similar to the β-iminoketone molecule, acetimidoylacetone, as the ketamine tautomer where the methine group is replaced formally by the *cis*-tetracarbonylrhenium moiety. The formation of this complex can be regarded as a Schiff-base condensation reaction of the "metalla-acetylacetone" molecule.

The title complex represents the parent compound of several known "metalla-β-iminoketone" molecules.[2] In each case a "metalla-β-diketone" molecule is treated with a primary amine, affording the corresponding "metalla-β-iminoketone" derivative. The structure of the *cis*-tetracarbonylrhenium metalla analogue of N-phenylacetimidoylacetone has been determined.[3] The structure of this complex confirms the formulation of these complexes as metalla analogues of the ketamine tautomer of β-iminoketone molecules. However, the electronic structure is described best by the zwitterionic resonance form shown below.

*Department of Chemistry, Vanderbilt University, Nashville, TN 37235.
†Department of Chemistry, University of California, Los Angeles, CA 90024.

Procedure

A 50-mL, two-necked flask is removed from a 130° drying oven and is flushed well with prepurified nitrogen while a gas inlet is attached to the side neck. This flask is charged with 0.30 g (0.78 mmole) of cis-[acetyltetracarbonyl-(1-hydroxyethylidene)rhenium],[1] a magnetic stirring bar, and 10 mL of dichloromethane that has been dried over and freshly distilled from P_4O_{10}. (■ **Caution**. *Metal carbonyl compounds are toxic chemicals and should be handled in the hood.*) The clear, yellow solution is cooled to 0° and ammonia is bubbled into the reaction solution by the following procedure. A 14.5-cm disposable glass pipette is fitted with a 2-cm cork. This assembly is attached to a cylinder of anhydrous ammonia by means of tygon tubing, and the cork is placed loosely in a joint (19/22) of the two-necked flask. The position of the pipette is adjusted until the tip of the pipette is well under the level of the reaction solution. While a mild nitrogen flush is continued through the gas inlet, gaseous ammonia is bubbled into the solution at a moderate rate. During the addition an off-white precipitate forms, and after 1 minute of bubbling the pipette is removed and the flask is stoppered and then removed from the 0° bath. Within 5 minutes of stirring at 25°, the precipitate dissolves, affording a clear pale lime-yellow solution. If the precipitate does not dissolve after this time, then more ammonia should be added until the solid begins to dissolve (approximately 15 sec more at a moderate bubbling rate). After the solution is at 25° for an additional 45 minutes, the solvent is removed at reduced pressure (5 torr, 20°), affording 0.23 g (77%) of the product as a pale lime-yellow solid. This solid is of sufficient purity for most purposes, although it may be crystallized from hexane solution as an off-white crystalline solid. *Anal.* Calcd. for $C_8H_8NO_5Re$: C, 25.00; H, 2.10; N, 3.64. Found: C, 24.65; H, 2.25; N, 3.69.

Properties

cis-[Acetyl(1-aminoethylidene)tetracarbonylrhenium] is crystallized from hexane solution as an off-white solid that melts at 91–93° and is air-stable for at least 3 days. It has good to excellent solubility in most organic solvents. The infrared spectrum in cyclohexane solution shows $\nu(C\equiv O)$ bands at 2080 (m), 1985 (s, sh), 1978 (vs), and 1945 (s) cm^{-1} and a $\nu(C\cdots O, C\cdots N)$ band at 1555 (m) cm^{-1}. The ^1H nmr spectrum (CDCl$_3$ versus TMS) shows a singlet at τ 7.32 for the acetyl methyl resonance, a singlet at τ 7.14 for the imino methyl resonance, a broad singlet at τ 0.89 for the exo-N—H proton resonance, and a broad singlet at τ -3.34 for the enolic N—H proton resonance. Both N—H resonances disappear when the spectrum is recorded in methanol-d_4 with no decomposition of the sample. When a THF solution of the title complex or any metalla-β-iminoketone molecule is treated with sodium hydride, a gas, presumably hydrogen, is readily evolved, apparently with the formation of the corresponding "metalla-β-iminoketonato" anion as characterized from the infrared spectrum. These anions may coordinate to a variety of metal ions as do the "metalla-β-diketonate" anions.[4,5]

References

1. C. M. Lukehart and J. V. Zeile, *J. Am. Chem. Soc.*, **98**, 2365 (1976). See synthesis number 44.
2. C. M. Lukehart and J. V. Zeile, *Inorg. Chem.*, **17**, 2369 (1978).
3. C. M. Lukehart and J. V. Zeile, *J. Am. Chem. Soc.*, **100**, 2774 (1978).
4. C. M. Lukehart, G. P. Torrence, and J. V. Zeile, *Inorg. Chem.*, **15**, 2393 (1976).
5. C. M. Lukehart and G. P. Torrence, *Chem. Commun.*, 1978, 183.

46. HYDRIDO COMPLEXES OF COBALT WITH BIS(PHOSPHINES)

$$CoBr_2 \cdot 4H_2O + 2\text{diphos} \xrightarrow{\text{EtOH}} CoBr_2(\text{diphos})_2 + 4H_2O$$

$$CoBr_2(\text{diphos})_2 \xrightarrow[\text{NaBH}_4]{\text{EtOH}} CoH(\text{diphos})_2$$

Submitted by V. D. BIANCO,* S. DORONZO,* and N. GALLO*
Checked by H. BONNEMANN† and M. SAMSON†

The preparation of the hydrido complexes of cobalt with 1,2-bis(diphenylphosphino)ethane (ethylenebis[diphenylphosphine], diphos), [CoH{C_2H_4-$(Ph_2P)_2$}$_2$], by the reaction of $CoBr_2$ with the bis(phosphine) and successive reduction of $CoBr_2(\text{diphos})_2$ with sodium tetrahydridoborate or lithium tetrahydridoaluminate in tetrahydrofuran or ethanol has been reported,[1,2] This complex also can be prepared in good yields starting from $Co(acac)_3$,‡ using dibutylaluminum hydride as a reducing agent.[3] Here, we give the preparation and properties of the hydrido complex of cobalt with *cis*-vinylenebis(diphenylphosphine) $Ph_2PCH=CHPPh_2$ (dp), which is an interesting chelating agent, together with the complex of the saturated bis(phosphine), (diphos).

---
*Istituto di Chimica Generale ed Inorganica, Università degli Studi di Bari, Bari, Italy.
†Max-Planck-Institut f. Kohlenforschung, 4330 Mülheim a. d. Ruhr, West Germany.
‡acac = 2,4-pentanedionato.

A. HYDRIDOBIS[cis-VINYLENEBIS(DIPHENYLPHOSPHINE)]-COBALT(I)

Procedure

The preparation is carried out under dry, oxygen-free nitrogen and the product is manipulated in the absence of air.

The bis(phosphine) is prepared as reported previously.[4] A 250-mL, round bottomed flask equipped with a stopcock is flushed with nitrogen by alternately applying a vacuum and refilling with nitrogen. A magnetic stirring bar and the bis(phosphine)* (4.9 g, 12.4 mmole) are introduced. Next, 70 mL of deoxygenated ethanol is added and the flask is equipped with a reflux condenser. The mixture is heated to 70° (water bath) and a solution of $CoBr_2 \cdot 4H_2O$ (1.8 g, 6.2 mmole)† in ethanol (30 mL) is added slowly, over a period of 1 hour with stirring, by means of a syringe inserted through a rubber septum placed on the stopcock. The solution becomes yellow-green. At this time, the heating is stopped and, after cooling is completed, a solution of $Na[BH_4]$ (0.71 g, 18.7 mmole) in aqueous (1:20, v/v) ethanol (20 mL) is added in the same manner, slowly and with stirring. The red reaction mixture is stirred for a few hours (overnight) and then the red precipitate that forms is filtered under nitrogen, washed with ethanol (3 × 20 mL), and dried under vacuum. The crude product (5 g, 6 mmole) is crystallized from deoxygenated benzene (700 mL) by slow addition of oxygen-free ethanol (700 mL) or pentane. After 24 hours, the crystals are isolated by filtration and dried. Yield of red crystals: 4.21 g (80%); mp 230°. *Anal.* Calcd. for $C_{52}H_{45}P_4Co$: C, 73.25; H, 5.28; Co, 6.92; P, 14.54. Found: C, 73.12, 73.16; H, 5.32, 5.33; Co. 6.89; 6.81; P, 14.44, 14.46. Mass spectrum: m/e 852 (±1) M⁺; 640; 455; 396; 262; 212; 183; 108.

Properties

Hydridobis[cis-vinylenebis(diphenylphosphine)]cobalt(I) is a red crystalline solid that is unstable in air both in the solid state and in solution. The complex is soluble in tetrahydrofuran (9.2×10^{-3} mole/L, at 20°), toluene (10.2×10^{-3} mole/L), chloroform (6.3×10^{-3} mole/L), dichloromethane (1.6×10^{-3} mole/L), and benzene (8.4×10^{-3} mole/L). It is insoluble in ethanol, diethyl ether, acetone, and pentane. The infrared spectrum in Nujol mull shows a band at 1885 (m) cm^{-1}, attributable to the Co-H stretching vibration.

---

*The commercially available cis-diphosphine contains up to 15% of the *trans* compound (Strem product). The mixture can be used as such.

†The use of $CoBr_2 \cdot 6H_2O$ has no influence on the reaction.

B. HYDRIDOBIS[ETHYLENEBIS(DIPHENYLPHOSPHINE)]COBALT(I)

Procedure

This procedure is the same as the preceding one, starting with 3.3 g (8.3 mmole) of the bis(phosphine) (diphos) (in 50 mL of deoxygenated ethanol) and 1.24 g (4.1 mmole) of $CoBr_2 \cdot 4H_2O$ (in 30 mL of ethanol). The reduction is carried out with $Na[BH_4]$ (0.5 g, 12.4 mmole) in aqueous ethanol (30 mL). After crystallization of the crude product from benzene by addition of ethanol or pentane, 3.40 g (96% yield) of complex (red crystals) is obtained (mp 265°). *Anal.* Calcd. for $C_{52}H_{49}P_4Co$: C, 72.91; H, 5.72; Co, 6.88; P, 14.47. Found: C, 72.86, 73.00; H, 5.70, 5.77; Co, 6.91, 6.73; P, 14.50, 14.36. Mass spectrum: m/e 856 (±1) M$^+$; 642, 457, 429, 398, 183.

Properties

The solubility (mole/L, at 20°) of the complex in common organic solvents is as follows: benzene, 5.8×10^{-3}; dichloromethane, 3.9×10^{-3}; chloroform, 5.4×10^{-3}; pyridine, 2.0×10^{-3}; tetrahydrofuran, 6.0×10^{-3}; toluene, 2.4×10^{-3}. The infrared spectrum in Nujol mull shows a band at 1885 (m) cm^{-1} due to ν(Co-H).

References

1. F. Zingales, F. Canziani, and A. Chiesa, *Inorg. Chem.*, **2**, 1303 (1963).
2. A. Sacco and R. Ugo, *J. Chem. Soc.*, **1964**, 3274.
3. J. Lorberth, H. Noth, and P. V. Rinze, *J. Organomet. Chem.*, **16**, P1 (1969).
4. A. M. Aguiar and D. J. Daigle, *J. Am. Chem. Soc.*, **86**, 2299 (1964).

Chapter Seven

METAL CLUSTER COMPLEXES

47. TRI-μ-CARBONYL-NONACARBONYLTETRARHODIUM, $Rh_4(CO)_9(\mu\text{-}CO)_3$

$RhCl_3 + 2Cu + 4CO + NaCl \longrightarrow Na[Rh(CO)_2Cl_2] + 2Cu(CO)Cl$

$4Na[Rh(CO)_2Cl_2] + 6CO + 2H_2O \longrightarrow Rh_4(CO)_{12} + 2CO_2 + 4NaCl + 4HCl$

Submitted by S. MARTINENGO,* G. GIORDANO,* and P. CHINI*
Checked by G. W. PARSHALL† and E. R. WONCHOBA†

Two general methods have been given for the preparation of $Rh_4(CO)_{12}$: high-pressure methods that employ $RhCl_3$ or $Rh_2(CO)_4Cl_2$ with metal powders as halogen acceptors,[1] and atmospheric pressure syntheses starting with $K_3[RhCl_6]\cdot H_2O$ and copper powder in water,[2] or with $Rh_2(CO)_4Cl_2$ and $NaHCO_3$ in hexane.[3] The synthesis described here is conducted at atmospheric pressure with $RhCl_3\cdot 3H_2O$ as the starting material rather than $K_3[RhCl_6]\cdot H_2O$. The overall procedure requires about 30 hours, with a yield of 80-90%.

The compound $Rh_4(CO)_{12}$ has a large number of applications. It is used in catalysis directly or as a catalyst precursor, and it is the starting material both for substitution reactions and for the synthesis of other rhodium carbonyl

---

*Centro del C.N.R. per lo studio dei composti di coordinazione nei bassi stati di ossidazione. Via G. Venezian, 21, 20133 Milano, Italy.

†Central Research and Development Department, E. I. Du Pont Co., Wilmington, DE 19898.

clusters. In particular it is easily converted into $Rh_6(CO)_{16}$, thus providing a method for the synthesis of this carbonyl in a highly pure state.[4]

Procedure

The starting material, $RhCl_3 \cdot 3H_2O$, may be obtained commercially,* or prepared as described in the literature.[5] Copper powder is first activated by washing with a mixture of concentrated hydrochloric acid and acetone and then rinsing with acetone and drying *in vacuo*. The solvent CH_2Cl_2 may be laboratory grade and is distilled from anhydrous sodium carbonate before use. (■ **Caution.** *The operations must be carried out in a well ventilated hood because carbon monoxide is used.*) A two-necked, 1-L flask, equipped with a stopcock on the side neck and a magnetic stirring bar, is charged with activated copper powder (1.5 g, 24 mmole), sodium chloride (0.6 g, 10 mmole), and water (200 mL).

A 100-mL dropping funnel with a pressure-equalizing tube is then placed on the central neck. The funnel is stoppered, and the entire apparatus is pumped *in vacuo* until the water is well degassed. It then is filled through the side stopcock with carbon monoxide from a CO line, including a mineral oil bubbler and a pressure-release bubbler with an 80-mm mercury column. The dropping funnel is opened while CO is passed through it, and it is charged with a degassed solution of $RhCl_3 \cdot 3H_2O$ (2.6 g, Rh 39.6%, 9.0 mmole) in water (50 mL). The funnel is closed, and the solution is dropped into the flask at a rate of about 1 mL/min while the mixture is stirred vigorously. During the addition, care must be taken that the pressure in the flask never lowers, for the lack of CO can cause the partial formation of rhodium metal. The best conditions for the reaction are achieved by adding the carbon monoxide at such a rate that a small portion of the gas continuously escapes from the pressure-release bubbler. For the same reason, the stirring must be vigorous enough to ensure good contact between the gaseous and aqueous phases. The rate of absorption of the CO is initially slow, but gradually increases, reaching a maximum when about half the $RhCl_3$ solution has been added.

At the end of the addition, the last traces of the $RhCl_3$ solution in the funnel are washed down into the flask with 10 mL of degassed water, and the mixture is stirred for 2 hours. During this time, some $Rh_4(CO)_{12}$ begins to separate as an orange powder. The dropping funnel is then opened in a CO stream and charged with a degassed solution of disodium citrate (0.4 M, 50 mL). The funnel is stoppered, and the solution of citrate is dropped into the yellow mixture during 50 minutes. The mixture is then stirred for 20 hours, and the resulting orange suspension of the carbonyl is filtered under carbon monoxide atmosphere.†

*Available from Alfa Products, Ventron Corp., P.O. Box 299, Danvers, MA 92923.
†The CO atmosphere is required to prevent the precipitation of the insoluble CuCl by decomposition of Cu(CO)Cl.

The solid is washed with water to remove all the mother liquor and then dried *in vacuo* (~10^{-1} torr) at room temperature.

The product is extracted on the filter under CO with the minimum amount of CH_2Cl_2 (about 25 mL, in 5-mL portions). The solution is quickly evaporated to dryness *in vacuo* on a water bath at room temperature, and the resulting crystalline mass is dried *in vacuo* for 2 hours to give 1.55-1.75 g of $Rh_4(CO)_{12}$ with a yield of 80-90%.* *Anal.* Calcd. for $Rh_4(CO)_{12}$: Rh, 55.08%. Found: Rh, 54.74%. This product is sufficiently pure for most purposes, the impurity being a trace of $Rh_6(CO)_{16}$. A $Rh_6(CO)_{16}$-free product can be obtained by extraction, under carbon monoxide with pentane, followed by crystallization at $-70°C$, filtration at this temperature, and drying in a stream of CO.

Properties

The compound $Rh_4(CO)_{12}$ is a dark-red solid when crystalline and orange when powdered. It decomposes under nitrogen at 130-140°, giving $Rh_6(CO)_{16}$ (lit. 150°).[1a, 1c] The same decompositon also occurs at room temperature, but it is very slow. For this reason $Rh_4(CO)_{12}$ is best stored under a CO atmosphere. It is stable in air, although on prolonged standing decomposition occurs. It is readily soluble in CH_2Cl_2 and $CHCl_3$, soluble in pentane (~12 g/L at 25°), toluene and tetrahydrofuran (THF) (~10 g/L), and acetone, and less soluble in cyclohexane and methanol.

Under nitrogen the solutions slowly decompose to $Rh_6(CO)_{16}$. This decomposition is faster at higher temperatures and provides a method for the synthesis of $Rh_6(CO)_{16}$. Very pure crystalline $Rh_6(CO)_{16}$ is formed in high yields on heating a saturated solution of $Rh_4(CO)_{12}$, or a suspension, in heptane at 80-90° under nitrogen until the orange-red color has disappeared.

The solutions in polar solvents (THF, acetone) are unstable toward water, which causes transformation into $Rh_6(CO)_{16}$ through the violet $[H_3O]_2$-$[Rh_{12}(CO)_{30}]$.[6]

The compound $Rh_4(CO)_{12}$ does not react with CO at room temperature at 300 atm. At higher temperature it is transformed into $Rh_6(CO)_{16}$, even under high CO pressures (80-120°, 360-420 atm).[7] More recently IR evidence has been reported for the formation of $Rh_2(CO)_8$ by the reaction of $Rh_4(CO)_{12}$ with CO at 430 atm and $-19°$.[8]

The IR spectrum in heptane solution shows bands at 2101 (w), 2074.5 (s), 2069 (s), 2059 (w, sh), 2044.5 (m), 2041.5 (m, sh), 2001 (w), 1918.4 (w), 1882 (s), 1848 (w) cm^{-1}.[9]

Tri-μ-carbonyl-nonacarbonyltetrarhodium is the starting material for the synthesis of a large number of rhodium carbonyl cluster compounds.[10]

---

*The checkers report that a ninefold scale-up of this synthesis proceeds satisfactorily.

References

1. (a) W. Hieber and H. Lagally, *Z. anorg. allg. Chem.*, **251**, 96 (1943). (b) S. H. Chaston and F. G. A. Stone, *J. Chem. Soc. (A)*, **1969**, 500. (c) B. L. Booth, M. J. Else, R. Fields, H. Goldwhite, and R. N. Haszeldine, *J. Organomet. Chem.*, **14**, 417 (1968).
2. S. Martinengo, P. Chini, and G. Giordano, *J. Organomet. Chem.*, **27**, 389 (1971).
3. G. Giordano, S. Martinengo, D. Strumolo, and P. Chini, *Gazzetta*, **105**, 613 (1975).
4. P. Chini and S. Martinengo, *Inorg. Chim. Acta*, **3**, 315 (1969).
5. S. N. Anderson and F. Basolo, *Inorg. Synth.*, **7**, 214 (1963).
6. P. Chini, S. Martinengo, and G. Garlaschelli, *Chem. Commun.*, **1972**, 709.
7. P. Chini and S. Martinengo, *Inorg. Chim. Acta*, **3**, 21 (1969).
8. R. Whyman, *J. Chem. Soc., Dalton Trans.*, **1972**, 1375.
9. F. Cariati, P. Fantucci, V. Valenti, and P. Barone, *Rend. ist. Lomb., Sci.*, **105**, 122 (1971).
10. P. Chini, G. Longoni, and V. G. Albano, *Adv. Organomet. Chem.*, **XIV**, 285 (1976).

48. DIPOTASSIUM μ_6-CARBIDO-NONA-μ-CARBONYL-HEXACARBONYLHEXARHODATE(2-)

$$K_2[Rh_6(CO)_6(\mu\text{-}CO)_9\text{-}\mu_6\text{-}C]$$

$$6[RhCl_6]^{3-} + 45OH^- + 26CO + CHCl_3 \longrightarrow$$
$$[Rh_6(CO)_{15}C]^{2-} + 39Cl^- + 11CO_3^{2-} + 23H_2O$$

Submitted by S. MARTINENGO,* D. STRUMOLO,* and P. CHINI*
Checked by G. W. PARSHALL† and E. R. WONCHOBA†

The title compound, $K_2[Rh_6(CO)_{15}C]$, has been prepared in about 70% yield by the reaction of $Rh_4(CO)_{12}$ with sodium hydroxide under carbon monoxide in methanol containing some chloroform.[1] The benzyltrimethylammonium and μ-nitrido-bis(triphenylphosphorus)(1+) [bis(triphenylphosphine)imminium] salts have been prepared by reaction of the corresponding $[Rh(CO)_4]^-$ salts with CCl_4.[2]

This anion provides both the first example of a trigonal prismatic hexanuclear cluster and one of the more easily prepared carbido-carbonyl cluster compounds.

It is the starting material for the preparation of many other unusual carbido carbonyl compounds, some of which, for example, $Rh_8(CO)_{19}C$, $Rh_{12}(CO)_{25}C_2$, and $[H_3O][Rh_{15}(CO)_{28}C_2]$, have been structurally characterized.[3]

The synthesis described here is a modification that starts directly from

---

*Centro del C.N.R. per lo studio dei composti di coordinazione nei bassi stati di ossidazione. Via G. Venezian 21, 20133 Milano, Italy.

†Central Research and Development Department, E. I. Du Pont Co., Wilmington, DE 19898.

$K_3[RhCl_6] \cdot H_2O$ and an excess of KOH in a mixture of methanol and chloroform under carbon monoxide.
The overall procedure requires about 50–60 hours with about 70% yield.

Procedure

All the operations must be carried out in the rigorous absence of oxygen under carbon monoxide and nitrogen containing less than 2 ppm of oxygen by volume.

The solvents are laboratory grade. Methanol and chloroform are distilled and stored under nitrogen before use. Sodium-dried tetrahydrofuran (THF) is redistilled over $LiAlH_4$ under nitrogen before use. (■ *Caution. To minimize the chance of explosion, peroxide-free THF must be employed. See Reference 8 for details.*) The aqueous solutions are carefully degassed *in vacuo* and saturated with nitrogen before use. Potassium hexachlororhodate(3-) monohydrate is prepared as described in the literature.[4] The operations are carried out with the usual Schlenk apparatus.[5]

■ *Caution. The first stage of the synthesis must be carried out under a well-ventilated hood because carbon monoxide is used.*

A 250-mL Schlenk flask is charged with $K_3[RhCl_6] \cdot H_2O$ (3 g, Rh 23.15%), and a magnetic stirring bar is introduced. The flask is stoppered, carefully evacuated to 10^{-2} torr, and filled with carbon monoxide from a CO line including a mineral oil bubbler and a pressure-release bubbler with an 80 mm mercury column.[5] The purge cycle should be repeated twice. The flask is opened while a stream of CO is passed through it, and a solution of KOH (4.7 g) in methanol (80 mL), previously prepared under nitrogen, and chloroform (5 mL) are introduced into the flask. The flask is restoppered, and the mixture is stirred vigorously for 42 hours, while the CO pressure in the flask is maintained by continuous addition. The absorption of carbon monoxide is fast during the first 4 hours but subsequently goes more slowly.* At the end of the reaction the CO is pumped off *in vacuo*, and the flask is filled with nitrogen for the subsequent operations, which are all carried out under nitrogen. The mixture is filtered, and the solid is washed with 10-mL portions of methanol until the washings are nearly colorless. The combined solution and washings are evaporated to dryness *in vacuo* with stirring, and the solid mass is treated with water (20 mL). A yellow crystalline powder separates from the brown mother liquor. The mixture is briefly stirred *in vacuo* to eliminate the last traces of methanol and then filtered under nitrogen. The residual traces of the product in the reaction flask are washed out with 5 mL of air-free saturated aqueous KCl. The product is washed with an air-free saturated aqueous solution of KCl in 5-mL portions until the washings are no longer brown and is dried at room temperature at 10^{-1} torr.

*Care must be taken that the CO pressure never decreases in the flask. This is achieved by adding the CO at such a rate that some of the gas continuously escapes from the pressure-release bubbler.

Under a nitrogen atmosphere, the product is extracted from the filter with THF (about 12 mL), which leaves the KCl undissolved, and the solution is cautiously evaporated to dryness *in vacuo*, giving first a thick yellow-brown glass, which, on heating on a water bath at 45° and 10^{-1} torr, slowly loses THF to give a yellow powder (about 2.5 hr are required). This product still contains variable amounts of THF, which can be evaluated from the rhodium content.* The product obtained under the above conditions weighs 1.0-1.1 g and contains about 3 moles of THF; the yield is about 70%. *Anal.* Calcd. for $K_2[Rh_6(CO)_{15}C] \cdot 3THF$: Rh, 45.9%. Found: Rh, 46.5%. The reaction has been carried out with up to 6 g of $K_3[RhCl_6] \cdot H_2O$ without complications. A smaller scale also is possible, but to avoid problems in handling the product we never use less than 1 g of $K_3[RhCl_6] \cdot H_2O$. The purity of the product is best judged from the color, which must be yellow, and from the IR spectrum. The product can be purified by precipitation from the aqueous solution on saturation with KCl, followed by filtration and extraction as above.

A highly pure compound can be obtained by transformation into the benzyltrimethylammonium salt (see below), which can be crystallized from THF-isopropyl alcohol.

Properties

The compound $K_2[Rh_6(CO)_{15}C]$ is a yellow powder. It is sensitive to air both in the solid state and in solution and is quite soluble in water, methanol, ethanol, acetone, THF, and acetonitrile. The salts of other cations can be obtained by metathesis, in water for the cesium salt and in methanol for the larger tetraalkylammonium or phosphonium cations. The tetraethylammonium salt is sparingly soluble in THF, whereas the benzyltrimethylammonium and bis-(triphenylphosphine)imminium salts are soluble. All of these salts are soluble in acetone and acetonitrile. The yellow solution of the potassium salt in THF shows characteristic IR bands at 2040 (vw), 1990 (vs), 1885 (vw), 1845 (s), 1830 (sh, m) 1815 (sh, br) and 1775 (vw, br) cm^{-1}. The IR spectral band shapes depend on solvents and cations. The oxidation of $K_2[Rh_6(CO)_{15}C]$ with iron(III) ammonium sulfate in water under carbon monoxide leads to the octanuclear carbido carbonyl cluster $Rh_8(CO)_{19}C$,[6] whereas under nitrogen $Rh_{12}(CO)_{25}C_2$[7] or $[H_3O][Rh_{15}(CO)_{28}C_2]$[8] is obtained.

References

1. V. G. Albano, M. Sansoni, P. Chini, and S. Martinengo, *J. Chem. Soc., Dalton Trans.*, 1973, 651.

*Prolonged heating *in vacuo*, to eliminate all the THF, is not convenient, as some decomposition occurs. The THF content must be evaluated in every preparation, as the drying conditions are difficult to reproduce.

2. V. G. Albano, P. Chini, S. Martinengo, D. J. A. McCaffrey, D. Strumolo, and B. T. Heaton, *J. Am. Chem. Soc.*, **96**, 8106 (1974).
3. P. Chini, G. Longoni, and V. G. Albano, *Adv. Organomet. Chem.*, **XIV**, 285 (1976) and references therein.
4. G. P. Kauffman and J. Hwa-san Tsai, *Inorg. Synth.*, **8**, 217 (1966).
5. D. F. Shriver, *The Manipulation of Air-Sensitive Compounds*, McGraw-Hill, 1969, Chaps. 7 and 9, and references threin.
6. V. G. Albano, M. Sansoni, P. Chini, S. Martinengo, and D. Strumolo, *J. Chem. Soc., Dalton Trans.*, **1975**, 305.
7. V. G. Albano, P. Chini, S. Martinengo, M. Sansoni, and D. Strumolo, *J. Chem. Soc., Dalton Trans.*, **1978**, 459.
8. V. G. Albano, M. Sansoni, P. Chini, S. Martinengo, and D. Strumolo, *J. Chem. Soc., Dalton Trans.*, **1976**, 970.

49. DISODIUM DI-μ-CARBONYL-OCTA-μ_3-CARBONYL-ICOSACARBONYLDODECARHODATE(2−), Na$_2$[Rh$_{12}$(CO)$_{20}$(μ-CO)$_2$(μ_3-CO)$_8$]

$$3Rh_4(CO)_{12} + 2CH_3COONa + H_2O \longrightarrow Na_2[Rh_{12}(CO)_{30}] + CO_2 + 2CH_3COOH + 5CO$$

Submitted by S. MARTINENGO* and P. CHINI*
Checked by G. W. PARSHALL† and E. R. WONCHOBA†

The compound Na$_2$[Rh$_{12}$(CO)$_{30}$] can be prepared by reaction of Rh$_2$(CO)$_4$Cl$_2$ with sodium acetate in methanol under an atmosphere of carbon monoxide.[1] It contains one of the first polynuclear anions to be formed when the rhodium carbonyls or carbonyl halides are reduced by the action of alkaline reagents in alcohols or by alkali metals in tetrahydrofuran (THF). It provides a unique example of a double octahedral cluster carbonyl anion in which the "noble gas rule" is not obeyed,[1,2] and it is a starting material for the preparation of other polynuclear rhodium carbonyl anions.[1,3-5] The synthesis reported here is a modification of the original method. The starting material is Rh$_4$(CO)$_{12}$, now easily prepared at atmospheric pressure.[6-8] The reaction is fast, and the overall procedure requires about 6–7 hours with 80–85% yields.

*Centro del C.N.R. per lo studio dei composti di coordinazione nei bassi stati di ossidazione. Via G. Venezian 21, 20133 Milano, Italy.
†Central Research and Development Department, E.I. Du Pont Co., Wilmington, DE 19898.

Procedure

The synthesis must be carried out under a very pure nitrogen atmosphere ($O_2 \leqslant$ 2 ppm in volume) using the Schlenk-apparatus technique.[9] All the solvents and solutions are thoroughly degassed *in vacuo* and stored under nitrogen before use. The solvents are laboratory grade. Acetone and methyl formate are simply distilled, whereas benzene is refluxed and distilled over sodium.

■ **Caution.** *The first stage of the synthesis must be carried out in a well-ventilated hood because some carbon monoxide is evolved.*

A 100-mL Schlenk tube equipped with a magnetic stirring bar is charged with $Rh_4(CO)_{12}$ (1 g, Rh 53.7%) and $CH_3COONa \cdot 3H_2O$ (1 g). (An excess of sodium acetate is used to buffer the acid that is formed during the reaction.) The tube is stoppered and then is carefully evacuated at 10^{-2} torr and filled with nitrogen. The purge cycle can be repeated. The tube is opened while a good nitrogen stream is passed through it, and acetone (20 mL) and water (2 mL) are introduced with a syringe. The tube is restoppered and the mixture is vigorously stirred at room temperature (about 25°) until all the solid $Rh_4(CO)_{12}$ has disappeared (about 45 min). The tube is then opened while nitrogen is passed through it, and an NaCl solution in water (5%, 25 mL) is added to the mixture. The tube is restoppered, and the mixture is concentrated *in vacuo*, while it is stirred on a water bath at about 30°, until all the acetone has evaporated. The evaporation of the acetone after the addition of the aqueous sodium chloride solution and the subsequent filtration should be carried out as soon as possible, since prolonged contact of the product with the acid mother liquor lowers the yields.

The final volume is about 25 mL. The sodium salt separates initially as black crystals (probably containing coordinated acetone), which subsequently transform into a dark-violet powder when all the acetone is eliminated. Sometimes the product separates out first as an oil, which crystallizes as the acetone is evaporated. When the concentration *in vacuo* is completed, nitrogen is reintroduced, and the product is immediately filtered from the brown mother liquor on a fritted-glass filter under nitrogen. It is then washed with an NaCl aqueous solution (5%) until all the brown mother liquor is eliminated, and the washings are only weakly reddish-violet. Generally, four washings with 5 mL of solution are sufficient. The product is then dried *in vacuo* (about 10^{-1} torr) at room temperature. The dry product is usually contaminated by minor amounts of NaCl and $Rh_6(CO)_{16}$. It can be purified by extraction with methyl formate (15 mL). The filtered extract is then treated with benzene (30 mL) and concentrated to about 15 mL *in vacuo* until the sodium salt crystallizes. The product is filtered, washed with benzene (3 × 5 mL) and dried *in vacuo* for approximately 3 hours at about 10^{-1} torr and 45°, until the IR band of the methyl formate at 1730 cm^{-1} has nearly disappeared. Yield: 0.8–0.85 g, 82–88% based on Rh. The product strongly retains the last traces of solvents, which are not convenient to remove. *Anal.* Calcd. for $Na_2Rh_{12}(CO)_{30}$: Rh, 58.2. Found: Rh, 55.5. This product is sufficiently pure (about 95%) for most purposes, the impurities being essentially traces of solvents. The IR spectrum in acetonitrile solution shows bands at 2080 (mw), 2055 (vs), 2020 (w), 1815 (m, br), 1802 (vw), and 1763 (s) cm^{-1}. Nearly pure compounds are obtained by conversion into the substituted ammonium cation salts.

The synthesis also has been carried out on a larger scale [2 g of $Rh_4(CO)_{12}$] without appreciable differences, but still larger quantities are not recommended because of the longer times required for the operations, which cause slightly lower yields. No problems arise for smaller scale syntheses, although for handling reasons it is not convenient to use less than 0.2-0.3 g of $Rh_4(CO)_{12}$.

Properties

The compound $Na_2[Rh_{12}(CO)_{30}]$ is violet when finely powdered and black when crystalline. It is soluble in methanol, ethanol, acetone, and acetonitrile with a dark-violet, permanganate-like color. It is moderately soluble in water, from which it precipitates as a hydrated salt on addition of NaCl. This salt loses the water on prolonged pumping under vacuum. The hydrated salt is soluble in THF, whereas the anhydrous salt is only sparingly soluble. The aqueous solutions are not stable for long and decompose slowly to unidentified brown products. The decomposition is very fast on heating. Solutions in organic solvents are more stable, and decomposition requires several days or prolonged heating. Salts of other cations can be obtained from the sodium salt by metathesis reactions with the corresponding chlorides. For the potassium and cesium salts, the exchange is best carried out by addition of 5% aqueous KCl or CsCl to the acetone solution of the sodium salt and concentration *in vacuo*.

Double exchange with the bulky ammonium cations is simply carried out in alcoholic solution. The tetramethyl- and tetraethylammonium salts are insoluble in THF, while salts of larger cations, such as tetrabutyl- and benzyltrimethylammonium and μ-nitrido-bis(triphenylphosphorus)(1+), are soluble. All these salts are soluble in acetone and acetonitrile.

The sodium salt is briefly stable in air in the solid state, whereas in solution it is quickly oxidized. The $[Rh_{12}(CO)_{30}]^{2-}$ anion reacts with carbon monoxide in THF solution, giving slow disproportionation to the $[Rh_5(CO)_{15}]^-$ anion and $Rh_6(CO)_{16}$.[10]

Under a hydrogen atmosphere there is slow transformation into the $[Rh_{13}(CO)_{24}H_3]^{2-}$ anion.[3] The $[Rh_{12}(CO)_{30}]^{2-}$ anion is unstable toward bromide and iodide ions, which therefore cannot be used in the double-exchange reactions.[4] It reacts with HCl or I_2 to give derivatives of the $[Rh_6(CO)_{15}X]^-$ (X = Cl, I) species.[5]

References

1. P. Chini and S. Martinengo, *Inorg. Chim. Acta*, **3**, 299 (1969).
2. V. G. Albano and P. L. Bellon, *J. Organomet. Chem.*, **29**, 405 (1969).
3. V. G. Albano, A. Ceriotti, P. Chini, G. Ciani, S. Martinengo, and W. M. Anker, *Chem. Commun.*, 1975, 859.
4. S. Martinengo, P. Chini, G. Giordano, A. Ceriotti, V. G. Albano, and G. Ciani, *J. Organomet. Chem.*, **88**, 375 (1975).
5. P. Chini, S. Martinengo, and G. Giordano, *Gazzetta*, **102**, 330 (1972).
6. S. Martinengo, P. Chini, and G. Giordano, *J. Organomet. Chem.*, **27**, 389 (1971).
7. G. Giordano, S. Martinengo, D. Strumolo, and P. Chini, *Gazzetta*, **105**, 613 (1975).
8. S. Martinengo, G. Giordano, and P. Chini, *Inorg. Synth.*, this volume p. 209.
9. D. F. Shriver, *The Manipulation of Air-Sensitive Compounds*, McGraw-Hill, New York, 1969, Chaps. 7 and 9, and references therein.
10. A. Fumagalli, T. E. Koetzle, F. Takusagawa, P. Chini, and S. Martinengo, unpublished results.

50. μ-NITRIDO-BIS(TRIPHENYLPHOSPHORUS)(1+) UNDECACARBONYLHYDRIDOTRIFERRATE(1−), [(Ph$_3$P)$_2$N] [Fe$_3$H(CO)$_{11}$]

(A) $Fe(CO)_5 \xrightarrow[(2) \ CH_3COOH/CH_3OH]{(1) \ NaBH_4/THF} Na[Fe_3H(CO)_{11}] + CO$

$Na[Fe_3H(CO)_{11}] + [(Ph_3P)_2N]Cl \xrightarrow{CH_3OH} [(Ph_3P)_2N][Fe_3H(CO)_{11}] + NaCl$

(B) $Fe_3(CO)_{12} + 4KOH \xrightarrow{CH_3OH} K_2[Fe_3(CO)_{11}] + K_2CO_3 + 2H_2O$

$K_2[Fe_3(CO)_{11}] + CH_3COOH \xrightarrow{CH_3OH} K[Fe_3H(CO)_{11}] + CH_3COOK$

$K[HFe_3(CO)_{11}] + [(Ph_3P)_2N]Cl \xrightarrow{CH_3OH} [(Ph_3P)_2N][Fe_3H(CO)_{11}] + KCl$

Submitted by H. A. HODALI,* C. ARCUS,* and D. F. SHRIVER*
Checked by D. P. JONES† and P. J. KRUSIC†

The anion [Fe$_3$H(CO)$_{11}$]$^-$ was first isolated by Hieber and Brendle[1] in 1957, who employed a base disproportionation of Fe$_3$(CO)$_{12}$ followed by protonation.

*Department of Chemistry, Northwestern University, Evanston, IL 60201.
†Central Research and Development Department, E. I. du Pont de Nemours and Co., Wilmington, DE 19898.

It has been isolated with several different counter cations: $[Ni(phenanthroline)_3]^{2+}$, $[Ni(NH_3)_6]^{2+}$, $[(CH_3CH_2)_4N^+]$,[2] and $[(CH_3CH_2)_3HN^+]$.[3,4] Procedure B, which is a modification of Hieber and Brendle's synthesis, provides a convenient method of preparing $[HFe_3(CO)_{11}]^-$, if $Fe_3(CO)_{12}$ is available. Another method, given in Procedure A, is based on the reduction of $Fe(CO)_5$ with $NaBH_4$. This synthesis, which may be preferred because of the more readily available starting materials, is based on previous fragmentary evidence that $[HFe_3(CO)_{11}]^-$ is produced from $Fe(CO)_5$ and BH_4^-.[5] In both procedures the product is obtained in crystalline form and in high yield by use of the μ-nitrido-bis(triphenylphosphorus)(1+) cation.

General Procedure

■ **Caution.** *Regardless of the procedure chosen, the reactions must be run in a hood owing to the high toxicity of iron carbonyls and carbon monoxide.*

As is described here the transfer of solids is carried out in a nitrogen-flushed dry box, and the syntheses are performed in standard Schlenk apparatus under a nitrogen atmosphere.[6] Alternatively, the entire synthesis may be performed in an inert-atmosphere dry box.

Tetrahydrofuran (THF) is stored over KOH, then distilled under N_2 from calcium hydride, and stored in an inert atmosphere over calcium hydride. Methanol is purified by fractional distillation; methanolic KOH is prepared from reagent grade KOH and is purged with nitrogen before use; glacial acetic acid (reagent grade) and distilled water are purged with nitrogen before use. Iron pentacarbonyl* is degassed by three cycles of freezing, pumping, and thawing. μ-Nitrodo-bis(triphenylphosphorus)(1+) chloride, $[(Ph_3P)_2N]Cl$,† is recrystallized from water and dried under high vacuum. Commercial triiron dodecacarbonyl* contains 5–10% methanol stabilizer, which is removed by subjecting the sample to high vacuum before use.

Procedure A

A sample of 0.78 g (20.0 mmole) of sodium tetrahydroborate‡ is placed in a 100-mL, round-bottomed flask containing a magnetic stirring bar. The flask is attached to an inlet tube for N_2 purge or vacuum, a 50-mL dropping funnel, and a bubbler. The flask and dropping funnel are evacuated and flushed with nitrogen several times. Under a flow of nitrogen, the flask is charged with 30 mL of THF. Into the dropping funnel under a flow of nitrogen are added 2.69

---

*Available from Strem Chemicals, 150 Andover St., Danvers, MA 01923.
†Available from Alfa Products, 152 Andover St., Danvers, MA 01923.
‡Available from Alfa Products, Ventron Corp., P.O. Box 299, Danvers, MA 01923.

mL (20.0 mmole) of previously degassed $Fe(CO)_5$ and 20 mL of THF. The $Fe(CO)_5$ solution is added dropwise with continuous stirring over a period of 1 hour. Within the first 10 minutes of addition, the solution in the flask undergoes a series of color changes from colorless to light yellow, light brown, dark brown, and finally dark red. When the addition of $Fe(CO)_5$ is complete, the solution is stirred for 6 hours and then filtered through a medium-porosity Schlenk-ware frit. While the solution is being rapidly stirred, the solvent is stripped from the filtrate under reduced pressure and residual THF is removed from the product by keeping it at reduced pressure (about 0.1 torr) for several hours. Under a nitrogen atmosphere, the resulting red gummy residue is dissolved in 20 mL of methanol, which has been purged with nitrogen. To the methanolic solution are added under nitrogen flush 11.4 mL (200 mmole) of glacial acetic acid and then a solution of 5.7 g (10 mmole) of μ-nitrido-bis-(triphenylphosphorus)(1+) chloride ($[Ph_3P)_2N]$Cl) in 20 mL of air-free methanol. A dark-red crystalline substance separates (if crystals have not formed at this point, 20 mL of degassed water should be added to precipitate the product). Using inert gas Schlenk techniques, the mixture is then filtered (medium-porosity frit), washed first with three 20-mL portions of air-free water and then with three 20-mL portions of air-free methanol. The product is dried under vacuum overnight. For purification, a sample (1.0 g) of the product is dissolved in 100 mL of methanol. After filtration, distilled water (25 mL) is added gradually. Dark-red crystals separate and these are filtered, washed with three 20-mL portions of water, and dried under vacuum overnight. *Anal.* Calcd. for $C_{47}H_{31}NFe_3P_2O_{11}$: C, 55.60; H, 3.08; N, 1.38. Found: C, 56.35; H, 3.28; N, 1.35. Yield of purified product: 5.0 g [about 74% based on $Fe(CO)_5$].

Procedure B

Inside a dry box, a sample of 1.5 g (2.7–3.0 mmole) of $Fe_3(CO)_{12}$ is placed in a 100-mL, round-bottomed flask. Under a stream of nitrogen, 15 mL of 2 *M* methanolic KOH is introduced into the flask by syringe. The reaction mixture is stirred for 30 minutes, during which time the solution changes from dark green to dark red-brown. The solution is filtered through a medium-porosity frit. Under nitrogen flow, 3.5 mL (60 mmole) of glacial acetic acid is added to the filtrate. A deep reddish-purple solution results. At this point, 2.9 g (5.0 mmole) of $[(Ph_3P)_2N]$Cl dissolved in 20 mL of methanol is introduced by syringe into the reaction flask. A dark-red crystalline solid separates and is collected on a medium-porosity Schlenk-type frit. The solid product is first washed with three portions of 20 mL of air-free water and then with three 20-mL portions of air-free methanol. The product is dried under vacuum (0.1 torr) overnight. This material is sufficiently pure for most synthetic work. However, further purification can be achieved by crystallization from methanol/water, as described in procedure A. *Anal.* Calcd. for $C_{47}H_{31}NFe_3P_2O_{11}$: C, 55.60; H, 3.08; N, 1.38; Fe, 16.50. Found: C, 55.39; H, 3.30; N, 1.44; Fe, 15.76. Yield of purified product: 2.3 g [about 77% based on $Fe_3(CO)_{12}$].

Properties

μ-Nitrido-bis(triphenylphosphorus) (1+) undecacarbonylhydridotriferrate (1-), $[(Ph_3P)_2N][HFe_3(CO)_{11}]$, is obtained as shiny dark red, platelike crystals. The compound is moderately air sensitive in the solid form, but it can be handled in air for short periods of time. It decomposes slowly, in solution when exposed to air. It is soluble in CH_2Cl_2, diethyl ether, and THF, slightly soluble in methanol, and insoluble in water. It can be stored for long periods of time in an inert atmosphere (e.g., N_2) at room temperature. The IR spectrum of the compound contains the following CO stretching vibrations (Nujol or Fluorolube mull): 2066 (w), 2088 (sh), 1996 (s), 1980 (s), 1961 (s), 1945 (sh), 1937 (s), 1921 (m), 1725 (m), cm^{-1}. The nmr spectrum (deuterated acetone solvent and acetone internal standard) has a hydride band at δ - 14.53 ppm.

The crystal structure of the hydrido anion, $[HFe_3(CO)_{11}]^-$, has been established by X-ray analysis[3] and its Mössbauer spectrum has been recorded.[2] On acidification at low temperature, $[HFe_3(CO)_{11}]^-$ is converted to $H_2Fe_3(CO)_{11}$, which decomposes above $-40°$.[7] Reaction of this anion with CH_3SO_3F gives the compound $HFe_3(CO)_{10}(COCH_3)$.[8] Contrary to previous proposals, recent evidence indicates that $[HFe_3(CO)_{11}]^-$ is not an important catalytic species in the Reppe conversion of CO and olefins to alcohols.[9]

Recently, it has been demonstrated that this anion is the active species in the reduction of the carbon-nitrogen triple bond in acetonitrile[10] and it is proposed to be a key intermediate in the reduction of nitro arenes with triiron dodecacarbonyl.[11]

References

1. W. Hieber and G. Brendel, *Z. anorg. Chem.*, **289**, 324 (1957).
2. K. Farmery, M. Kilner, R. Greatrex, and N. N. Greenwood, *J. Chem. Soc. (A)*, **1969**, 2339.
3. L. F. Dahl and J. F. Blount, *Inorg. Chem.*, **4**, 1373 (1965).
4. J. R. Wilkinson and L. J. Todd, *J. Organomet. Chem.*, **118**, 199 (1976).
5. D. T. Haworth and J. R. Ruff, *J. Inorg. Nucl. Chem.*, **17**, 184 (1961).
6. D. F. Shriver, *The Manipulation of Air-Sensitive Compounds*, McGraw-Hill, New York, 1969.
7. H. A. Hodali, D. F. Shriver, and C. A. Ammlung, *J. Am. Chem. Soc.*, **100**, 5239 (1978).
8. D. F. Shriver, D. Lehman, and D. Strope, *J. Am. Chem. Soc.*, **97**, 1594 (1975).
9. H. C. Kang, C. H. Mauldin, T. Cole, W. Slegeir, K. Cann, and R. Pettit, *J. Am. Chem. Soc.*, **99**, 8323 (1977).
10. M. A. Andrews and H. D. Kaesz, *J. Am. Chem. Soc.*, **99**, 6763 (1977).
11. H. des Abbayes and H. Alper, *J. Am. Chem. Soc.*, **99**, 98 (1977).

51. BIS[μ-NITRIDO-BIS(TRIPHENYLPHOSPHORUS)(1+)] UNDECACARBONYLTRIFERRATE, [(PH$_3$P)$_2$N]$_2$[Fe$_3$(CO)$_{11}$]

$$Fe_3(CO)_{12} + 4KOH \xrightarrow{MeOH} K_2Fe_3(CO)_{11} + K_2CO_3 + 2H_2O$$

$$K_2Fe_3(CO)_{11} + 2[Ph_3P)_2N]Cl \longrightarrow [(Ph_3P)_2N]_2[Fe_3(CO)_{11}] + 2KCl$$

Submitted by H. A. HODALI* and D. F. SHRIVER*
Checked by D. J. JONES† and P. J. KRUSIC†

The dianion [Fe$_3$(CO)$_{11}$]$^{2-}$ was first isolated by Hieber[1] as the [Fe(en)$_3$]$^{2+}$ (en = ethylenediamine) salt. Attempts to prepare the tetraethylammonium salt were unsuccessful.[2] In the synthesis reported here, this dianion is isolated as the μ-nitrido-bis(triphenylphosphorus)(1+) salt, (Ph$_3$P)$_2$N$^+$, which affords high yields of a pure product.

Procedure

■ **Caution.** *The reaction must be run in a hood owing to the highly toxic nature of iron carbonyls and carbon monoxide.*

All manipulations are carried out on a high-vacuum line, in a nitrogen-flushed dry box, or in standard Schlenk apparatus under a nitrogen atmosphere. Anhydrous diethyl ether is further dried by distillation under nitrogen from sodium benzophenone ketyl. Methanol is purified by fractional distillation and purged with nitrogen before use. Dichloromethane is distilled from phosphorus pentoxide and purged with nitrogen before use. Methanolic KOH is prepared from reagent grade KOH and is purged with nitrogen before use. Glacial acetic acid, reagent grade, is purged with nitrogen before use. Commercial triiron dodecacarbonyl‡ contains 5-10% methanol stabilizer, which is removed by subjecting the sample to high vacuum before use. Commercial μ-nitrido-bis(triphenylphosphorus)(1+) chloride ([Ph$_3$P)$_2$N]Cl)§ is recrystallized from water (using charcoal) and dried under high vacuum (0.001 torr) overnight.

Into a 200-mL, round-bottomed flask, containing a magnetic stirring bar and a 5.1-g (9-10 mmole) sample of Fe$_3$(CO)$_{12}$[3], is added 50 mL of 2 *M* methano-

---

*Department of Chemistry, Northwestern University, Evanston, IL 60201.
†Central Research and Development Department, E.I. du Pont de Nemours and Co., Inc., Wilmington, DE 19898.
‡Available from Strem Chemicals, 150 Andover St., Danvers, MA 01923.
§Available from Alfa Products, 152 Andover St., Danvers, MA 01923.

lic KOH. After a few minutes of stirring, the solution changes from dark green to dark red-brown. The mixture is stirred for 30 minutes and then filtered through a medium-porosity Schlenk-type frit. A solution of 12.6 g (22.0 mmole) of $[(Ph_3P)_2N]Cl$ in 50 mL of methanol is added to the filtrate. A thick red-brown precipitate separates, and the mixture is filtered under a nitrogen atmosphere with a medium-porosity frit. The dark-red solid that is collected is then dried under vacuum. Crystals are obtained by dissolving the product in 100 mL of CH_2Cl_2 and filtering the resulting solution with a medium-porosity Schlenk-type fritted filter. To the filtrate is added 60 mL of diethyl ether to precipitate a crystalline red-brown product. The crystalline product is filtered, washed with three 20-mL portions of diethyl ether, and then dried under vacuum overnight. *Anal.* Calcd. for $C_{83}H_{60}N_2Fe_3P_4O_{11}$: C, 64.19; H, 3.90; N, 1.80; Fe, 10.79. Found: C, 63.76; H, 3.81; N, 1.60; Fe, 11.03. The yield of purified product is 11.9 g (about 86% based on $Fe_3(CO)_{12}$).

Properties

Bis[μ-nitrido-bis(triphenylphosphorus)(1+)] undecacarbonyltriferrate(2-) is obtained as shiny dark red-brown rectangular crystals that are moderately air sensitive but can be handled in air for short periods of time. When in solution, the compound slowly decomposes upon exposure to air. It is soluble in CH_2Cl_2, slightly soluble in acetone, and insoluble in diethyl ether. It can be stored for long periods of time in an inert atmosphere (e.g. N_2) at room temperature. The IR spectrum shows the following CO stretching frequencies (Fluorolube or Nujol mull) 2010 (vw), 1935 (s), 1900 (s), 1883 (s), 1878 (s, sh), 1965 (sh), 1667 (m) cm^{-1}.

The $[Fe_3(CO)_{11}]^{2-}$ anion is a very useful starting material and its reaction with methylating agents has been described.[4] Upon reaction with hydroxide, it disproportionates to give other iron carbonyl anions.[5,6] Acidification of $[Fe_3(CO)_{11}]^{2-}$ at low temperatures yields $H_2Fe_3(CO)_{11}$.[7] Interconversion of $[Fe_3(CO)_{11}]^{2-}$ into other iron carbonyl anions is possible and under certain conditions the conversion is quantitative.[2]

References

1. W. Heiber, J. Sedlemeir, and R. Werner, *Chem. Ber.*, **90**, 278 (1957).
2. K. Farmery, M. Kilner, R. Greatrex, and N. N. Greenwood, *J. Chem. Soc. (A)*, **1969**, 2339.
3. W. McFarlane and G. Wilkinson, *Inorg. Synth.*, **8**, 181 (1965).
4. D. F. Shriver, D. Lehman, and D. Strope, *J. Am. Chem. Soc.*, **97**, 1594 (1975).
5. W. Hieber and D. H. Schubert, *Z. anorg. allg. Chem.*, **338**, 37 (1965).
6. W. Hieber and H. Beutner, *Z. Naturforsch.*, **17b**, 211 (1962).
7. H. A. Hodali, D. F. Shriver, and C. A. Ammlung, *J. Am. Chem. Soc.*, **100**, 5239 (1978).

52. μ_3-ALKYLIDYNE-TRIS(TRICARBONYLCOBALT) COMPOUNDS: ORGANOCOBALT CLUSTER COMPLEXES

Submitted by DIETMAR SEYFERTH,* MARA O. NESTLE,* JOHN S. HALLGREN,* JOSEPH S. MEROLA,* GARY H. WILLIAMS,* and CYNTHIA L. NIVERT*

The μ_3-alkylidyne-tris(tricarbonylcobalt) complexes, I, are relative newcomers to organotransition metal chemistry. The first member of this class to be prepared (1, R = CH_3) was reported in 1958.[1] During the next 10 years more complexes of this type were prepared, but the chemistry of this novel class of organocobalt carbonyls remained relatively unexplored. Since 1969, several research groups have been actively investigating $RCCo_3(CO)_9$ complexes, and much new and interesting chemistry has been uncovered. The organocobalt carbonyl cluster complexes have been reviewed at various stages of the development of their chemistry.[2-5]

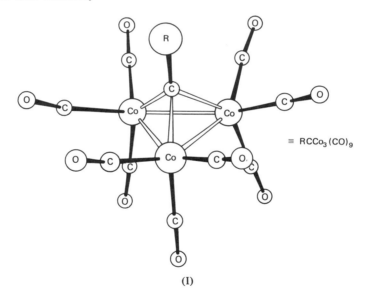

(I)

A wide variety of $RCCo_3(CO)_9$ complexes may be prepared (R = H, halogen, alkyl, aryl, alkenyl, acyl, alkoxycarbonyl, aminocarbonyl, triorganosilyl, dialkyl phosphino, alkoxy, dialkylamino, etc.), either by direct reactions of octacarbonyl dicobalt with organic tri- and dihalides or by acid-induced reactions of

*Department of Chemistry, Massachusetts Institute of Technology, Cambridge, MA 02139.

cobalt carbonyl complexes of terminal acetylenes, $(RC_2H)Co_2(CO)_6$.[2] Further interconversions of cluster complexes thus prepared have added many new members to this class.[2] Replacement of one to three of the nine carbon monoxide ligands with other donor molecules (tertiary phosphines, arenes, dienes, isocyanides) also is possible.[3]

Oxidation of these complexes releases the apical carbon atom and its substituent in the form of an organic derivative.[2] Thus ceric ammonium nitrate oxidation of an $RCCo_3(CO)_9$ complex produces the carboxylic acid, RCO_2H, when the reaction is carried out in aqueous acetone. Thermolysis of $RCCo_3(CO)_9$ produces acetylenes or acetylene hexacarbonyl dicobalt complexes, depending on the reaction conditions. Useful applications of these cluster complexes in organic synthesis remain to be developed.

A general description of the physical and chemical properties of the μ_3-alkylidyne-tris(tricarbonylcobalt) complexes may be helpful at this point. All are highly colored, with colors ranging from red to purple to brown to black, depending on the apical substituent. Some can be heated to 100-185° without decomposition and many of them may be sublimed in vacuum at 50-80°. Some others, however, are stable at room temperature only for limited periods of time. Most $RCCo_3(CO)_9$ complexes, but not all, are air-stable in the solid state and in solution. Most of the complexes with which we have worked are crystalline solids, usually (but not always) with well-defined melting and/or decomposition points. However, when the apical carbon atom bears a long carbon chain substituent, the melting points are lowered, and oils, rather than crystalline solids, can result. Most of the $RCCo_3(CO)_9$ complexes are soluble in the common organic solvents. In view of their color, stability, and solubility properties, thin layer and column chromatography are applicable to their detection, isolation, and purification. Their combustion analysis has presented no difficulties. The general chemical reactivity of the $RCCo_3(CO)_9$ complexes includes instability toward attack by oxidizing agents, many bases, and many nucleophiles. Many, however, are stable toward protonic and Lewis acids. Mass spectroscopy is very useful in their identification and characterization. Initial fragmentation results in loss of the nine carbon monoxide ligands, leaving the $[RCCo_3]^+$ fragment.

In general, the infrared spectra of μ_3-alkylidyne-tris(tricarbonylcobalt) complexes show five bands in the terminal carbonyl region, in the range 2150-1950 cm^{-1}, with the intensity pattern medium, very strong, strong, weak, very weak. The last and weakest of these is a ^{13}C E-mode-derived band. Since these complexes are diamagnetic, completely normal proton nmr spectra are obtained for the organic substituents on the apical carbon atom of the cluster. The ^{13}C nmr spectra of μ_3-alkylidyne-tris(tricarbonylcobalt) compounds have been obtained; the signal due to the apical carbon atom of the cluster can be seen only with some difficulty.[6,7]

Acknowledgment

The preparations reported here were developed during the course of our research in the area of organocobalt carbonyl chemistry, which was supported by the National Science Foundation. We acknowledge with thanks this generous assistance.

References

1. R. Markby, I. Wender, R. A. Friedel, F. A. Cotton, and H. W. Sternberg, *J. Am. Chem. Soc.*, **80**, 6529 (1958).
2. D. Seyferth, *Adv. Organomet. Chem.*, **14**, 97 (1976).
3. B. R. Penfold and B. H. Robinson, *Acc. Chem. Res.*, **6**, 73 (1973).
4. G. Pályi, F. Piacenti, and L. Markó, *Inorg. Chim. Acta Rev.*, **4**, 109 (1970).
5. D. Seyferth, J. E. Hallgren, R. J. Spohn, A. T. Wehman, and G. H. Williams, *Special Lectures, XXXIIIrd International Congress of Pure and Applied Chemistry*, July 26-30, 1971, Vol. 6, Butterworths, London, pp. 133-149.
6. D. Seyferth, G. H. Williams, and D. D. Traficante, *J. Am. Chem. Soc.*, **96**, 604 (1974).
7. D. Seyferth, C. S. Eschbach, and M. O. Nestle, *J. Organomet. Chem.*, **97**, C11 (1975).

53. μ_3-METHYLIDYNE AND μ_3-BENZYLIDYNE-TRIS(TRICARBONYLCOBALT)

$$\tfrac{9}{2} Co_2(CO)_8 + 2HCBr_3 \longrightarrow 2HCCo_3(CO)_9 + 18CO + 3CoBr_2 *$$

$$HCCo_3(CO)_9 + (C_6H_5)_2Hg \longrightarrow C_6H_5CCo_3(CO)_9 + Hg + C_6H_6$$

Submitted by MARA O. NESTLE,† JOHN E. HALLGREN,† and DIETMAR SEYFERTH†
Checked by PETER DAWSON‡ and BRIAN H. ROBINSON‡

μ_3-Methylidyne-tris(tricarbonylcobalt) is a useful starting material for further conversions to other μ_3-alkylidyne-tris(tricarbonylcobalt) complexes. Its reactions with diarylmercury compounds (as described here for diphenylmercury), arylmercuric halides, and α-(haloalkyl)mercurials are very useful in the preparation of aryl and alkyl derivatives of the $CCo_3(CO)_9$ cluster.[2] μ_3-Methylidyne-

---

*This balanced equation has no connection with the actual course of the reaction, nor does it necessarily give a complete picture of the reaction in terms of products. For a study concerning this question, see Reference 1.

†Department of Chemistry, Massachusetts Institute of Technology, Cambridge, MA 02139.

‡Department of Chemistry, University of Otago, P.O. Box 56, Dunedin, New Zealand.

tris(tricarbonylcobalt) also reacts with olefins and dienes, adding its C—H bond to the C=C bond.[3] Its reactions with silicon and germanium hydrides provide a very effective route to μ_3-alkylidyne-tris(tricarbonylcobalt) complexes bearing silyl and germyl substituents at the apical carbon atom.[4] μ_3-Benzylidyne-tris-(tricarbonylcobalt) complexes are of interest because they undergo facile Friedel-Crafts acylation and arylation by $RC(O)Cl/AlCl_3$ systems.[5]

A. μ_3-METHYLIDYNE-TRIS(TRICARBONYLCOBALT)

Procedure

■ **Caution.** *This reaction should be carried out in an efficient fume hood since both octacarbonyldicobalt and the evolved carbon monoxide are toxic. Octacarbonyldicobalt, as supplied, is pyrophoric and thus must be handled under an inert atmosphere at all times.*

A 1-L, three-necked, round-bottomed Pyrex flask is equipped with a serum cap, a solids-transfer tube (in the center neck), and a T-tube, one side of which is connected to a prepurified nitrogen gas supply and the other to a Nujol bubbler. The vessel is evacuated, flamed out, and flushed with dry nitrogen. It is then charged with octacarbonyldicobalt* (64.4 g, 0.188 mole) by means of the transfer tube. The transfer tube is removed and replaced with an adapter tube holding a stirrer sleeve and shaft that is attached to an overhead stirring motor. Tetrahydrofuran (500 mL, freshly distilled from sodium/benzophenone) is added by syringe through the serum cap, and the mixture is stirred under nitrogen until the cobalt carbonyl has dissolved. Bromoform (tribromomethane) (28.9 g, 0.114 mole) then is added by syringe, and the resulting mixture is warmed gently to 50°. The reaction temperature should not be permitted to exceed 50°, since decomposition occurs above that temperature. At 50° carbon monoxide evolution begins. This temperature is maintained while the solution is being stirred, until gas evolution ceases (about 6 hr). The reaction mixture is cooled to room temperature and then poured into 1500 mL of 10% aqueous hydrochloric acid. All manipulations from this point on may be carried out in the air, with no further precautions to exclude moisture and oxygen. The hydrolysis mixture is stirred well. Subsequently, the product is extracted with 500-mL portions of pentane until the extracts are only faintly colored. The organic layer is washed 8 to 10 times with equal volumes of water, until the water layer no longer is yellow. The organic phase is dried over anhydrous magnesium sulfate and filtered with suction. The solvent is removed under reduced pressure (rotary evaporator), and the residue is sublimed at 50° and 0.1

---

*Strem Chemicals, Inc., P.O. Box 108, Newbury Port, MA 01950. The material was used as supplied; further purification was found to be unnecessary.

torr. (■ **Caution.** *The vacuum must be released by admission of nitrogen or argon since the residue from the sublimation is pyrophoric. The residue is rendered harmless by treatment with water.*) The sublimate is resublimed to yield 18.88 g (34% yield based on the cobalt charged) of μ_3-methylidyne-tris-(tricarbonylcobalt), mp 105-107° (dec.).[6] *Anal.* Calcd. for $C_{10}HO_9Co_3$: C, 27.18; H, 0.23. Found: C, 27.30; H, 0.38.

Properties

μ_3-Methylidyne-tris(tricarbonylcobalt) is an air-stable, crystalline, purple solid.[6] It is soluble in common organic solvents, giving intensely purple solutions that are stable in air for several hours. The infrared spectrum of $HCCo_3(CO)_9$ in carbon tetrachloride solution shows the C—H stretch at 2980 (w) cm^{-1}, four carbonyl bands at 2105 (m), 2060 (vs), 2045 (s), and 2025 (w) cm^{-1}, and bands at 865 (m) and 723 (w) cm^{-1}. Its proton nmr spectrum (in $CDCl_3$) shows a singlet at δ 12.08 ppm downfield from internal tetramethylsilane.

B. μ_3-BENZYLIDYNE-TRIS(TRICARBONYLCOBALT)

Procedure

■ **Caution.** *This reaction is carried out under an atmosphere of carbon monoxide and therefore should be performed in an efficient fume hood. Diphenylmercury is a toxic chemical and care should be taken that its solutions are not spilled on the skin or ingested. Its vapor pressure, however, is low, and so it is not as dangerous as the dialkylmercurials, nor is it as dangerously toxic as the dialkylmercurials.*[7]

A 100-mL, three-necked, round-bottomed flask is equipped with a reflux condenser topped with a gas inlet tube that is connected to a T-tube, one side of which leads to a prepurified nitrogen gas supply and the other to a Nujol bubbler, a magnetic stirring bar, and a glass tube to be used for bubbling carbon monoxide through the reaction mixture. The vessel is flame dried, purged with dry nitrogen, and charged with μ_3-methylidyne-tris(tricarbonylcobalt) (0.88 g, 2.0 mmole), diphenylmercury* (0.70 g, 2.0 mmole), and 50 mL of dry benzene that has been distilled from sodium/benzophenone. Carbon monoxide is bubbled through the solution for 30 minutes, and then the mixture is heated at reflux, with stirring, while carbon monoxide is bubbled slowly through the solution. Heating is continued until thin layer chromatography† shows only one brown

---

*Alfa Products, Ventron Corporation, P.O. Box 299, Danvers, MA 01923. If the material as purchased is of good quality it may be used without further purification.

†Eastman Chromagram Sheet No. 6061 (silica gel), using hexane as eluent. The $HCCo_3(CO)_9$ produces a purple spot of high R_f.

spot. A reaction time of about 2 hours is required. The mixture is cooled to room temperature and filtered in air. The residue is washed with acetone to yield 0.15 g (38%) of mercury. The filtrate is evaporated to dryness under reduced pressure. The solid that remains is purified by column chromatography (40 × 600 mm column containing 100 mesh Mallinckrodt reagent grade silicic acid, with hexane as eluent). A faint purple band (starting material) may elute first. A brown band follows. The eluate containing the material in the latter is evaporated at reduced pressure to give 0.99 g (95%) of μ_3-benzylidyne-tris(tricarbonylcobalt), mp 103–106°. This material is sublimed at 50° and 0.1 torr to give 0.95 g (91%) of pure product, mp 107°. *Anal.* Calcd. for $C_{16}H_5O_9Co_3$: C, 37.10; H, 0.97. Found: C, 37.10; H, 1.00.

Properties

μ_3-Benzylidyne-tris(tricarbonylcobalt) is an air-stable, brown, crystalline solid. It is soluble in organic solvents to give intensely brown solutions that are stable in air for several hours. Its infrared spectrum (in carbon tetrachloride solution) shows five bands in the terminal carbonyl region: 2109 (m), 2062 (vs), 2047 (s), 2028 (w) and 1980 (vw) cm^{-1}. Other bands observed include: 3080 (w), 1480 (w), 1435 (w), 1255 (w), 900 (w), 685 (m), 655 (m), 625 (w), and 605 (s) cm^{-1}. The nmr spectrum of $C_6H_5CCo_3(CO)_9$ (in carbon tetrachloride) shows the phenyl proton signals as a multiplet at δ 7.13–7.63 ppm downfield from internal tetramethylsilane.

References

1. B. L. Booth, G. C. Casey and R. N. Haszeldine, *J. Chem. Soc., Dalton Trans.*, **1975**, 1850.
2. D. Seyferth, J. E. Hallgren, R. J. Spohn, G. H. Williams, M. O. Nestle, and P. L. K. Hung, *J. Organomet. Chem.*, **65**, 99 (1974).
3. (a) D. Seyferth and J. E. Hallgren, *J. Organomet. Chem.*, **49**, C41 (1973).
   (b) N. Sakamoto, T. Kitamura, and T. Joh, *Chem. Lett.*, **1973**, 583.
   (c) T. Kamijo, T. Kitamura, N. Sakamoto, and T. Joh, *J. Organomet. Chem.*, **54**, 265 (1973).
4. D. Seyferth and C. L. Nivert, *J. Am. Chem. Soc.*, **99**, 5209 (1977).
5. D. Seyferth, G. H. Williams, A. T. Wehman, and M. O. Nestle, *J. Am. Chem. Soc.*, **97**, 2107 (1975).
6. G. Pályi, F. Piacenti, and L. Markó, *Inorg. Chim. Acta Rev.*, **4**, 109 (1970).
7. L. Friberg and J. Vostal, Eds., *Mercury in the Environment*, CRC Press, Cleveland, OH 1972.

54. μ_3-[(ETHOXYCARBONYL)METHYLIDYNE]- and μ_3-{[(METHYLAMINO)CARBONYL]METHYLIDYNE}-TRIS(TRICARBONYLCOBALT)

$$\tfrac{9}{2} Co_2(CO)_8 + 2CCl_3CO_2C_2H_5 \longrightarrow 2C_2H_5O_2CCCo_3(CO)_9 + 18CO + 3CoCl_2$$

$$C_2H_5O_2CCCo_3(CO)_9 + (C_2H_5CO)_2O + HPF_6 \longrightarrow$$
$$[(OC)_9Co_3CCO]^+ [PF_6]^- (s) + 2C_2H_5CO_2H$$

$$[(OC)_9Co_3CCO]^+ [PF_6]^- + 2CH_3NH_2(g) \longrightarrow$$
$$CH_3NHC(O)CCo_3(CO)_9 + [CH_3NH_3]PF_6$$

Submitted by JOSEPH S. MEROLA,* JOHN E. HALLGREN* and DIETMAR SEYFERTH*
Checked by PETER DAWSON† and BRIAN H. ROBINSON†

The ethyl ester of μ_3-(carboxymethylidyne)-tris(tricarbonylcobalt) is a useful precursor to a large variety of organocobalt cluster complexes of type $RC(O)CCo_3(CO)_9$.[1,2] It is easily converted to the decacarbonyl-*tetrahedro*-carbontricobalt cation, $[(OC)_9Co_3CCO]^+$, by treatment with concentrated sulfuric acid or with hexafluorophosphoric acid in propionic anhydride medium. The $[(OC)_9Co_3CCO]^+$ cation can react with many nucleophiles to give new, functionally substituted μ_3-alkylidyne-tris(tricarbonylcobalt) complexes.[1] One example of many is given below.

A. μ_3-[(ETHOXYCARBONYL)METHYLIDYNE]-TRIS(TRICARBONYLCOBALT)

Procedure

■ **Caution.** *The cautionary remarks made with respect to the synthesis of $HCCo_3(CO)_9$ apply here as well.*

A 1-L, three-necked, round-bottomed flask is equipped with a serum cap, a nitrogen inlet tube connected to a T-piece that leads to a prepurified nitrogen supply and to a Nujol bubbler, and a solids-transfer tube (in the center neck). The flask is evacuated, flamed out, and flushed with dry nitrogen. It then is charged with octacarbonyldicobalt (73.4 g, 0.215 mole) and 440 mL of dry tetrahydrofuran (THF). The transfer tube is removed and replaced with an

---

*Department of Chemistry, Massachusetts Institute of Technology, Cambridge, MA 02139.
†Department of Chemistry, University of Otago, P.O. Box 56, Dunedin, New Zealand.

adapter tube holding a stirrer sleeve and shaft that is attached to an overhead stirring motor. The mixture is stirred to effect solution. Then 22.9 g (0.12 mole) of ethyl trichloroacetate is added with a syringe through the serum cap. The resulting brown-yellow solution is stirred at room temperature for 1 hour, during which time a slow evolution of gas takes place. The solution then is warmed with a heating mantle to about 30°. This results in vigorous gas evolution, and the solution slowly becomes purple. Gas evolution is complete after about 2 hours at 30°. The purple solution then is heated at 50° for an additional hour.

The mixture is allowed to cool to room temperature and then is suction filtered in air to obtain a bright-blue residue (cobalt salts, which are discarded) and an intensely purple-colored filtrate. The latter is treated with 1500 mL of 10% aqueous hydrochloric acid to destroy remaining noncluster cobalt carbonyl. The product ester is extracted into 500 mL of pentane. The extracts are washed several times with equal portions of water. The organic layer is dried over anhydrous magnesium sulfate. Filtration is followed by evaporation of the filtrate at reduced pressure (rotary evaporator). The solid that remains is sublimed at 50° and 0.1 torr to yield 38.79 g (53%) of purple crystals of μ_3-[(ethoxycarbonyl)methylidyne]-tris(tricarbonylcobalt), mp 45–46°.[2] *Anal.* Calcd. for $C_{13}H_5O_{11}Co_3$: C, 30.38; H, 0.98. Found: C, 30.30; H, 0.96.

μ_3-[(Ethoxycarbonyl)methylidyne]-tris(tricarbonylcobalt) is a low-melting solid and has a tendency to remain in the form of an oil at this point. In this event, crystallization can be induced by redissolving the oil in pentane, removing the pentane at reduced pressure, and repeating this procedure until a solid is obtained.

■ **Caution.** *The sublimation residue is pyrophoric and therefore nitrogen or argon must be used to break the vacuum after the sublimation is complete. The sublimation residue is rendered harmless by treatment with water.*

Properties

μ_3-[(Ethoxycarbonyl)methylidyne]-tris(tricarbonylcobalt) is stable in air and dissolves in most organic solvents to form deep-purple solutions that are stable in air for long periods of time.[2] The terminal carbonyl absorptions are seen in the infrared spectrum of $C_2H_5O_2CCCo_3(CO)_9$ (in CCl_4) at 2115 (m), 2065 (vs), 2040 (s), 2020 (w) and 1980 (vw) cm^{-1} and the ester carbonyl absorption at 1685 (s) cm^{-1}. The other bands observed include 2990 (w), 2950 (w), 1250 (s), (w), 2880 (w), 2540 (w), 2500 (w), 1445 (w), 1390 (w), 1318 (w), 1250 (s), 1185 (w), 1098 (w), 1060 (s), 870 (w), and 705 (m) cm^{-1}. The proton nmr spectrum (in CCl_4) exhibits a triplet at δ 1.35 ppm (3H, J = 7 Hz, CH_3) and a quartet at δ 4.35 ppm (2H, J = 7 Hz, CH_2).

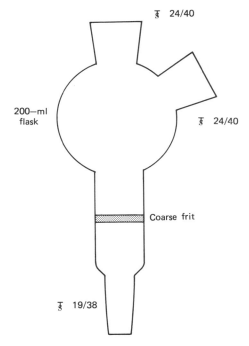

Fig. 1. Standard cation formation apparatus.

B. μ_3-{[(METHYLAMINO)CARBONYL]METHYLIDYNE}-TRIS(TRICARBONYLCOBALT)

Procedure

The apparatus used is shown in Fig. 1. It consists of a 200-mL, round-bottomed flask modified by attachment of a straight 30-mm glass tube with a coarse glass frit and a 19/38 standard taper joint colinear with the neck of the flask. A 24/40 standard taper joint is attached at an angle of 45° to the neck. To the apparatus are added, at neck A, a nitrogen inlet tube connected to a T-piece that leads to a prepurified nitrogen supply and to a Nujol bubbler, at neck B, an air-tight, foldover rubber stopper ("no-air stopper"), and, below the glass frit at the 19/38 joint, a 100-mL, three-necked, round-bottomed flask equipped with a nitrogen inlet tube connected to a T-piece that leads to a prepurified nitrogen supply and a Nujol bubbler and whose third neck is closed with a glass stopper.

The apparatus is equipped with a magnetic stirring bar, dried, flushed with nitrogen and charged with 1.04 g (2.0 mmole) of μ_3-[(ethoxycarbonyl)-methylidyne]-tris(tricarbonylcobalt) and 15 mL of propionic anhydride. The mixture is stirred under nitrogen (apparatus horizontal) until the cobalt complex has dissolved and 0.6 g (2.6 mmole) of 65% aqueous hexafluorophosphoric acid* is added by means of a disposable pipette. A black precipitate forms immediately. The mixture is filtered under positive nitrogen pressure (apparatus vertical). The black solid thus obtained is washed four times with 10-mL portions of reagent grade dichloromethane. In a separate experiment[1] it has been identified by analysis as the salt $[(OC)_9Co_3CCO]^+ [PF_6]^-$.

The black solid then is slurried in 10 mL of reagent grade dichloromethane (apparatus horizontal), and gaseous methylamine is passed slowly over the surface of the stirred mixture until all the solid has dissolved. Manipulations from this point on may be carried out without exclusion of air. The solution is poured into 100 mL of 10% aqueous hydrochloric acid and the product is extracted with dichloromethane. The extracts are dried over anhydrous magnesium sulfate. Filtration is followed by removal of solvent from the filtrate at reduced pressure (rotary evaporator). The red solid that remains is recrystallized from hexane to give 0.93 g (93%) of $CH_3NHC(O)CCo_3(CO)_9$, mp 124-126° (dec.), as black platelets. *Anal.* Calcd. for $C_{12}H_4NO_{10}Co_3$: C, 28.88; H, 0.81; N, 2.81. Found: C, 28.66; H, 0.94; N, 2.65.

Properties

μ_3-{[(Methylamino)carbonyl]methylidyne}-tris(tricarbonylcobalt) forms black, air-stable crystals that dissolve in most organic solvents to form purple solutions that are not stable for long periods of time. The infrared spectrum of $CH_3NHC(O)CCo_3(CO)_9$ (in CCl_4) shows bands in the terminal carbonyl region at 2114 (m), 2065 (vs), 2044 (s), 2018 (w), and 1985 (vw) cm^{-1}. The following other bands are observed in the ir spectrum: 3470 (w), 2960-2840 (w), 1633 (s), 1487 (s), 1415 (m), and 690 (m) cm^{-1}. Its proton nmr spectrum (in CCl_4) consists of a doublet at δ 3.07 ppm (3H, $J = 5$ Hz, CH_3) and a multiplet at δ 6.0 ppm (1H, NH).

References

1. D. Seyferth, J. E. Hallgren, and C. S. Eschbach, *J. Am. Chem. Soc.*, **96**, 1730 (1974).
2. G. Pályi, F. Piacenti, and L. Markó, *Inorg. Chim. Acta Rev.*, **4**, 109 (1970).

*Alfa Products, Ventron Corporation, P.O. Box 299, Danvers, MA 01923.

55. μ_3-(CHLOROMETHYLIDYNE)- AND μ_3-[(tert-BUTOXYCARBONYL)METHYLIDYNE]-TRIS(TRICARBONYLCOBALT)

$$\tfrac{9}{2} Co_2(CO)_8 + 2CCl_4 \longrightarrow 2ClCCo_3(CO)_9 + 18CO + 3CoCl_2$$

$$ClCCo_3(CO)_9 + 2AlCl_3 \longrightarrow [(OC)_9Co_3CCO]^+ [AlCl_4]^- \cdot AlCl_3$$

$$[(OC)_9Co_3CCO]^+ [AlCl_4]^- \cdot AlCl_3 + (CH_3)_3COH \longrightarrow$$
$$(CH_3)_3CO_2CCCo_3(CO)_9 + 2AlCl_3 + Cl^-$$

Submitted by CYNTHIA L. NIVERT,* GARY H. WILLIAMS,* and DIETMAR SEYFERTH*
Checked by STEVEN H. STRAUSS†

μ_3-(Chloromethylidyne)-tris(tricarbonylcobalt) finds useful application in the preparation of other derivatives of the $CCo_3(CO)_9$ cluster. Its reactions with aluminum chloride in the presence of diverse arenes give μ_3-benzylidyne-tris-(tricarbonylcobalt) complexes, $ArCCo_3(CO)_9$, generally in good yield.[1] On the other hand, its reaction with aluminum chloride under the conditions described in this synthesis provides an alternate route to the decacarbonyl-*tetrahedro*-carbon-tricobalt cation.[2] Treatment of $ClCCo_3(CO)_9$ with an alkali metal gives the esr-observable radical anion $[ClCCo_3(CO)_9]^{\overline{\cdot}},$[3] and its reaction with triphenylarsine results in formation of the dimer $(OC)_9Co_3C-CCo_3(CO)_9$.[4] Reactions of $ClCCo_3(CO)_9$ with aryl Grignard reagents in large excess produce the respective μ_3-benzylidyne-tris(tricarbonylcobalt) complexes.[5] (This reaction, however, proceeds better with $BrCCo_3(CO)_9$).

A. μ_3-(CHLOROMETHYLIDYNE)-TRIS(TRICARBONYLCOBALT)

Procedure

■ **Caution.** *The cautionary remarks made with respect to the synthesis of* $HCCo_3(CO)_9$ *apply here as well.*

A 2-L, three-necked, round-bottomed flask is equipped with a nitrogen inlet tube connected to a T-piece that leads to a prepurified nitrogen supply and to a Nujol bubbler, a solids-transfer tube, and an adapter tube holding a stirrer sleeve and shaft that is connected to an overhead stirring motor. The apparatus is evacuated, flamed out, and flushed with dry nitrogen. Octacarbonyldicobalt

---
*Department of Chemistry, Massachusetts Institute of Technology, Cambridge, MA 02139.
†Department of Chemistry, Northwestern University, Evanston, IL 60201.

(60.0 g, 0.175 mole) then is added under nitrogen by means of the solids-transfer tube. The latter is replaced by a rubber serum cap, and 600 mL of reagent grade carbon tetrachloride, used without further purification, is added by means of a syringe. The rubber serum cap is replaced by a thermometer. The reaction mixture is stirred under nitrogen and heated to 50°. This temperature is maintained (but not exceeded) until evolution of carbon monoxide ceases (within 4 hr). During this time the color of the mixture changes from yellow-brown to red-purple. The mixture is allowed to cool to room temperature and then is filtered, in air, through Celite (diatomaceous earth). The solvent is evaporated at reduced pressure (rotary evaporator). The resulting residue is dissolved in 1500 mL of hexane and the solution thus formed is washed with equal volumes of 10% hydrochloric acid and water until the aqueous layer is colorless. The hexane layer is dried over anhydrous magnesium sulfate and evaporated to dryness at reduced pressure. The solid that remains is sublimed at 45° and 0.1 torr to give 23.5 g (42%, based on cobalt charged) of purple, crystalline $ClCCo_3(CO)_9$, mp 131–133°. (■ *Caution. The vacuum must be released by admitting argon or nitrogen, since the sublimation residue is pyrophoric. The sublimation residue is rendered harmless by treatment with water.*) *Anal.* Calcd. for $C_{10}O_9ClCo_3$: C, 25.21. Found: C, 25.15.

Properties

μ_3-(Chloromethylidyne)-tris(tricarbonylcobalt) is an air-stable, purple, crystalline solid, mp 131–133°.[6] It is soluble in common organic solvents to give deep-purple solutions that are stable in air. Its infrared spectrum (in CCl_4) shows bands at 900 (s) and 595 (s) cm^{-1}, in addition to terminal carbonyl absorptions at 2112 (m), 2065 (vs), 2045 (s), 2020 (w), and 1985 (vw) cm^{-1}.

B. μ_3-[(*tert*-BUTOXYCARBONYL)METHYLIDYNE]- TRIS(TRICARBONYLCOBALT)

Procedure

A flame-dried 100-mL, three-necked, round-bottomed flask is equipped with a nitrogen inlet tube connected to a T-piece that leads to a prepurified nitrogen supply and to a Nujol bubbler, a rubber serum cap, and a magnetic stirring bar. It then is evacuated, flamed out again, and flushed with dry nitrogen. μ_3-(Chloromethylidyne)-tris(tricarbonylcobalt) (2.00 g, 4.4 mmole) and 50 mL of reagent grade dichloromethane are added. To the resulting purple solution is added, using a transfer vial, 1.75 g (13.2 mmole) of reagent grade aluminum chloride (used without further purification) as rapidly as possible, since it is hygroscopic. The mixture is stirred at room temperature under nitrogen for 30 minutes, during which time the initially purple solution becomes yellow-brown. The yellow-brown solution usually appears to contain some undissolved solids. However, control experiments have established that most, if not all, of the active $[(OC)_9Co_3CCO]^+$ reagent is in solution. Such solutions are stable at room temperature in the absence of moisture for at least 3 days. The insolubles very likely are derived for the most part from the aluminum chloride. The formation of the tricobaltcarbon decacarbonyl cation may be monitored by thin layer chromatography [Eastman Chromagram Sheet No. 6061 (silica gel); hexane eluent]. Upon completed reagent cation formation, tlc shows a brown spot of zero R_f and complete disappearance of the purple spot of high R_f, $ClCCo_3(CO)_9$. The yellow-brown solution is very sensitive to moisture and must be kept under a dry nitrogen atmosphere at all times.

At this point the conversion of μ_3-(chloromethylidyne)-tris-(tricarbonylcobalt) is complete, and its reactions with a wide variety of nucleophiles, such as alcohols, thiols, amines, alkylating agents, and the more nucleophilic aromatic compounds, may be carried out. The reaction with *tert*-butanol is described here.

To the solution of $[(OC)_9Co_3CCO]^+$ $[AlCl_4]^- \cdot AlCl_3$, thus prepared, is added by syringe 5 mL (53 mmole) of *tert*-butyl alcohol that has been purified by distillation from calcium hydride. The reaction is complete within 5 minutes, as indicated by a color change from yellow-brown to purple. [Tlc shows a purple spot of high R_f (conditions as above) when dichloromethane is used as eluent.] The reaction mixture is poured in air into 200 mL of cold 10% hydrochloric acid. The dichloromethane layer is extracted several times with 100-mL portions of 10% hydrochloric acid. The organic layer is washed with water, dried over a mixture of anhydrous calcium chloride and magnesium sulfate, and evaporated at reduced pressure to leave the crude product as a solid. (The calcium chloride removes any *tert*-butyl alcohol that remains.) The solid product thus obtained is dissolved in 5 mL of hexane and poured onto the top of a silicic acid pad (4 in. deep). Elution with hexane yields a small red-brown band that may be discarded. Elution with 70:30 v/v hexane/dichloromethane and evaporation of the solvent yields the desired product. Finally, sublimation at 50° at 0.1 torr gives 1.67 g (70%) of pure, purple μ_3-[(*tert*-butoxycarbonyl)methylidyne]-tris(tricarbonylcobalt), mp 48-49°.[7] (The yield is based on the $ClCCo_3(CO)_9$ charged.) *Anal.* Calcd. for $C_{15}H_9O_{11}Co_3$: C, 33.24; H, 1.67. Found: C, 33.01; H, 1.54.

Properties

μ_3-[(*tert*-Butoxycarbonyl)methylidyne]-tris(tricarbonylcobalt) is air stable and is best stored in a refrigerator. Its infrared spectrum (in CCl_4) shows terminal carbonyl absorptions at 2111 (m), 2065 (vs), 2045 (s), 2020 (w), and 1985 (vw) cm^{-1}, and the ester carbonyl absorption at 1675 cm^{-1}. Other bands in its IR spectrum include: 2990 (w), 2940 (w), 1460 (w), 1398 (w), 1373 (m), 1265 (sh), 1250 (m), 1163 (s), 1061 (m), and 1037 (m) cm^{-1}. Its nmr spectrum (in CDCl$_3$) shows a singlet at δ 1.58 ppm downfield from internal tetramethylsilane.

References

1. R. Dolby and B. H. Robinson, *J. Chem. Soc., Dalton Trans.*, 1972, 2046.
2. D. Seyferth and G. H. Williams, *J. Organomet. Chem.*, 38, C11 (1972).
3. T. W. Matheson, B. M. Peake, B. H. Robinson, J. Simpson, and D. J. Watson, *Chem. Commun.*, 1973, 894.
4. T. W. Matheson and B. H. Robinson, *J. Chem. Soc. (A)*, 1971, 1457.
5. R. Dolby and B. H. Robinson, *J. Chem. Soc., Dalton Trans.*, 1973, 1794.
6. R. Ercoli, E. Santambrogio, and G. T. Casagrande, *Chim. Ind. (Milano)*, 44, 1344 (1962).
7. D. Seyferth, J. E. Hallgren, and P. L. K. Hung, *J. Organomet. Chem.*, 50, 265 (1973).

56. TETRACARBONYLBIS-µ-(2-METHYL-2-PROPANE-THIOLATO)-DIIRIDIUM(I)

$$\{IrCl(C_8H_{12})\}_2 + 2Li(SC(CH_3)_3) \longrightarrow \{Ir[SC(CH_3)_3](C_8H_{12})\}_2 + 2LiCl$$
$$\{Ir[SC(CH_3)_3](C_8H_{12})\}_2 + 4CO \longrightarrow \{Ir[SC(CH_3)_3](CO_2\}_2 + 2C_8H_{12}$$

Submitted by D. de MONTAUZON* and R. POILBLANC*
Checked by R. H. BALLAN†, C. S. CHIN†, and L. VASKA†

A material suitable for studies of the fundamental reactions of dinuclear complexes has been sought as a companion to Vaska's complex [Ir(CO)(PA$_3$)$_2$X]. This newly described carbonyl compound allows synthesis of several series of polynuclear compounds of iridium in formal oxidation states +I, +II, +III and plays the role of a typical dinuclear species.

Procedure

■ **Caution.** *Carbon monoxide is a highly toxic, colorless, and odorless gas and the reaction should be performed only in an efficient fume hood. 2-Methyl-2-propanethiol is highly volatile and toxic and has a very obnoxious odor. Upon contact with air butyllithium is spontaneously combustible. All reactions and operations must be carried out under nitrogen.*

---

*Laboratoire de Chimie de Coordination du CNRS, B.P. 4142, 31030 Toulouse Cedex, France.
†Department of Chemistry, Clarkson College of Technology, Potsdam, NY 13676.

Di-μ-chloro-bis(1,5-cyclooctadiene)diiridium(I), $[IrCl(C_8H_{12})]_2$, has been prepared in very good yield by previously reported methods[1] from H_2IrCl_6 or $IrCl_3 \cdot 3H_2O$.*

To a solution of 6.71 g (10 mmole) of $[IrCl(C_8H_{12})]_2$ in 100 mL of toluene is added 20 mmole of $t\text{-}C_4H_9SLi$ (obtained by mixing of 8.60 mL 2.3 M butyllithium[†] and 2.30 mL of 2-methyl-2-propanethiol[†]) in 60 mL of toluene-diethyl ether (1:1) mixture. The solution, which becomes instantaneously violet, is stirred magnetically for 2 hours. The solvent is evaporated under reduced pressure. The solid residue is redissolved in 150 mL of pentane and filtered in an inert atmosphere.[‡] $\{Ir(SC(CH_3)_3)(C_8H_{12})\}_2$ can be recrystallized from the concentrated solution.[§]

Carbon monoxide is bubbled through the pentane solution of $\{Ir[SC(CH_3)_3]\text{-}(C_8H_{12})\}_2$ for 15 minutes with stirring. The initially violet solution becomes red-brown. After concentrating to 20 mL, crystallization occurs at $-40°$, giving black crystals. Red crystals can also be obtained under mild conditions ($-20°$) from dilute solutions. Concentration of the mother liquor to 5 mL yields additional product. Overall yield: 5.70 g (85% based on $[IrCl(C_8H_{12})]_2$). *Anal.* Calcd. for $C_{12}H_{18}S_2O_4Ir_2$: C, 21.36; H, 2.67; S, 9.50. Found: C, 21.30; H, 2.80; S, 9.20. *MW by tonometry in C_6H_6.* Found: 686. Calcd.: 675. mp 127°.

Properties

Tetracarbonyl bis-μ-(2-methyl-2-propanethiolato)diiridium(I) is stable to air and moisture and is soluble in many organic solvents. The infrared spectra of the black and red species in hexadecane solution exhibit three CO stretching bands [2061 (s), 2040 (vs), 1986 (vs) cm^{-1}] analogous to those of the previously reported rhodium complexes.[3] The analogous compound $[Ir(\mu\text{-}SC_6H_5)(CO)_2]_2$ can be obtained in the same way (with $LiSC_6H_5$) in high yield (70%) [ν_{CO} = 2072 (m), 2052 (vs), 2003 (vs) cm^{-1}].[4] Moreover, the analogous compound $[Ir(\mu\text{-}SCF_3)(CO)_2]_2$ [ν_{CO} = 2096 (m), 2076 (vs), 2029 (vs) in hexadecane solution] and the tetrahedral carbonyl $\{Ir[\mu\text{-}P(C_6H_5)_2](CO)_2\}_2$ [ν_{CO} = 2045 (s), 1985 cm^{-1} in hexadecane solution] can also be prepared.[5] The crystal

---

*H_2IrCl_6, K and K, Rare and Fine Chemicals, 26201 Miles Road, Cleveland, OH 44128. $IrCl_3 \cdot 3H_2O$, Alfa Division, Ventron Corp., 16207 South Carmenita Road, Cerrites, CA 90701. Available in Europe from: "Compagnie des Métaux Précieux" 74 bd Pierre Vaillant Couturier 94 IVRY, France.

[†]Butyllithium, Aldrich Chemical Co., 940 W. Saint Paul Avenue, Milwaukee, WI 53223. *tert*-Butyl mercaptan, Aldrich Chemical Co. Available in Europe from: Fluka AG., Chemische Fabrik, CH-9470 Buchs, Switzerland.

[‡]Various types of glass apparatus designed for the purpose of filtering in an inert atmosphere can be used.[2] The authors used the apparatus shown in Fig. 1.

[§]Crystallization at $-40°$ from pentane gives 7.0 g (90%) of $\{Ir[SC(CH_3)_3](C_8H_{12})\}_2$, mp 189° *Anal.* Calcd. for $C_{24}H_{42}S_2Ir_2$: C, 37.01; H, 5.39; S, 8.23. Found: C, 36.95; H, 5.40; S, 8.35. *MW by tonometry in C_6H_6.* Found: 800. Calcd.: 779.

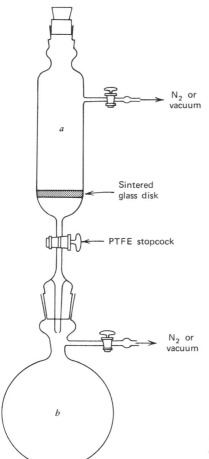

Fig. 1. Filtration apparatus: (a) 250-mL dropping funnel with glass frit, (b) 250-mL, round-bottomed flask.

structure of the black form in the analogous compound $[Ir(\mu\text{-}SC_6H_5)(CO)_2]_2$ suggests the existence of intra- and intermolecular metal–metal interactions,[4] whereas in the red form only the independent dinuclear unit exists.[5]

This complex does not react with H_2, CO, mineral acids (HCl, $HClO_4$), or amines.[5] It forms adducts with alkynes (C_4F_6[6]) and phosphines[4]; these adducts are examples of trinuclear crownlike complexes. Addition of 2 moles of a tertiary phosphine ligand L to $[Ir(\mu\text{-}S\text{-}tert\text{-}C_4H_9)(CO)_2]_2$ affords two types of complexes: the crown-like $[Ir(\mu\text{-}S\text{-}tert\text{-}C_4H_9)]_3(CO)_5L_2$ [L = $P(CH_3)_3$][6] and the disubstituted derivatives $[Ir(\mu\text{-}S\text{-}tert\text{-}C_4H_9)(CO)L]_2$.[4] Interestingly, these last disubstituted compounds react with H_2,[7] tetracyanoethylene, O_2, and SO_2[5] and can be referred to as "dinuclear Vaska's complexes."

References

1. J. L. Herde, J. C. Lambert, and C. V. Senoff, *Inorg. Synth.*, **15**, 18 (1974).
   R. H. Crabtree and G. E. Morris, *J. Organomet. Chem.*, **135**, 395 (1977).
   G. Pannetier, R. Bonnaire, and P. Fougeroux, *J. Less-Common. Met.*, **21**, 437 (1970).
2. R. B. King, *Organometallic Syntheses*, Vol. 1, J. J. Eisch and R. B. King, Ed., Academic, New York, 1965, p. 55.
   W. L. Jolly, *Inorg. Synth.*, **11**, 116 (1968).
3. P. Kalck and R. Poilblanc, *Inorg. Chem.*, **14**, 2779 (1975).
4. J. J. Bonnet, J. Galy, D. de Montauzon, and R. Poilblanc, *J. Chem. Soc. Chem. Commun.*, **1977**, 47.
5. R. Poilblanc, J. J. Bonnet, A. Maisonnat, D. de Montauzon and A. Thorez, unpublished results.
6. R. Poilblanc, *Nouv. J. Chim.*, **2**, 145 (1978) and references therein.
7. A Thorez, A. Maisonnat, and R. Poilblanc, *Chem. Commun.*, **1977**, 518. J. J. Bonnet, A. Thorez, A. Maisonnat, J. Galz and R. Poilblanc, *J. Am. Chem. Soc.*, **101**, 5940 (1979).

57. DECACARBONYL-μ-[DIBORANE(6)]-μ-HYDRIDO-TRIMANGANESE

$$Mn_2(CO)_{10} + Na[BH_4] \xrightarrow[\Delta]{THF} \text{red anions} \xrightarrow[\text{cyclohexane}]{H_3PO_4} Mn_3(B_2H_6)(H)(CO)_{10} + \text{other products}$$

Submitted by H. D. KAESZ* and S. W. KIRTLEY†
Checked by DONALD F. GAINES‡ and MARK B. FISCHER‡

Considerable variation is encountered in the hydridometal carbonyl cluster chemistry of manganese and rhenium. While treatment of $Re_2(CO)_{10}$ with $Na[BH_4]$ and subsequent acidification yields $Re_3H_3(CO)_{12}$ as the principal product,[1] similar treatment of $Mn_2(CO)_{10}$ yields principally the title compound[2] whose rhenium analogue has not been observed. The trimer $Mn_3H_3(CO)_{12}$ is also obtained in this preparation, but in low yields. A good method for obtaining $Mn_3H_3(CO)_{12}$, using KOH/EtOH treatment of $Mn_2(CO)_{10}$ followed by acidification, has been reported in a previous volume of this series.[3]

---

*Department of Chemistry, University of California, Los Angeles, CA 90024.
†Department of Chemistry, Smith College, Northampton, MA.
‡Department of Chemistry, University of Wisconsin–Madison, Madison, WI 53706.

Procedure

■ **Caution.** *Poisonous carbon monoxide may be evolved in this reaction and must be properly vented. All metal carbonyls should also be considered to be toxic compounds.*

All operations should be carried out under an atmosphere of nitrogen unless otherwise noted. Manganese carbonyl can be purchased from Strem Chemicals (Beverly, MA) or Pressure Chemicals (Pittsburgh, PA). Tetrahydrofuran (THF) and cyclohexane solvents should be freshly distilled under nitrogen from CaH_2. Phosphoric acid is stripped of dissolved oxygen by bubbling purified nitrogen through it by way of a glass frit for a period of 30 minutes.

Dimanganese decacarbonyl (5.0 g, 12.8 mmole) is placed in a dry three-necked, 250-mL flask equipped with a reflux condenser, an N_2-inlet and a powder addition tube. The dimanganese decacarbonyl is dissolved in 50 mL of THF, and sodium tetrahydroborate(1-) (1.0 g, 26.5 mmole, previously vacuum dried for 24 hr) is slowly added through the powder addition tube. The mixture is heated at reflux for 6 hours. Under this treatment, the color of the solution quickly changes from yellow to dark red. Excess sodium tetrahydroborate(1-) and other insoluble materials are removed by filtration and the THF is evaporated under vacuum for 2 hours to produce a dark-red tar. Cyclohexane (50 mL) and oxygen-free H_3PO_4 (10 mL, 85%) are added and the mixture is stirred vigorously for 1 hour at room temperature. The cyclohexane layer is transferred with a syringe into a 250-mL Schlenk flask and the acid layer is extracted with two additional 50-mL portions of cyclohexane. The collected cyclohexane layers are evaporated under vacuum to produce a brick-red powder. A water-cooled cold finger is fitted into the flask and the solid is pumped at 25° and 0.1 torr for 24 hours to sublime away the yellow $Mn_2(CO)_{10}$ and light-red $Mn_3H_3(CO)_{12}$ impurities.[3] The resultant dark-red powder is pure but can be crystallized from a mixture of methylcyclohexane and cyclohexane (10:90). Yield of powder is about 1 g (15-25%).

Anal. Calcd. for $H_7Mn_3B_2(CO)_{10}$: Mn, 34.80; C, 25.36; H, 1.49; B, 4.57; MW 473.6. Found[4]: Mn, 34.60; C, 25.51; H, 1.41; B, 4.69; MW 440 (vapor pressure osmometer, in cyclohexane solution under argon).

Properties

The compound exists as a red powder that is moderately stable in air, but it should be stored under nitrogen if it is to be kept for prolonged periods. The crystals take the shape of dark-red needles and are moderately stable in air. The compound is soluble in organic solvents but insoluble in water. The structure has been determined by single crystal X-ray diffraction (see Fig. 1.)[2b] Infrared spectral bands in the carbonyl region appear at 1959.0 (w), 1964.8 (m), 1979.5 (s),

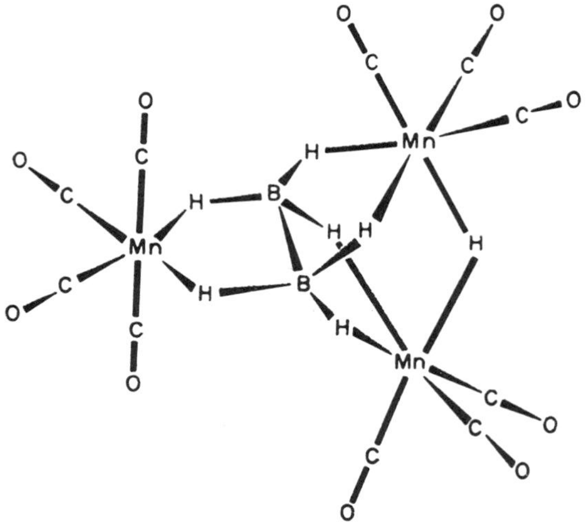

Fig. 1. Molecular configuration of $Mn_3(B_2H_6)(H)(CO)_{10}$. Reprinted with permission from H. D. Kaesz, W. Fellmann, G. R. Wilkes, and L. F. Dahl, J. Am. Chem. Soc., **89**, 2753 (1965). Copyright 1965 by the American Chemical Society.

2000.5 (s), 2031.7 (s), 2045.3 (s), 2060.1 (s), and 2116.7 (m) cm^{-1}. The lower energy regions show maxima at 688 (m), 675 (m), 662 (s), 629 (s), 608 (m), 536 (m), 504 (m), 492 (m), 464 (m), 442 (m), and 416 (m) cm^{-1}. The only other infrared absorptions in the regions scanned are two weak and broad bands at 995 and 935 cm^{-1}. The corresponding perdeutero compound has been prepared and its spectrum has been reported.[2] Magnetic resonances[5] at 270 MHz, ^{1}H nmr in C_6D_6 versus tetramethylsilane: two broad peaks of relative areas 1:2 at δ 10.53 ppm and δ 12.65 ppm, respectively, and a sharp spike at δ 17.20 ppm; at 86.7 MHz,[11] B nmr in C_6D_6 versus $BF_3 \cdot O(C_2H_5)_2$, one broad peak at -40.91 ppm.

References and Notes

1. D. K. Huggins, W. Fellmann, J. M. Smith, and H. D. Kaesz, J. Am. Chem. Soc., **86**, 4841 (1964).
2. (a) W. Fellmann, D. K. Huggins, and H. D. Kaesz, Proceedings of the VIIIth International Conference on Coordination Chemistry, V. Gutmann, Ed., Springer-Verlag, Berlin, 1964, pp. 225-257. (b) H. D. Kaesz, W. Fellmann, G. R. Wilkes, and L. F. Dahl, J. Am. Chem. Soc., **89**, 2753 (1965).
3. B. F. G. Johnson, R. D. Johnston, J. Lewis, and B. H. Robinson, Inorg. Synth., **12**, 43 (1970).
4. Schwartzkopf Microanalytical Laboratory, Woodside, NY.
5. The authors gratefully acknowledge the checkers for these data.

CORRECTION

BARIUM TETRACYANOPLATINATE(II) TETRAHYDRATE*

Submitted by SARAH KURTZ,[†] LEAH E. TRUITT,[†]
RICHARD C. SUNDELL,[†] and JACK M. WILLIAMS[‡]

With slight modification the synthesis of $Ba[Pt(CN)_4] \cdot 4H_2O$[1] is improved considerably by replacing the last paragraph with the text given below.

The murky solution is heated to boiling with stirring and immediately filtered (medium-grit glass filter). The filtrate is transferred to a 1000-mL beaker, then cooled to room temperature, and finally placed in an ice-bath overnight. The crystals are collected by filtration and the filtrate is disposed of in a waste-platinum receptacle (or treated as below). The material is purified by adding 3 mL of H_2O per gram of solid and heating with stirring in a boiling water bath until dissolution of the yellow crystals is complete. If any other solid is present at this point, the solution should be filtered hot. The $Ba[Pt(CN)_4] \cdot 4H_2O$ is then allowed to recrystallize slowly by cooling as above. The recrystallization is repeated until the desired purity is achieved. A measure of the quantity of barium hydroxide and barium cyanide removed during each step can be determined by measuring the solution pH. When the filtrate reaches pH 7, the product is quite pure. Two recrystallizations are usually adequate. A second collection of crystals as previously recommended[1] may not be advantageous

---

*Research performed under the auspices of the Office of Basic Energy Sciences of the U.S. Department of Energy.

[†]Research participants sponsored by the Argonne Center for Educational Affairs; from Manchester College, North Manchester, IN, Washington College, Chestertown, MD, and Bethany College, Lindsborg, KA, respectively.

[‡]Chemistry Division, Argonne National Laboratory, Argonne, IL 66045.

because the additional amount of product collected will be small, in addition to being contaminated with barium hydroxide and barium cyanide. The solubility of $Ba[Pt(CN)_4] \cdot 4H_2O$ is approximately 2 g/100 mL H_2O at 3°.

References

1. R. L. Maffly and J. M. Williams, *Inorg. Synth.*, **19**, 112 (1979).

INDEX OF CONTRIBUTORS

Adler, A. D., 20:97
Angelici, R. J., 20:188
Anton, D. L., 20:134
Arcus, C., 20:218
Aresta, M., 20:69
Armington, A. F., 20:1, 9
Arnáiz Garciá, F. J., 20:82

Bailar, J. C., Jr., 20:18
Ballan, R. H., 20:237
Barefield, E. K., 20:108
Bartlett, N., 20:34
Behrman, E. C., 20:50
Besinger, R., 20:20
Bianco, V. D., 20:206
Bonnemann, H., 20:206
Bulkowski, J., 20:127
Bump, C. M., 20:161
Busse, P. J., 20:181

Chalilpoyil, P., 20:101
Chang, C. K., 20:147
Chen, G. J. J., 20:119
Chin, C. S., 20:237
Chini, P., 20:209, 212, 215
Clark, S. F., 20:57
Coffey, C., 20:25
Coleman, W. M., 20:127
Cuellar, E. A., 20:97
Cushing, M. A., Jr., 20:76, 81, 90, 143, 155
Cutler, A., 20:127

Dabrowiak, J., 20:115
Darby, D., 20:53
Darst, K. P., 20:200
Dawson, P., 20:226, 230
Del Donno, T. A., 20:48, 85, 106, 109, 111
de Montauzon, D., 20:237
Desmarteau, D. D., 20:36, 38
Dilts, J. A., 20:53
Dilworth, J. R., 20:119

Di Nello, R. K., 20:147
Dolphin, D., 20:134, 143, 147, 155
Doronzo, S., 20:206
Dose, E. V., 20:87
Doyle, J., 20:181
Drake, J. E., 20:171, 176
Durand, B., 20:50

Ellis, G. D., 20:82
English, D. R., 20:106, 108, 109, 111

Fackler, J. P., Jr., 20:185
Fenton, D. E., 20:90
Fernando, Q., 20:53
Fischer, M. B., 20:240
Flood, M., 20:61
Freeman, G., 20:108

Gaines, D. F., 20:240
Gallo, N., 20:206
Gayda, S. E., 20:90
Gerrity, D. P., 20:20, 28
Gillard, R. D., 20:57
Giordano, G., 20:209
Giordano, R. S., 20:97
Gladysz, J. A., 20:200, 204
Glavinčevski, B. M., 20:171, 176
Glicksman, H. D., 20:46
Goedken, V. L., 20:87
Greene, B., 20:181

Haight, G. P., Jr., 20:63
Halko, D. J., 20:134
Hall, J. R., 20:185
Hallgren, J. S., 20:224, 226, 230
Harrowfield, J. MacB., 20:85
Hemmings, R. T., 20:171, 176
Henderson, H. E., 20:171, 176
Herlt, A. J., 20:85
Hirons, D. A., 20:185
Hobbs, D. T., 20:204

Index of Contributors

Hodali, H. A., 20:218, 222
Hogenkamp, H. P. C., 20:134
Howe-Grant, M., 20:101
Huang, A., 20:50
Huber, W., 20:11
Hullinger, F., 20:11
Hung, Y., 106, 111

Ittel, S. D., 20:76, 81, 90, 143, 155

Jesson, J. P., 20:76
Johnson, E. C., 20:143
Jones, D. P., 20:218, 222
Jordan, J., 20:161

Kaesz, H. D., 20:240
Karsch, H. H., 20:69
Kawaguchi, S., 20:65
Kerr, G. T., 20:48
Kirchner, S. J., 20:53
Kirtley, S. W., 20:240
Krumpolc, M., 20:63
Krusic, P. J., 20:218, 222
Kuhlmann, E., 20:65
Kurtz, S., 20:243

Lawlor, L., 20:33
Lee, G. C., 20:29
Lee, S. R., 20:11
Lever, A. B. P., 20:147
Lien, C. A., 20:11
Lincoln, R., 20:41, 44, 46
Lippard, S. J., 20:101
Luckehart, C. M., 20:200, 204

McDonald, J. W., 20:119
Mackay, K. M., 20:171
Malin, J. M., 20:61
Marks, T. J., 20:97
Martinengo, S., 20:209, 212, 215
Marzilli, L. G., 20:101
Merola, J. S., 20:224, 230
Merrill, P., 20:63
Mertes, K. B., 20:20, 24, 25, 26, 28, 29
Michalczyk, M. J., 20:26

Nestle, M. O., 20:224, 226
Nivert, C. L., 20:224, 234

O'Connor, J. J., 20:1, 9

Okeya, S., 20:65
Orchin, M., 20:65, 181

Pan, W. H., 20:185
Pâris, J. M., 20:50
Parshall, G. W., 20:209, 212, 215
Passmore, J., 20:33
Pedrosa De Jesus, J., 20:57
Petersen, J. D., 20:57
Place, D., 20:115
Poilblanc, R., 20:237
Preble, J. C., 20:11

Quick, M. H., 20:188

Ramaswamy, B. S., 20:147
Raymond, K. N., 20:109
Regan, C. M., 20:90
Reimer, K. J., 20:188
Richards, R. L., 20:19
Robinson, B. H., 20:226, 230
Roček, J., 20:63

Sams, J. R., 20:155
Samson, M., 20:206
Sargeson, A. M., 20:85
Schram, E. P., 20:82
Schweizer, A. E., 20:48
Segal, J. A., 20:196
Selover, J. C., 20:200, 204
Seppelt, K., 20:33, 34, 36, 38
Seyferth, D., 20:224, 226, 230, 234
Shaver, A., 20:188
Sheridan, P. S., 20:57
Shriver, D. F., 20:218, 222
Silverman, R. B., 20:127, 134
Silverthorn, W. E., 20:196
Smith, W. L., 20:109
Sorrell, T. N., 20:161
Stearley, K. L., 20:24, 29
Stojakovic, D. R., 20:97
Strauss, S. H., 20:234
Strumolo, D., 20:212
Sundell, R. C., 20:243
Swile, G. A., 20:185

Tanzella, F., 20:34
Taylor, L. T., 20:127
Torrence, G. P., 20:204
Truitt, L. E., 20:243

Tsin, T. B., 20:155

Van Dyke, C. H., 20:176
Vaska, L., 20:237
Verkade, J. G., 20:76, 81
Vetsch, H., 20:11
Vidusek, D. A., 20:20, 26
Viswanathan, N., 20:176

Walton, R. A., 20:41, 44, 46
Warfield, L. T., 20:200
Whitmore, A., 20:61

Wilkinson, G., 20:41, 44, 46
Williams, G. H., 20:224, 234
Williams, J. M., 20:20, 24-26, 28, 29, 243
Wold, A., 20:9, 11, 18, 50
Wonchoba, E. R., 20:209, 212, 215
Wood, T. E., 20:41, 44

Yarbrough, L. W., II, 20:76, 81

Zahray, R., 20:181
Zeile, J. V., 20:200, 204

SUBJECT INDEX

Names used in this Subject Index for Volume XX, as well as in the text, are based for the most part upon the "Definitive Rules for Nomenclature of Inorganic Chemistry," 1957 Report of the Commission on the Nomenclature of Inorganic Chemistry of the International Union of Pure and Applied Chemistry, Butterworths Scientific Publications, London, 1959; American version, *J. Am. Chem. Soc.*, 82, 5523-5544 (1960); and the latest revisions [Second Edition (1970) of the Definitive Rules for Nomenclature of Inorganic Chemistry]; also on the Tentative Rules of Organic Chemistry–Section D; and "The Nomenclature of Boron Compounds" [Committee on Inorganic Nomenclature, Division of Inorganic Chemistry, American Chemical Society, published in *Inorganic Chemistry*, 7, 1945 (1968) as tentative rules following approval by the Council of the ACS]. All of these rules have been approved by the ACS Committee on Nomenclature. Conformity with approved organic usage is also one of the aims of the nomenclature used here.

In line, to some extent, with *Chemical Abstracts* practice, more or less inverted forms are used for many entries, with the substituents or ligands given in alphabetical order (even though they may not be in the text); for example, derivatives of arsine, phosphine, silane, germane, and the like; organic compounds; metal alkyls, aryls, 1,3-diketone and other derivatives and relatively simple specific coordination complexes: *Iron, cyclopentadienyl- (also as Ferrocene); Cobalt(II), bis(2,4-pentanedionato)-* [instead of *Cobalt (II) acetylacetonate*]. In this way, or by the use of formulas, many entries beginning with numerical prefixes are avoided; thus *Vanadate (III), tetrachloro-*. Numerical and some other prefixes are also avoided by restricting entries to group headings where possible: *Sulfur imides*, with formulas; *Molybdenum carbonyl*, $Mo(CO)_6$; both *Perxenate*, $HXeO_6^{3-}$, and *Xenate(VIII)*, $HXeO_6^{3-}$. In cases where the cation (or anion) is of little or no significance in comparison with the emphasis given to the anion (or cation), one ion has been omitted; e.g., also with less well-known complex anions (or cations): $CsB_{10}H_{12}CH$ is entered only as *Carbaundecaborate*$(1-)$, *tridecahydro-* (and as $B_{10}CH_{13}^-$ in the Formula Index).

Under general headings such as *Cobalt(III) complexes* and *Ammines,* used for grouping coordination complexes of similar types having names considered unsuitable for individual headings, formulas or names of specific compounds are not usually given. Hence it is imperative to consult the Formula Index for entries for specific complexes.

As in *Chemical Abstracts* indexes, headings that are phrases are alphabetized straight through, letter by letter, not word by word, whereas inverted headings are alphabetized first as far as the comma and then by the inverted part of the name. Stock Roman numerals and Ewens-Bassett Arabic numbers with charges are ignored in alphabetizing unless two or more names are otherwise the same. Footnotes are indicated by *n,* following the page number.

Subject Index

Acetaldehyde:
rhenium complexes, 20:201-205
Acetamide, N-methyl cobalt complex, 20:230, 232
Acetato(triphenylphosphine) complexes of platinum metals, 17:124-132
Acetic acid, ethylenediamine-N,N'-di-, cobalt complexes, 18:100, 103
—, trifluoro-:
osmium complexes, 17:131
rhenium complexes, 17:127
Acetonitrile:
copper complex, 19:90
iron(II) complex, 18:6, 15
molybdenum complex, 20:120
Acetylene complexes:
of nickel(O), 17:117-124
Alkanes, cyclo-, see Cycloalkanes
β-Aluminas:
Al, Ag, Ga, K, Li, NH$_4$, NO, Rb, and Tl, 19:51-58
see also Aluminum oxide
Aluminate(1−), dihydridobis(2-methoxyethoxy)-:
sodium, 18:149
—, tetrachloro-:
pentathiazyl, 17:190
Aluminum, bis(diethylamino)hydrido-, 17:41
—, (diethylamino)dihydro-, 17:40
—, hexa-μ-acetyl-tris(tetracarbonylmanganese)-, 18:56, 58
—, trihydrido(trimethylamine)-, 17:37
—, triiodotris(pyridine)-, 20:83
—, tris[cis-(diacetyltetracarbonylmanganese)]-, 18:56, 58
Aluminum ammonium oxide (Al$_{11}$(NH$_4$)O$_{17}$), 19:56
Aluminum compounds, 18:145
Aluminum gallium oxide (Al$_{11}$GaO$_{17}$), 19:56
Aluminum hydrido complexes, 17:36-42
Aluminum lithium oxide (Al$_{11}$LiO$_{17}$), 19:54
Aluminum nitrosyl oxide (Al$_{11}$(NO)O$_{17}$), 19:57
Aluminum potassium oxide (Al$_{11}$KO$_{17}$), 19:55
Aluminum rubidium oxide (Al$_{11}$RbO$_{17}$), 19:55

Aluminum thallium oxide (Al$_{11}$TlO$_{17}$), 19:53
Aluminum trihydride, 17:6
Amine, tris[2-(diphenylarsino)ethyl]-, 16:177
—, tris[2-(diphenylphosphino)ethyl]-, 16:176
—, tris[2-(methylthio)ethyl]-, 16:177
Aminoboranes, 17:30-36, 159-164
Amine complexes:
of cobalt(III), 16:93-96; 17:152-155
of iridium(III) and rhodium(III), 17:152-155
of osmium, 16:9-12
of ruthenium, 16:13-15, 75
Ammines:
chromium, 18:75
cobalt, 18:67, 75
Ammonia:
compd. with TaS$_2$ (1:1), 19:42
Ammonium, tetrabutyl-:
heptahydrodiborate(1-), 17:25
tetrahydroborate(1-), 17:23
Ammonium, tetraethyl-:
cyanate, 16:132
cyanide, 16:133
Ammonium compounds, quaternary, 17:21-26
Ammonium cyanate, 16:136
Ammonium decavanadate(V) (NH$_4$)$_6$-[V$_{10}$O$_{28}$] hexahydrate, 19:140, 143
Ammonium $catena$-polyphosphate (NH$_4$)$_n$(P$_n$O$_{(3n+1)}$H$_2$), 19:278
Ammonium periodate ((NH$_4$)$_2$H$_3$IO$_6$):
crystal growth of by double infusion, 20:15
[13]aneN$_4$, see 1,4,7,10-Tetraazacyclotridecane
[15]aneN$_4$, see 1,4,8,12-Tetraazacyclopentadecane
[16]aneN$_4$, see 1,5,9,13-Tetraazacyclohexadecane
Aniline, N,N-dimethyl-:
chromium complex, 19:157
molybdenum complex, 19:84
—, N-(ethoxymethyl)-:
platinum complex, 19:175
—, N-[(ethylamino)methyl]-:
platinum complex, 19:176

—, N-methyl-:
 borane deriv., 17:162
—, N,N'-methylenedi-:
 platinum complex, 19:176
—, 2-(methylthio)-, 16:169
Anilinoboranes, 17:162
Anion exchange resin, Bio-Rad AG l-X8,
 in resolution of cobalt edda
 complexes, 18:109
Anisole:
 chromium complex, 19:155
Antimonate, hexafluoro, mercury
 $(Hg_{2.91}[SbF_6])$, 19:26
Antimonate(1-), hexachloro-, pentathiazyl, 17:191
Antimonate(V), μ-fluoro-bis(pentafluoro-, :
 mercury $(Hg_3[Sb_2F_{11}]_2)$, 19:23
Apparatus:
 reaction tube for $Ir(CO)_3Cl_{1.10}$, 19:19
 reactor for metal atom syntheses, 19:59-62
Arsenate, hexafluoro, mercury
 $(Hg_{2.86}[AsF_6])$, 19:25
Arsenate(V), hexafluoro, mercury
 $(Hg_3[AsF_6]_2)$, 19:24
Arsine, see Amine, tris[2-(diphenylarsino)ethyl] –
Arsine, (2-bromophenyl)dimethyl-, 16:185
—, dimethyl(pentafluorophenyl)-, 16:183
—, 1,2-phenylenebis(dimethyl-:
 nickel(O)complex, 17:121
 platinum complex, 19:100
—, (stibylidynetri-1,2-phenylene)tris-(dimethyl-, 16:187
—, triphenyl-:
 copper complex, 19:95
 nickel(O) complex, 17:121
—, tris[2-(dimethylarsino)phenyl]-, 16:186
Arsonium, tetraphenyl-:
 cyanate, 16:134
 cyanide, 16:135
 hexacarbonylniobate(1-), 16:72
 hexacarbonyltantalate(1-), 16:72
Aza-2-boracycloalkanes, 1,3-di-, see
 Cycloalkanes
5-Aza-2,8,9-trioxagermatricyclo[3.3.3.0]-

undecane, 1-ethyl-, see Germanium,
 ethyl{{2,2',2''-nitrilotris[ethanolato]}(3-)-N,O,O',O''}
5-Aza-2,8,9-trioxastannatricyclo[3.3.3.0]-undecane, 1-ethyl-, see Tin, ethyl
 {{2,2',2''-nitrilotris[ethanolato]}-(3-)-N,O,O',O''}-
Azobenzene complexes of nickel(O), 17:121-123

Barium tetracyanoplatinate(II), tetrahydrate (correction), 20:243
Benzaldehyde, O-amino-macrocyclic
 ligands from, 18:31
Benzene:
 chromium complex, 19:157
 molybdenum complexes, 20:196-199
—, 1,3-bis(trifluoromethyl), Cr complex, 19:70
—, 1-bromo-2-(methylthio)-, 16:169
—, chloro-:
 chromium complex, 19:157
 molybdenum, 19:82
—, O-dimethyl, chromium complex, 19:198
—, fluoro-chromium, 19:157
—, (methoxymethyl), tungsten complex, 19:165
—, methylenebis-, see Methane, diphenyl-
—, pentafluoro-:
 nickel complex, 19:72
Benzene-d_5, bromo-, 16:164
1,2-Benzenedicarbonitrile:
 macrocyclic ligands from, 18:47
Benzenethiol:
 iridium complex, 20:238
Benzoic acid:
 molybdenum complex, 19:133
—, methyl ester:
 chromium complex, 19:157
1,5,8,12-Benzotetraazacyclotetradecine,
 3,10-dibromo-1,6,7,12-tetrahydro-:
 copper complex, 18:50
Benzylamine, N,N-dimethyl-, tungsten, 19:169
Benzylideneacetone, 16:104; 17:135
Bicyclo[2.2.1]heptane-7-methanesulfonate, 3-bromo-1,7-dimethyl-2-oxo-, [1R(endo,anti)]-, see (+)-

Camphor-π-sulfonate, α-bromo-
Bicyclo[6.6.6]ane-1,3,6,8,10,13,16,19-N$_8$, see 1,3,6,8,10,13,16,19-Octaazatricyclo[6.6.6]eicosane
2,2'-Bi-1,3-dithiolylidene:
  radical cation:
    bis[2,3-dimercapto-2-butenedinitrilato-(2-)] cuprate(2-) (2:1), 19:31
    bis[2,3-dimercapto-2-butenedinitrilato-(2-)] nickelate(2-) (2:1), 19:31
    bis[2,3-dimercapto-2-butenedinitrilato-(2-)] platinate(1-) (1:1), 19:31
    bis[2,3-dimercapto-2-butenedinitrilato-(2-)] platinate(2-) (2:1), 19:31
    iodide (8:15), 19:31
    iodide (11:8), 19:31
    iodide (24:63), 19:31
    salt with 2,2'-(2,5-cyclohexadiene-1,4-diylidene)bis[propanedinitrile] (1:1), 19:32
    selenocyanate (14:8), 19:31
    tetracyanoplatinate(II) (2:1), 19:31
    tetrafluoroborate(1-) (3:2), 19:30
    thiocyanate (14:8), 19:31
2,2'-Bi-1,3-dithiolylidene (TTF), 19:28
Bio-Rad AG 1-X8:
  anion exchange resin, in resolution of cobalt edda complexes, 18:109
2,2'-Bipyridine:
  molybdenum complex, 19:135
  nickel complex, 17:121
  niobium complex, 17:78
2,2'-Bipyridinium(2+) pentachlorooxomolybdate(V), 19:135
Bis(guanidinium) dibromotetracyanoplatinate(IV):
  hydrate, 19:11
Bis(guanidinium) tetracyanoplatinate(II), 19:11
Bis(trimethylsilyl)amine:
  transition metal complexes, 18:112
2-Boracycloalkanes, 1,3-diaza-, see Cycloalkanes
Borane, bis(dimethylamino)-, 17:30
—, chlorodiethyl-, 17:195
—, tri-sec-butyl-, 17:27
—, tris(diethylamino)-, 17:159
—, tris(N-methylanilino)-, 17:162
—, tris(1-methylpropyl)-, see Borane, tri-sec-butyl-
—, tris[(1-methylpropyl)amino]-, 17:160
Borane(6), di-, see Diborane(6)
Boranes, 18:145
—, trisamino-, 17:159-164
Borate(1), heptahydrodi-:
  methyltriphenylphosphonium, 17:24
  tetrabutylammonium, 17:25
—, tetrafluoro-,
  [[tris[μ-[(2,3-butanedione dioximato)]-(2-)-O;O']-difluorodiborato(2-)]-N,N',N'',N''',N'''',N''''']cobalt(III), 17:140
—, tetrahydro-:
  calcium, 17:17
  purification of, 17:19
  copper complex, 19:96
  methyltriphenylphosphonium, 17:22
  tetrabutylammonium, 17:23
—, tri-sec-butylhydro-
  potassium, 17:26
Borato complexes:
  tetrahydro-:
    of hydridonickel and platinum complexes, 17:88-91
    of titanium, 17:91
  tris(dioximato)di-, of cobalt(III) and iron(II), 17:139-147
8-Bornanesulfonic acid, 3-bromo-2-oxo-(+)-, 16:93; 18:106
Boron:
  compd. with TaS$_2$ (1:2), 19:42
—, dihydro(isocyano)(trimethylamine)-, 19:233, 234
Boron chloride (B$_2$Cl$_4$), see Diborane(4), tetrachloro-
Bpy, see 2,2'-Bipyridine
(+)-Brcamsul, see (+)-α-bromocamphor-π-sulfonate
Bromide:
  guanidinium tetracyanoplatinate (0.25:2:1) monohydrate, 19:19, 12
  potassium tetracyanoplatinate (0.3:2:1) trihydrate, 19:1, 4, 15
2-Bromophenyl methyl sulfide, 17:169
2,3-Butanedione:
  macrocyclic ligands from, 18:23
2,3-Butanedione dioxime:
  cobalt complexes, 20:128-132
  diborato complexes with cobalt(III) and iron(II), 17:140-142, 144-145

2,4-Butanedione, dihydrazone, 20:88
1-Butanol, 4-(ethylphenylphosphino)-, 18:189, 190
2-Butanone, 1,1,1-trifluoro-4-(2-thienyl)-4-thioxo-, *see* 3-Buten-2-one, 1,1,1-trifluoro-4-mercapto-4-(2-thienyl)-
2-Butene, 2,3-dimethyl-:
  rhodium complex, 19:219
Butenedinitrile:
  nickel(O) complex, 17:23
2-Butenedinitrile, 2,3-dimercapto-, copper, nickel, and platinum complexes, 19:31
3-Buten-2-one, 4-phenyl-, *see* Benzylideneacetone
—, 1,1,1-trifluoro-4-mercapto-4-(2-thienyl)-, 16:206
  metal derivs., 16:210
sec-Butylamine:
  boron deriv., 17:160
Butyric acid:
  molybdenum complex, 19:133
Butyric acid, 2-ethyl-2-hydroxy, chromium complex, 20:63
tert-Butyl acetate:
  cobalt complex, 20:234, 235
2,3:9,10-Bzo$_2$[14]hexeneN$_4$, 5,7,12,14-Me$_4$, *see* Dibenzo[*b,i*][1,4,8,11]-tetraazacyclotetradecine, 7,16-dihydro-6,8,15,17-tetramethyl-

Cadmium:
  toluene slurry, 19:78
—, ethyliodo-, 19:78
Calcium bis[tetrahydroborate(1-)], 17:17
Calcium tartrate:
  crystal growth of, in silica gel, 20:9
Camphor-π-sulfonate, α-bromo-(+)-, resolving agent, 16:93; 18:106
Carbene complexes:
  of chromium, platinum, and tungsten, 19:164-180
  of chromium and tungsten, 17:95-100
  of tungsten, 19:182
Carbodiimide, digermyl-, 18:163
Carbonates:
  crystal growth of, in silica gel, 20:4
Carbon dioxide:
  iron trimethylphosphine complex, 20:73
Carbonyl complexes, 18:53
  of chromium, 19:155-157; 178, 179

Subject Index 253

  of chromium and tungsten, 19:164-171
  of chromium and tungsten carbene complexes, 17:95-100
  cobalt, 20:224, 226, 230, 234
  of cobalt(III), rhodium(III), and iridium(III), 17:152-155
  of Cr and Mn, 19:188-203
  of η^5-cyclopentadienyliron, chromium, molybdenum, and tungsten complexes, 17:100-109
  of iridium, 19:19-21; 20:237, 238
  iron, 20:218, 222
  of manganese, 19:158-163, 227, 228; 20:24
  of manganese and rhodium, 20:192-195
  rhenium, 20:201-205
  of rhodium, 17:115; 20:209, 212, 215
  of ruthenium, 17:126-127
  of tungsten, 19:172, 181-187
Carboxylato complexes:
  of platinum metals, 17:124-132
Carbyne complexes:
  cobalt, 20:224-235
  tungsten, 19:172
Cesium tetracyanoplatinate (1.75:1) dihydrate, 19:6, 7
Cesium tetracyanoplatinate fluoride (2:1:0.19), 20:29
Cesium tetracyanoplatinate (hydrogen difluoride) (2:1:0.23), 20:26
Cesium tetracyanoplatinate (hydrogen difluoride) (2:1:0.38), 20:28
Chloride:
  metal, volatile, 20:41
  potassium tetracyanoplatinate (0.3:2:1) trihydrate, 19:15
Chloro complexes:
  iridium carbonyl one dimensional elec. conductors, 19:19
Chromate(III), diaquabis(malonato)-potassium, trihydrate, *cis-* and *trans-*, 16:81
—, diaquabis(oxalato)-:
  potassium:
    dihydrate, *cis-*, 17:148
    trihydrate, *trans-*, 17:149
—, tris(malonato)-:
  potassium trihydrate, 16:80
—, tris(oxalato)-:
  tripotassium, trihydrate, 19:127

254 Subject Index

Chromate(V), bis[2-ethyl-2-hydroxybutrato(2-)]oxosodium, 20:63
Chromium:
vaporization of, 19:64
—, (η^6-anisole)tricarbonyl-, 19:155
—, (η^6-benzene)tricarbonyl-, 19:157
—, bis[1,3-bis(trifluoromethyl)-η^6-benzene]-, 19:70
—, bis[(η^5-cyclopentadienyl)dinitrosyl-, 19:211
—, carbonyl(methyl-η^6-m-methylbenzoate)(thiocarbonyl)(triphenyl phosphite)-, 19:202
—, chloro(η^5-cyclopentadienyl)dinitrosyl-, 18:129
—, (η^5-cyclopentadienyl)isobutyldinitrosyl-, 19:209
—, dicarbonyl(η^5-cyclopentadienyl)nitrosyl-, 18:127
—, dicarbonyl(η^6-o-dimethylbenzene)-(thiocarbonyl)-, 19:197, 198
—, dicarbonyl (methyl-η^6-benzoate)-(thiocarbonyl)-, 19:200
—, dicarbonyl (methyl-η^6-m-methylbenzoate)(thiocarbonyl)-, 19:201
—, pentacarbonyl[(diethylamino)ethoxymethylene]-, 19:168
—, tetranitrosyl-, 16:2
—, tricarbonyl(η^6-chlorobenzene)-, 19:157
—, tricarbonyl(η^5-cyclopentadienyl)silyl-, 17:104, 109
—, tricarbonyl(N,N-dimethyl-η^6-aniline)-, 19:157
—, tricarbonyl(η^6-fluorobenzene)-, 19:157
—, tricarbonyl(methyl-η^6-benzoate)-, 19:157
Chromium(O), pentacarbonyl(dihydro-2(3H)-furanylidene)-, 19:178, 179
—, pentacarbonyl(methoxymethylcarbene)-, 17:96
—, pentacarbonyl[1-(phenylthio)ethylidene]-, 17:98
Chromium(III), aquabis(dioctylphosphinato)hydroxo-:
poly-, 16:90
—, aquabis(diphenylphosphinato)hydroxo-:
poly-, 16:90
—, aquabis(ethylenediamine)hydroxodithionate, cis-, 18:84
—, aquahydroxobis(methylphenylphosphinato)-:
poly-, 16:90
—, aquahydroxobis(phosphinato)-polymers, 16:91
—, bis(dioctylphosphinato)hydroxopoly-, 16:91
—, bis(diphenylphosphinato)hydroxopoly-, 16:91
—, bis(phosphinato)-:
complexes, polymers, 16:89-92
—, diaquabis(ethylenediamine)-, tribromide, 18:85
—, di-μ-hydroxo-bis[bis(ethylenediamine)-:
bis(dithionate), 18:90
tetrabromide, dihydrate, 18:90
tetrachloride, dihydrate, 18:91
tetraperchlorate, 18:91
—, di-μ-hydroxo-bis[tetraammine-:
tetrabromide, tetrahydrate, 18:86
tetraperchlorate, dihydrate, 18:87
—, hexakis(dimethyl sulfoxide)-tribromide, 19:126
—, hydroxobis(methylphenylphosphinato)-:
poly-, 16:91
—, hydroxobis(phosphinato), polymers, 16:91
—, malonato complexes, 16:80-82
—, tetraammineaquachloro, cis-, sulfate, 18:78
—, tetraammineaquahydroxo-, cis-, dithionate, 18:80
—, tetraamminediaqua, cis-, triperchlorate,
—, 18:82
tris[bis(trimethylsilyl)amido]-, 18:118
—, tris(ethylenediamine), tribromide, 19:125
Chromium ammine and ethylenediamine complexes, 18:75
Chromium bromide (CrBr$_3$):
anhydrous, 19:123, 124
Chromium cobalt oxide (CoCr$_2$O$_4$), 20:52
Chromium lithium oxide (CrLiO$_2$), 20:50
Chromium magnesium oxide (Cr$_2$MgO$_4$), 20:52
Chromium manganese oxide (Cr$_2$MnO$_4$), 20:52

Chromium nickel oxide (Cr_2NiO_4),
  20:52
Chromium nitrosyl, $Cr(NO)_4$, 16:2
Chromium zinc oxide (Cr_2ZnO_4), 20:52
Clathrochelates from iron(II) and cobalt-
  (III) dioxime complexes, 17:139-
  147
  nomenclature and formulation of,
    17:140
Cobalamin, (2,2-diethoxyethyl)-,
  20:138
—, hydroxo-, 20:138
—, methyl-, 20:136
Cobalamines:
  cobalt 2,3-butanedione dioxime (co-
    baloximes) as model systems,
    20:127
Cobalt:
  vaporization of, 19:64
—, μ_3-benzyldynetris(tricarbonyl-,
    20:226, 228
—, bis(dimethylcarbamodithioate-S,S')-
    nitrosyl, see Cobalt, bis(dimethyl-
    dithiocarbamato-S,S')nitrosyl-
—, bis(dimethyldithiocarbamato-S,S')-
    nitrosyl, 16:7
—, [1,2-bis(diphenylphosphino)ethane]di-
    nitrosyl-:
  tetraphenylborate, 16:19
—, μ_3-[(tert-butoxycarbonyl)methyl-
    idyne]-tris(tricarbonyl-), 20:234,
    235
—, μ_3-(chloromethylidyne)-tris(tricar-
    bonyl, 20:234
—, [(1,2,5,6-η)-1,5-cyclooctadiene]-
    [(1,2,3-η)-2-cycloocten-1-yl]-,
    17:112
—, dichlorobis(methyldiphenylphos-
    phine)nitrosyl-, 16:29
—, dinitrosylbis(triphenylphosphine)-,
    tetraphenylborate, 16:18
—, dinitrosyl(N,N,N',N'-tetramethylethyl-
    enediamine)-:
  tetraphenylborate, 16:17
—, μ_3-[(ethoxycarbonyl)methyledyne]-
    tris(tricarbonyl-, 20:230
—, [ethylenebis(diphenylphosphine)]
    dinitrosyl-:
  tetraphenylborate, 16:19
—, μ_3-[[(methylamino)carbonyl]methyl-
    idyne]-tris(tricarbonyl-, 20:230, 232
—, μ_3-methylidyne-tris(tricarbonyl-,
    20:226, 227
—, nitrosyltris(triphenylphosphine)-,
    16:33
Cobalt(I), hydridobis[ethylenebis(diphenyl-
    phosphine)-, 20:208
—, hydridobis[cis-vinylenebis(diphenyl-
    phosphine)]-, 20:207
—, pentakis(trimethyl phosphite)-, tetra-
    phenylborate, 20:81
Cobalt(II), bromo(2,12-dimethyl-3,7,11,17-
    tetraazabicyclo[11.3.1]heptadeca-
    1(17),2,11,13,15-pentaene)-:
  bromide, monohydrate, 18:19
—, diaqua(ethylenediamine-N,N'-diace-
    tato)-, 18:100
—, [5,14-dihydrodibenzo[b,i][1,4,8,11]-
    tetraazacyclotetradecinato(2-)]-,
    18:46
—, [N,N'-ethylenebis(thioacetylacetoni-
    minato)]-, see Cobalt(II), {{4,4'-
    (ethylenedinitrilo)bis[2-pentane-
    thionato]}(2-)-S,N,N',S'}-
—, {{4,4'-(ethylenedinitrilo)bis[2-pentane-
    thionato]}(2-)-S,N,N',S'}-, 16:227
—, [5,6,14,15,20,21-hexamethyl-1,3,4,-
    7,8,10,12,13,16,17,19,22-dodeca-
    azatetracyclo[8.8.4.1^{3,7}.1^{8,12}]-
    tetracosa-4,6,13,15,19,21-hexaene-
    $N^4,N^7,N^{13},N^{16},N^{19},N^{22}$]-, bis-
    (tetrafluoroborate), 20:89
—, [1,4,8,11,15,18,22,25-octamethyl-29H,-
    31H-tetrabenzo[b,g,l,q]porphinato-
    (2-)]-, 20:156
—, tetrakis(selenourea)diperchlorate,
    16:84
—, tris(selenourea)sulfato-, 16:85
Cobalt(III), amminebromobis(ethylenedi-
    amine)-:
  (+)-cis-, (+)-α-bromocamphor-π-sulfonate,
    16:93
  chloride:
    (+)-cis-, 16:95
    (−)-cis-, 16:96
  dithionate, (−)-cis-, 16:94
  halides, (+) and (−)-, 16:93
  nitrate, (+) and (−)-, 16:93
—, ammine(carbonato)bis(ethylenedi-
    amine)-:

bromide, hemihydrate, *cis*-, 17:152-154
perchlorate, *trans*-, 17:152
—, bis[2,3-butanedione dioximato(1-)] (4-*tert*-butylpyridine)(2-ethoxyethyl)-, 20:131
—, bis[2,3-butanedione dioximato(1-)]-(4-*tert*-butylpyridine)(ethoxymethyl)-, 20:132
—, bis(ethylenediamine)oxalato-, as cationic resolving agent, 18:96
chloride, monohydrate, 18:97
(−)-:
bromide, monohydrate, 18:99
(−)-[(ethylenediaminetetraacetato)cobaltate(III)], trihydrate, 18:100
(−)-[(ethylenediamine-N,N'-diacetato)dinitrocobaltate(III), 18:101
hydrogen (+)-tartrate, 18:98
(+)-:
hydrogen (+)-tartrate, 18:98
iodide, 18:99
—, bromobis[2,3-butanedione dioximato(1-)] (4-*tert*-butylpyridine)-, 20:130
—, bromobis[2,3-butanedione dioximato(1-)] (dimethyl sulfide)-, 20:128
—, (carbonato)(tetraethylenepentamine)-: perchlorate, trihydrate, α,β-*sym*-, 17:153-154
—, dibromo[2,12-dimethyl-3,7,11,17-tetraazabicyclo[11.3.1]heptadeca-1(17),2,11,13,15-pentaene)-:
bromide, monohydrate, 18:21
—, dibromo(2,3-dimethyl-1,4,8,11-tetraazacyclotetradeca-1,3-diene)-perchlorate, 18:28
—, dibromo(5,5,7,12,12,14-hexamethyl-1,4,8,11-tetraazacyclotetradecane)-:
meso-trans-, diperchlorate, 18:14
—, dibromo(tetrabenzo[*b*,*f*,*j*,*n*][1,5,9,12]-tetraazacyclohexadecine)-, bromide, 18:34
—, dibromo(2,3,9,10-tetramethyl-1,4,8,11-tetraazacyclotetradeca-1,3,8,10-tetraene)-:
bromide, 18:25
—, dichlorobis(ethylenediamine)-*trans*-, nitrate, 18:73

—, dichloro(1,5,9,13-tetraazacyclohexadecane)-, perchlorate, *trans*-, isomers I and II, 20:113
—, dichloro(1,4,8,12-tetraazacyclopentadecane)-, chloride, *trans*-, I and II isomers, 20:112
—, dichloro(1,4,7,10-tetraazacyclotridecane)-, chloride, *trans*-, 20:111
—, di-μ-hydroxo-bis[bis(ethylenediamine)-: bis(dithionate), 18:92
tetrabromide, dihydrate, 18:92
tetrachloride, pentahydrate, 18:93
tetraperchlorate, 18:94
—, di-μ-hydroxo-bis[tetraamminetetrabromide, tetrahydrate, 18:88
tetraperchlorate, dihydrate, 18:88
—, (ethylenediamine)[ethylenediamine-N,N'-diacetato(2-)]-:
asym-cis-, chloride, 18:105
(−)-*asym-cis*-:
(+)-α-bromocamphor-π-sulfonate, 18:106
chloride, trihydrate, 18:106
(−)-*sym-cis*-:
hydrogen tartrate, 18:109
nitrate, 18:109
(+)-*asym-cis*-:
(+)-α-bromocamphor-π-sulfonate, 18:106
chloride, trihydrate, 18:106
(+)-*sym-cis*-:
hydrogen tartrate, 18:109
nitrate, 18:109
—, hexaammine-:
triacetate, 18:68
trichloride, 18:68
—, [5,6,14,15,20,21-hexamethyl-1,3,4,7,-8,10,12,13,16,17,19,22-dodecaazatetracyclo[8.8.4.1^{3,7}.1^{8,12}]tetracosa-4,6,13,15,19,21-hexaene-N^4-$N^7,N^{13},N^{16},N^{19},N^{22}$]-, tris[tetrafluoroborate], 20:89
—, (1,3,6,8,10,13,16,19-octaazabicyclo-[6.6.6]eicosane)-, trichloride, 20:85
—, pentaammine(carbonato)-perchlorate, monohydrate, 17:152-154
—, tetraammineaquahydroxodithionate, *cis*-, 18:81
—, tetraamminediaqua, triperchlorate, *cis*, 18:83

Subject Index 257

—, tetraamminedinitrato-
  nitrate, cis- and trans-, 18:70, 71
—, [[tris[μ-[(2,3-butanedione dioximato)-
  (1-)-O:O]]difluorodiborato](2-)-
  $N,N',N'',N''',N'''',N'''''$]-, 17:140
Cobalt ammine and ethylenediamine complexes, 18:67, 75
Cobaltate(1-), tetracarbonyldipyridinemagnesium(2+), 16:58
Cobaltate(II) tetrachloro-:
  dipotassium, 20:51
Cobaltate(III), (carbonato)[ethylenediamine-N,N'-diacetato(2-)]-
  sodium, 18:104
—, (ethylenediamine-N,N'-diacetato)dinitro-:
  (−)-, (−)-[bis(ethylenediamine)oxalatocobalt(III)], 18:101
  (−)-sym-cis-, potassium, 18:101
  sym-cis-, potassium, 18:100
—, (ethylenediaminetetraacetato)(−)-:
  (−)-[bis(ethylenediamine)oxalato:
    cobalt(III), trihydrate, 18:100
    potassium dihydrate, 18:100
  (+)-, potassium, dihydrate, 18:100
  potassium, dihydrate, resolution of, 18:100
Cobalt(III) clathrochelates with dioximes, 17:139-147
Cobalt chromium oxide (CoCr$_2$O$_4$), 20:52
Cobalt complexes:
  amminebromobis(ethylenediamine), 16:93-96
  dinitrosyl-, 16:16-21
  with selenourea, 16:83-85
Cobalt potassium chloride (CoK$_2$Cl$_4$), 20:51
Cobyrinic acid, aquacyano-, heptamethyl ester, perchlorate, 20:141
—, aquamethyl-, heptamethyl ester, perchlorate, 20:141
—, dicyano-, heptamethyl ester, 20:139
Cod, see 1,5-Cyclooctadiene
Configuration:
  of $meso$- and $racemic$-(5,7,7,12,14,14-hexamethyl-1,4,8,11-tetraazacyclotetradeca-4,11-diene)nickel(II), 18:5
Copper:
  vaporization of, 19:64

—, chlorotris(triphenylphosphine)-, 19:88
—, hydroxo(tris-p-tolylphosphine)-, 19:89
Copper(I), hydroxo(triphenylphosphine)-, 19:87, 88
—, nitratobis(triphenylphosphine)-, 19:93
—, nitratotris(triphenylarsine)-, 19:95
—, nitratotris(triphenylstibine)-, 19:94
—, (tetrahydroborato)bis(triphenylphosphine)-, 19:96
—, tetrakis(acetonitrile)-, hexafluorophosphate, 19:90
Copper(II), aqua-μ-[[6,6'-(ethylenedinitrilo)bis(2,4-heptanedionato)](4-)-
  N,N',O,O^4:$O^2,O^{2'},O^4,O^{4'}$]-[oxovanadium(IV)]-, 20:95
—, [3,10-dibromo-1,6,7,12-tetrahydro-1,5,8,12-benzotetraazacyclotetradecinato(2-)]-, 18:50
—, [[6,6'-(ethylenedinitrilo)bis(2,4-heptanedionato)](2-)-$N,N',O^4,O^{4'}$]-, 20:93
—, μ-[[6,6'-(ethylenedinitrilo)bis(2,4-heptanedionato)](4-)-$N,N',O^4,O^{4'}$:-$O^2,O^{2'},O^4,O^{4'}$]di-, 20:94
—, [tetrabenzo[b,f,j,n][1,5,9,13]tetraazacyclohexadecine] dinitrate, 18:32
—, [5,9,14,18-tetramethyl-1,4,10,13-tetraazacyclooctadeca-5,8,14,17-tetraene-7,16-dionato(2-)-$N^1,N^4,-O^7,O^{16}$]-, 20:92
—, [5,10,15,20-tetraphenylporphyrinato-(2-)-$N^{21},N^{22},N^{23},N^{24}$]-, 16:214
Copper acetate (Cu$_2$(C$_2$H$_3$O$_2$)$_2$), 20:53
Copper(I) chloride:
  crystal growth of, in silica gel, 20:10
Copper complexes:
  with 5,10,15,20-tetraphenylprophyrin, 15:214
Copper(I) nitrate:
  complexes, 19:92-97
Crystal growth:
  complex dilution method, in silica gel, 20:6
  double-infusion, 20:11
  in gels, 20:1
  of one dimensional elec. conductors, 19:4
  of TaS$_2$, 2H(a) phase single crystals, 19:39
  of (TTF)(TCNQ) single crystals, 19:33

258 Subject Index

Cuprate(2-), bis[2,3-dimercapto-2-butene-
  dinitrilato(2-)]-:
  salt with 2,2-bi-1,3-dithiolylidene
  (1:2), 19:31
Cyanate:
  ammonium, 16:136
  tetraethylammonium, 16:132
  tetraphenylarsonium, 16:134
Cyanide:
  tetraethylammonium, 16:133
  tetraphenylarsonium, 16:135
—, iso-, see Isocyanide
Cyclam, see 1,4,8,11-Tetraazacyclotetra-
  decane
Cycloalkanes, 1,3-diaza-2-bora-, 17:164-
  167
1,2-Cyclohexanedione dioxime:
  diborato complexes with iron(II),
  17:140, 143-144
Cyclohexasilane, dodecamethyl-, 19:265
1,5-Cyclooctadiene:
  platinum complexes, 19:213, 214
  rhodium complex, 19:218; 20:191, 194
Cyclooctatetraene:
  uranium complex, 19:149, 150
1,3,5,7-Cyclooctatetraene lithium complex, 19:214
Cycloolefin complexes, 16:117-119;
  17:112
1,3-Cyclopentadiene, 1-bromomanganese
  complex, 20:193, 195
—, 1-bromo-2,3,4,5-tetrachloro-:
  manganese complex, 20:194, 195
  rhodium complex, 20:192
—, 1-chloro-:
  manganese and rhodium complexes,
  20:192
—, 1-chloro-2,3,4,5-tetraphenylrhodium
  complexes, 20:191, 192
—, 5-diazo-, 20:191
—, 1-iodo-:
  manganese complex, 20:193
—, 1,2,3,4,5-pentachloro-:
  manganese and rhodium complexes,
  20:193, 194
—, 1,2,3,4-tetrachloro-5-diazo-, 20:189,
  190
Cyclopentadienide:
  lithium, 17:179
  potassium, 17:173, 176

—, methyl-:
  potassium, 17:175
2,4-Cyclopentadien-1-one, hydrazone,
  20:190
Cyclopentadienyl complexes, 16:237;
  17:91, 100-109
  Cr, Mo, and W, 19:208
  manganese, 19:188, 196
  molybdenum, 20:196-199
  Pd, 19:221
Cyclopentadienyl compounds of Group IV
  elements, 17:172-178
Cyclopentasilane, decamethyl-, 19:256
Cyclotriphosphazenes:
  bromo fluoro, 18:194
Cyclotrithiazene, trichloro-, 17:188

daaenH$_4$, see 2,4-Heptanedione, 6,6'-
  (ethylenedinitrilo)bis-
daenH$_4$, see 1,4,10,13-Tetraazacyclooctа-
  deca-5,8,14,17-tetraene-7,16-dione,
  5,9,14,18-tetramethyl-
Dendrites:
  of metals, growth of, in silica gel, 20:7
Detonation:
  of 1-BrB$_5$H$_8$, 19:247, 248
1,3-Diaza-2-boracyclohexane, 1,3-dimethyl-,
  17:166
1,3-Diaza-2-boracyclopentane, 1,3-di-
  methyl-, 17:165
1,3,2-Diazaborinane, 1,3-dimethyl-, 17:166
1,3,2-Diazaborolane, 1,3-dimethyl-,
  17:165
Diazene complexes of nickel(O), 17
  17:117-124
Dibenzo[b,i][1,4,8,11]tetraazacyclotetra-
  decine, 5,14-dihydro-:
  cobalt complex, 18:45
Dibenzo[b,i][1,4,8,11]tetraazacyclotetra-
  decine, 7,16-dihydro-6,8,15,17-tetra-
  methyl-(5,7,12,14-Me$_4$-2,3:9,10-
  Bzo$_2$[14]hexaeneN$_4$), 20:117
1,6-Diboracyclodecane (B$_2$H$_2$(C$_4$H$_8$)$_2$),
  19:239, 241
Diborane(4), tetrachloro-, 19:74
Diborane(6):
  manganese complex, 20:240
—, bis[µ-(dimethylamino)]-, 17:32
—, 1,2:1,2-bis(tetramethylene)-, see
  1,6-Diboracyl

—, bromo-, 18:146
—, μ-(dimethylamino)-, 17:34
—, iodo-, 18:147
—, 1-methyl-, 19:237
Diborane(6)-d_5, bromo-, 18:146
  iodo-, 18:147
Diborate(1-), μ-hydro-bis(μ-tetramethylene)-:
  tetrabutylammonium, 19:243
Diethylamine:
  aluminum complexes, 17:40-41
  boron complexes, 17:30, 32, 34, 159
—, N-(ethoxymethyl)-:
  chromium complex, 19:168
Diethyl ether:
  cobalt alkylated complex, 20:131
Digermaselenane, 20:175
Digermatellurane, 20:175
Digermathiane, see Digermyl sulfide
Digermoxane, 20:176, 178
—, 1,3-dimethyl-, 20:176, 179
—, hexamethyl-, 20:176, 179
—, 1,1,3,3-tetramethyl-, 20:176, 179
Digermylcarbodiimide, 18:63
Digermyl sulfide, 18:164
5,26:13,18-Diimino-7,11:20,24-dinitrilodibenzo[c,n] [1,6,12,17] tetraazacyclodocosine, 18:47
7,12:21,26-Diimino-19,14:28,33:35,5-trinitrilo-5H-pentabenzo[c,h,m,r,w]-[1,6,11,16,21] pentaazacyclopentacosin:
  uranium complex, 20:97
β-Diketone, see 2,4-Pentanedione
β-Diketonimine, thio-, see 2-Pentanethione, 4,4'-(ethylenedinitrilo)bis-
Dilithium phthatocyanine, 20:159
Dimethyl phosphite:
  platinum complex, 19:98, 100
Dimethyl sulfide:
  cobalt complex, 20:128
Dimethyl sulfoxide:
  chromium complex, 19:126
  indium complexes, 19:259, 260
Diop, see Phosphine, [(2,2-dimethyl-1,3-dioxolane-4,5-diylbis(methylene)]-bis(diphenyl-
Dioximes:
  diborato clathrochelate complexes of

iron(II) and cobalt(III), 17:139-147
1,3-Dioxolane, 4,5-bis[(diphenylphosphino)methyl]-2,2-dimethyl-
  (+)-, rhodium(I) complex, 17:81
diphos, see Phosphine, ethylenebis(diphenyl-
Dipropylamine, 3,3'-diamino-:
  macrocyclic ligands from, 18:18
Disilaselenane, hexamethyl-, 20:173
Disilathiane, 19:275
—, 1,3-dimethyl-, 19:276
—, hexamethyl-, 19:276
—, 1,1,3,3-tetramethyl-, 19:276
Disilatellurane, hexamethyl-, 20:173
Disilizane, 1,1,1,3,3,3-hexamethyl-, see Bis(trimethylsilyl)amine
Distannoxane, hexamethyl-, 17:181
1,3-Dithiolylium tetrafluoroborate, 19:28
dmso, see dimethyl sulfoxide
2,6,10-Dodecatriene:
  nickel complex, 19:85
1,3,4,7,8,10,12,13,16,17,19,22-Dodecaazatetracyclo[8.8.4.1^{3,17}.1^{8,12}]tetracosa-4,6,13,15,19,21-hexene, 5,6,14,-15,20,21-hexamethyl-:
  cobalt, iron, and nickel complexes, 20:88
1-Dodecene:
  platinum complex, 20:181
dp, see Phosphine, vinylenebis(diphenyl-

edda, see Ethylenediamine-N,N'-diacetato
Electric conductors:
  nonmetal solid state, 19:1-58; 20:20-31
Electrochemical syntheses:
  of indium(III) complexes, 19:257
  of one dimentional platinum conductors, 19:13
en, see Ethylenediamine
Ethane, 1,2-bis(diphenylphosphino)-, see Phosphine, ethylenebis(diphenyl-
—, 1,2-bis(methylthio)-:
  molybdenum complex, 19:131
1,2-Ethanediamine, see Ethylenediamine
1,2-Ethanedione, 1,2-diphenyl-:
  dibutoxydiborato complex with iron(II), 17:140, 145
  dioxime, 17:145
Ethanethiol, 2-amino-:
  platinum complex, 20:104

Subject Index

Ethanimine:
  rhenium complex, 20:204
Ethanol, 2-mercapto-:
  platinum complex, 20:103
—, 2-methoxy-:
  aluminum complex, 18:147
—, 2,2',2''-nitrilotri-:
  germanium and tin complexes, 16:229-234
Ethene, 1,2-diphenyl-, see Stilbene
Ethenetetracarbonitrile:
  nickel complex, 17:123
Ethyl acetate:
  cobalt complex, 20:230
Ethylamine, 2-(diphenylphosphino)-N,N-diethyl-, 16:160
Ethylene:
  platinum complex, 19:215, 216; 20:181
Ethylenediamine:
  chromium complexes, 19:125
  osmium complex, 20:61
  rhodium complexes, 20:58-60
  ruthenium complexes, 19:118, 119
—, N-(2-aminoethyl)-N'-[2-[(2-aminoethyl)amino]ethyl]-, see Tetraethylenepentamine
—, N,N-bis[(diphenylphosphino)methyl]-N',N'-dimethyl-, 16:199-200
—, N,N'-bis[(diphenylphosphino)methyl]-N,N'-dimethyl-, 16:199-200
—, N-[(diphenylphosphino)methyl]-N,N',N'-trimethyl-, 16:199-200
—, N,N,N',N'-tetrakis[(diphenylphosphino)methyl]-, 16:198
—, N,N,N'-tris[(diphenylphosphino)methyl]-N'-methyl-, 16:199-200
Ethylenediamine complexes:
  of (carbonato)cobalt(III) complexes, 17:152-154
  chromium and cobalt complexes, 18:75
  of cobalt(III), 16:93-96
  macrocyclic ligands from, 18:37, 51
Ethyl methyl ether:
  cobalt alkylated complex, 20:32
Ethyl phosphite:
  nickel complexes, 17:119

Ferrate(1), dicarbonylbis(η-cyclopentadienyl)-:
  bis[tetrahydrofuran]magnesium(2+), 16:56
—, tetrachloro-:
  pentathiazyl, 17:190
Ferrate(1-), undecacarbonylhydridotri-, μ-nitrido-bis(triphenylphosphorus)-(1+), 20:218
Ferrate(2-), undecacarbonyltri-, bis[μ-nitrido-bis(triphenylphosphorus)-(1+)], 20:222
Ferrate(III), hexacyano-, samarium(III), tetrahydrate:
  crystal growth of, by double infusion, 20:13
Fluoride, cesium tetracyanoplatinate (0.19:2:1), 20:29
Fluorides:
  penta-, of Mo, Os, Re, and U, 19:137-139
Furan, tetrahydro-:
  chromium complex, 19:178, 179
  molybdenum complexes, 20:121, 122

Gallate(1-), tetrahydro-:
  lithium, 17:45
  potassium, 17:48
  sodium, 17:48
Gallium, trihydrido(trimethylamine)-, 17:42
Gallium trichloride, 17:167
Germane:
  derivs., 18:153
—, bromodimethyl-, 18:157
—, bromotrimethyl-, 18:153
—, chlorodimethyl-, 18:157
—, cyclopentadienyl-, 17:176
—, dimethyl-, 18:156
  mono halo derivs., 18:154
—, fluorodimethyl-, 18:159
—, iodo-, 18:162
—, iododimethyl-, 18:158
—, (methylthio)-, 18:165
—, (phenylthio)-, 18:165
—, tetramethyl-:
  in preparation of GeBr(CH$_3$)$_3$, 18:153
Germanium, ethyl{{2,2',2''-nitrilotris[ethanolato]}(3-)-N,O,O',O''}-, 16:229
Germanium hydride derivatives, 18:153
Germatrane, see Germanium, ethyl{{2,2',2''-nitrilotris[ethanolato]}(3-)-N,O,O',O''}-

Glyoxime, dimethyl-, *see* 2,3-Butanedione dioxime
—, diphenyl-, *see* 1,2-Ethanedione, 1,2-diphenyl dioxime
Gold, methyl[trimethyl(methylene)phosphorane]-, 18:141
Guanidinium bromide tetracyanoplatinate (2:0.25:1):
 monohydrate, 19:10, 12

Hematoprophyrin(IX), *see* 2,18-Porphinedipropanoic acid, 7,12-bis(1-hydroxyethyl)-3,8,13,17-tetramethyl-, dimethyl ester
Hemes:
 iron-57 incorperation 20:153
2,4-Heptanedione, 6,6'-(ethylenedinitrilo)bis:
 copper complexes, 20:93
Heptasulfurimide, *see* Heptathiazocine
Heptathiazocine, 18:203, 204
Hexaborane(10), 19:247, 248
1,5-Hexadiene:
 rhodium complex, 19:219
H$_2$ oep, *see* 21H,23H-Porphine, 2,3,7,8,-12,13,17,18-octaethyl-
H$_2$ TAPP, *see* Porphyrin, 5,10,15,20-tetrakis(2-aminophenyl)-
H$_2$ TNPP, *see* Porphyrin, 5,10,15,20-tetrakis(2-nitrophenyl)-
H$_2$ TpivPP, *see* Porphine, 5,10,15,20-tetrakis[2-(2,2-dimethylpropionamido)phenyl]-, *all-cis-*
H$_2$ tpp, *see* 21H,23H-Porphine, 5,10,15,20-tetraphenyl-
Hydride complexes:
 summary of previous preparations, 17:53
 of transition metals, 17:52-94
Hydrides:
 of boron, 17:17-36
 of gallium, 17:42-51
 of magnesium, 17:1-5
 of zinc, 17:6-16
Hydroaluminum complexes, 17:36-42
Hydrogallium complexes, 17:42-43
Hydroiron complexes, 17:69-72
Hydromanganese complexes, 20:240
Hydromolybdenum complexes, 17:54-64; 19:129, 130
Hydronickel complexes, 17:83-91

Hydridopalladium complexes, 17:83-91
Hydridorhenium complexes, 17:64-68
Hydridorhodium complexes, 17:81-83
Hydridoruthenium complexes, 17:73-80
Hydridotitanium complexes, 17:91-94
Hydridozinc complexes, 17:13-14
Hydridozirconium complexes, 19:224-226
Hydroboron complexes, 17:26-36
Hydrogen, [*cis*-(acetimidoylacetyltetracarbonylrhenium)]-, *see* Rhenium, acetyl(1-aminoethylidene)tetracarbonyl-, *cis*-
—, [*cis*-(diacetyltetracarbonylrhenium)]-, *see* Rhenium, acetyltetracarbonyl(1-hydroxyethylidene)-, *cis*-
[Hydrogen bis(sulfate)], rubidium tetracyanoplatinate (0.45:1:3):
 monohydrate, 20:20
(Hydrogen difluoride), cesium tetracyanoplatinate (0.23:2:1), 20:26
—, cesium tetracyanoplatinate (0.38:2:1), 20:28
—, rubidium tetracyanoplatinate (0.29:-2:1):
 hydrate (1:1.67), 20:24
—, rubidium tetracyanoplatinate (0.38:-2:1), 20:25
Hydrogen pentafluoroselenate(VI), 20:38

Imidazole, *N*-methyl-:
 iron complex, 20:167, 168
Indate(III), pentachloro-:
 bis(tetraethylammonium), 19:260
Indium(III), tribromotris(dimethyl sulfoxide)-, 19:260
—, trichlorotris(dimethyl sulfoxide)-, 19:259
—, tris(2,4-pentanedionato)-, 19:261
Indium bromide (InBr$_3$), 19:259
Indium chloride (InCl$_3$), 19:258
Intercalation compounds, 19:35-48
Iodates:
 crystal growth of, in silica gel, 20:5
Iridate, dicarbonyldichloro-:
 potassium (1:0.6) hemihydrate, 19:20
Iridate(IV), hexachloro-:
 diammonium, in recovery of Ir from laboratory residues, 18:132
Iridium:
 recovery of, from laboratory residues, 18:131

262 Subject Index

Iridium, bis-μ-(benzenethiolato)-tetracarbonyldi-, 20:238
—, bromo(dinitrogen)bis(triphenylphosphine)-, 16:42
—, bromonitrosylbis(triphenylphosphine)-:
  tetrafluoroborate, *trans*-, 16:42
—, chloro(dinitrogen)bis(triphenylphosphine)-, 16:42
—, chloronitrosylbis(triphenylphosphine)-:
  tetrafluoroborate, *trans*-, 16:41
Iridium, chloro(thiocarbonyl)bis(triphenylphosphine)-:
  trans-, 19:206
—, tricarbonylchloro- [Ir(CO)$_3$Cl$_{1.10}$], 19:19
Iridium(I), carbonylchlorobis(trimethylphosphine)-, 18:64
—, carbonyltetrakis(trimethylphosphine)-:
  chloride, 18:63
—, pentakis(trimethylphosphite), tetraphenylborate, 20:79
—, tetracarbonylbis-μ-(2-methyl-2-propanethiolato)-di-, 20:237
Iridium(III), (acetato)dihydridotris(triphenylphosphine)-, 17:129
—, pentaammine(carbonato)-:
  perchlorate, 17:153
Iridium carbonyl halides:
  one dimensional elec. conductors, 19:19
Iridium carbonyl trimethylphosphine complexes, 18:62
Iron:
  vaporization of, 19:64
—, (benzylideneacetone)tricarbonyl-, 16:104
—, bis(diethyldithiocarbamato)nitrosyl-, 16:5
—, bis[ethylenebis(diphenylphosphine)]-hydrido-:
  complexes, 17:69-72
—, (carbon dioxide-*C,O*)tetrakis(trimethylphosphine)-, 20:73
—, [(dimethylphosphino)methyl-*C,P*]-hydridotris(trimethylphosphine)-, *cis*-, 20:71

—, pentakis(trimethyl phosphite)-, 20:79
—, tetrakis(trimethylphosphine)-, 20:71
—, carbonyl η-diene complexes, 16:103-105
—, tetracarbonyl(chlorodifluorophosphine)-, 16:66
—, tetracarbonyl[(diethylamino)difluorophosphine)]-, *see* Iron, tetracarbonyl(diethylphosphoramidous difluoride)-
—, tetracarbonyl(diethylphosphoramidous difluoride)-, 16:64
—, tetracarbonyl(trifluorophosphine)-, 16:67
Iron(1+), dicarbonyl(η^5-cyclopentadienyl)(thiocarbonyl)hexafluoro-, phosphate, 17:110
Iron-57:
  insertion into porphyrins, 20:157
Iron(I), bis[ethylenebis(diphenylphosphine)]hydrido-, 17:71
—, dicarbonyl(η^5-cyclopentadienyl)-[(methylthio)thiocarbonyl]-, 17:102
Iron(II), bis(acetonitrile)(5,7,7,12,14,14-hexamethyl-1,4,8,11-tetraazacyclotetradeca-4,-1-diene)-:
  bis(trifluoromethanesulfonate), 18:6
—, bis(acetonitrile)(5,5,7,12,12,14-hexamethyl-1,4,8,11-tetraazacyclotetradecane)-:
  bis(trifluoromethanesulfonate), *meso*-, 18:15
—, bis[ethylenebis(diphenylphosphine)]-hydrido-:
  tetraphenylborate, 17:70
—, bis(*N*-methylimidazole)[(*all-cis*)-5,10,15,20-tetrakis[2-(2,2-dimethylpropanamido)phenyl] porphyrinato(2-)]-, 20:167
—, chlorobis[ethylenebis(diphenylphosphine)]hydrido-, 17:69
—, dichlorobis(trimethylphosphine)-, 20:70
—, (dioxygen)(*N*-methylimidazole)[(*all-cis*)-5,10,15,20-tetrakis[2-(2,2-dimethylpropanamido)phenyl] porphyrinato(2-)]-, 20:161, 168
—, [5,6,14,15,20,21-hexamethyl-1,3,4,7,-8,10,12,13,16,17,19,22-dodecaaza-

Subject Index 263

tetracyclo[8.8.4.1^{3,17}.1^{8,12}]-
tetracosa-4,6,13,15,19,21-hexaene-
$N^4,N^7,N^{13},N^{16},N^{19},N^{22}$]-, bis-
(tetrafluoroborate), 20:88
—, [phthatocyaninato(2-)]-, 20:160
—, tris(2,4-butanedione dihydrazone)-,
 bis(tetrafluoroborate), 20:88
—, [[tris[μ-[(2,3-butanedione dioximato)(2-)-O:O]]-dibutoxydiborato]-
 (2-)-$N,N',N'',N''',N'''',N'''''$]-,
 17:144
—, [[tris[μ-(2,3-butanedione dioximato)(2-)-O:O']]-difluorodiborato]-
 (2-)-$N,N',N'',N''',N'''',N'''''$]-,
 17:142
—, [[tris[μ-[(1,2-cyclohexanedione
 dioximato)(2-)-O:O']]-difluorodi-
 borato] (2-)-$N,N',N'',N''',N'''',$-
 N''''']-, 17:143
—, [[tris[μ-[(1,2-cyclohexanedione
 dioximato)(2-)-O:O']-dihydroxy-
 diborato] (2-)-$N,N',N'',N''',N'''',$
 N''''']-, 17:144
—, [[tris[μ-[(1,2-diphenyl-1,2-ethane-
 dione dioximato)(2-)-O:O']-di-
 butoxydiborato] (2-)-N,N',N'',
 N''',N'''',N''''']-, 17:145
Iron(III), bromo[(all-cis)-5,10,15,20-,
 tetrakis[2-(2,2-dimethylpropion-
 amido)phenyl] porphyrinato(2-)]-,
 20:166
—, chloro[7,12-diethyl-3,8,13,17-tetra-
 methyl-21H,23H-porphine-2,18-
 dipropionato(2-)]-, 20:152
—, chloro[dimethyl 7,12-bis(1-hydroxy-
 ethyl)-3,8,13,17-tetramethyl-2,18-
 porphinedipropanoato(2-)-N^{21},
 N^{22},N^{23},N^{24}]-, 16:216
—, chloro[dimethyl 7,12-diethenyl-
 3,8,13,17-tetramethyl-21H,23H-
 porphine-2,18-dipropionato(2-)]-,
 20:148
—, chloro[hematoprophyrin(IX) dimethyl
 ester]-, see Iron(III), chloro[di-
 methyl 7,12-bis(1-hydroxyethyl)-
 3,8,13,17-tetramethyl-2,18-por-
 phinedipropanoato(2-)-N^{21},N^{22},
 N^{23},N^{24}-
—, chloro[2,3,7,8,12,13,17,18-octa-
 ethyl-21H,23H-porphinato(2-)]-,
 20:151
—, tris[bis(trimethylsilyl)amido]-, 18:119
Iron carbonyl, Fe(CO)$_4$:
 complexes with difluorophosphines,
 16:63-67
Iron complexes:
 with hematoporphyrin (XI), 16:216
Iron(III) diborato clathrochelates of
 dioximes, 17:139-147
Isocyanide complexes of nickel(O),
 17:117-124
Isocyano complexes:
 boron, 19:233, 234

Lithium, [bis(trimethylsilyl)amido]-:
 in preparation of transition metal com-
 plexes, 18:115
—, (1,3,5,7-cyclooctatetraene)di-, 19:214
—, [2-(methylthio)phenyl]-, 16:170
—, μ-[phthatocyaninato(2-)]di-, 20:159
Lithium chromium oxide (CrLiO$_2$), 20:50
Lithium cyclopentadienide, 17:179
Lithium diphenylphosphide, 17:186
Lithium tetrahydridogallate(1-), 17:45
Lithium tetrahydridozincate(2-), 17:9, 12
Lithium tetramethylzincate(2-), 17:12
Lithium trihydridozincate(1-), 17:9, 10

Macrocyclic ligands, 16:220-225; 18:1; 20, 158
 conjugated, 18:44
 tetraazatetraenato complexes, 18:36
Magnesate, tetrachloro-, dipotassium,
 20:51
Magnesium:
 vaporization of, 19:64
—, bis[(trimethylsilyl)methyl]-, 19:262
Magnesium(2+), bis(tetrahydrofuran)-:
 bis[dicarbonyl-η-cyclopentadienyl-
 ferrate(1-)], 16:56
—, diindenyl, 17:137
—, metal carbonyl salts, 16:56
—, tetrakis(pyridine)-:
 bis[tetracarbonylcobaltate(1-)], 16:58
—, tetrakis(tetrahydrofuran)-:
 bis[dicarbonyl(η-cyclopentadienyl)-
 (tributylphosphine)molybdate(1-)],
 16:59
 bis[tricarbonyl(tributylphosphine)-
 cobaltate(1-)], 16:58
—, transition metal carbonyl salts, 16:56-60

Magnesium(II), [2,3,7,8,12,13,17,18-octaethyl-21H,23H-porphinato(2-)]-, 20:145
—, [1,4,8,11,15,18,22,25-octamethyl-29H,31H-tetrabenzo[b,g,l,q]porphinato(2-)]bis(pyridine)-, 20:158
Magnesium chromium oxide (Cr$_2$MgO$_4$), 20:52
Magnesium dihydride, 17:2
Magnesium potassium chloride (K$_2$MgCl$_4$), 20:51
Maleonitriledithiol, *see* 2-Butenedinitrile, 2,3-dimercapto-
Malonaldehyde, bromo-:
macrocyclic ligands from, 18:50
Malonic acid:
chromium complexes, 16:80-82
Manganate(II), tetrachloro-, dipotassium, 20:51
Manganese, acetylpentacarbonyl-, 18:57
—, [1,2-bis(diphenylphosphino)ethane]-(η^5-cyclopentadienyl)(thiocarbonyl)-, 19:191
Manganese, (bromo-η^5-cyclopentadienyl)tricarbonyl-, 20:193
—, bromopentacarbonyl-, 19:160
—, (bromotetrachloro-η^1-cyclopentadienyl)pentacarbonyl-, 20:194
—, (bromotetrachloro-η^5-cyclopentadienyl)tricarbonyl-, 20:194, 195
—, carbonyl(η^5-cyclopentadienyl)(thiocarbonyl)(triphenylphosphine)-, 19:189
—, carbonyltrinitrosyl-, 16:4
—, decacarbonyl-μ-[diborane(6)]-μ-hydrido-tri-, 20:240
—, dicarbonyl(η^5-cyclopentadienyl)(selenocarbonyl)-, 19:193, 195
—, dicarbonyl(η^5-cyclopentadienyl)(thiocarbonyl)-, 16:53
—, pentacarbonyl-:
thallium complex, 16:61
—, pentacarbonylchloro-, 19:159
—, pentacarbonyliodo-, 19:161, 162
—, pentacarbonyl(pentachloro-η^1-cyclopentadienyl)-, 20:193
—, tetracarbonyl[octahydrotriborato(1-)]-, 19:227, 228
—, tricarbonyl(chloro-η^5-cyclopentadienyl)-, 20:192
—, tricarbonyl(iodo-η^5-cyclopentadienyl)-, 20:193
—, tricarbonyl(pentachloro-η^5-cyclopentadienyl)-, 20:194
Manganese(II), [5,26:13,18-diimino-7,11:20,24-dinitrilodibenzo[c,n]-[1,6,12,17]tetraazacyclodocosinato(2-)]-, 18:48
Manganese chromium oxide (Cr$_2$MnO$_4$), 20:52
Manganese diphosphate (Mn$_2$P$_2$O$_7$), 19:121
Manganese potassium chloride (K$_2$MnCl$_4$), 20:51
5,7,7,12,14,14-Me$_6$[14]-4,11-diene-1,4,8,11-N$_4$, *see* 1,4,8,11-Tetraazacyclotetradeca-4,11-diene, 5,7,7,12,14,14-hexamethyl-
Mercury, bis[trimethyl(methylene)phosphorane]-:
dichloride, 18:140
Mercury(II), bis(selenourea)-:
dihalides, 16:85
—, dibromobis(selenourea)-, 16:86
—, dichlorobis(selenourea)-, 16:86
—, di-μ-chloro-dichlorobis(selenourea)di-, 16:86
Mercury(II) complexes:
with selenourea, 16:85-86
Mercury(II) halides:
complexes with selenourea, 16:85
Mercury bis[hexafluoroarsenate(V)]:
Hg$_3$[AsF$_6$]$_2$, 19:24
Mercury [μ-fluoro-bis(pentafluoroantimonate(V))]:
Hg$_3$[Sb$_2$F$_{11}$]$_2$, 19:23
Mercury hexafluoroantimonate:
Hg$_{2.91}$[SbF$_6$], 19:26
Mercury hexafluoroarsenate:
Hg$_{2.86}$[AsF$_6$], 19:25
Mesoporphyrine, *see* 21H,23H-Porphine-2,18-dipropionic acid, 7,12-diethyl-3,8,13,17-tetramethyl-
Metal atom syntheses, 19:59-69
Metal(II) chromites, 20:50
Metal(II) chromium(III) oxides, 20:50
Metal cluster complexes, 20:209
Metallic crystals:
growth of, in silica gel, 20:7

Metallatranes, 16:229-234
Metalloporphines, 20:143
Metalloporphyrins, 16:213-220
Methane, chloro-:
 cobalt complex, 20:234
—, diphenyl-:
 tungsten complex, 19:182
Methanesulfonate, 3-bromo-1,7-dimethyl-2-oxobicyclo[2.2.1]heptan-7-yl-, [1R(endo,anti)]-, see (+)-Camphor-π-sulfonate, α-bromo-
Methyl benzoate:
 chromium complex, 19:200
Methyl difluorophosphite, 16:166
Methyl diphenylphosphinite:
 nickel complexes, 17:119
Methylenediamine, N,N'-diphenyl-, see Aniline, N,N'-methylenedi-
Methylidyne complexes, see Carbyne complexes
Methyl m-methylbenzoate:
 chromium complex, 19:201, 202
Methyl phosphite:
 nickel complexes, 17:119
Methyl phosphorodifluoridite, 16:166
Methylene complexes, see Carbene complexes
mnt, see 2-butenedinitrile, 2,3-dimercapto-
Molybdate(1-), dicarbonylbis(η^5-cyclopentadienyl)(triphenylphosphine): tetrakis(tetrahydrofuran)magnesium-(2+), 16:59
Molybdate(3-), di-μ-bromo-hexabromo-μ-hydrido-di-:
 tricesium, 19:130
—, di-μ-chloro-hexachloro-μ-hydrido-di-:
 tricesium, 19:129
Molybdate(4-), octachloro-, see Molybdate(3-), di-μ-chloro-hexachloro-μ-hydrido-di-
Molybdate(5-), nonachlorodi:
 pentaammonium, monohydrate, 19:129
Molybdate(V), pentachlorooxo-:
 2,2'-bipyridinium(2+), 19:135
Molybdenum:
 vaporization of, 19:64
—, (η^3-allyl)(η^6-benzene)chloro(triphenylphosphine)-, 17:57
—, (η^6-benzene)bromo(η^5-cyclopentadienyl)-, 20:199
—, (η^6-benzene)chloro(η^5-cyclopentadienyl)-, 20:198
—, (η^6-benzene)(η^5-cyclopentadienyl)-, 20:196, 197
—, (η^6-benzene)(η^5-cyclopentadienyl)iodo-, 20:199
—, (η^6-benzene)dihydridobis(triphenylphosphine)-, 17:57
—, (η^6-benzene)tris(dimethylphenylphosphine)-, 17:59
—, (η^6-benzene)tris(dimethylphenylphosphine)dihydro-:
 bis(hexafluorophosphate), 17:60
—, (η^6-benzene)tris(dimethylphenylphosphine)hydrido-:
 hexafluorophosphate, 17:58
—, bis(η^6-benzene)-, 17:54
—, bis-μ-(benzoato)-dibromobis(tributylphosphine)di-, 19:133
—, bis(η^6-chlorobenzene)-, 19:81, 82
—, bis[η^5-cyclopentadienyldihalonitrosyl-:
 derivatives, 16:24
—, bis[η^5-cyclopentadienyldiiodonitrosyl-, 16:28
—, bis[dibromo-η^5-cyclopentadienylnitrosyl-], 16:27
—, bis[dichloro-η^5-cyclopentadienylnitrosyl-], 16:26
—, bis(diethyldithiocarbamato)dinitrosyl-, 16:235
—, bis(η^6-N,N-dimethylaniline)-, 19:84
—, bis(dinitrogen)bis[ethylenebis(diphenylphosphine)]-, trans-, 20:122
—, chloro(η^5-cyclopentadienyl)dinitrosyl-, 18:129
—, (η^5-cyclopentadienyl)dinitrosylphenyl-, 19:209
—, (η^5-cyclopentadienyl)ethyldinitrosyl-, 19:210
—, dicarbonyl(η^5-cyclopentadienyl)nitrosyl-, 16:24; 18:127
—, tetrakis-μ-(butrato)-di-, 19:133
—, tricarbonyl(η^5-cyclopentadienyl)silyl-, 17:104
Molybdenum(O), bis(*tert*-butyldifluorophosphine)tetracarbonyl-:
 cis-, 18:175
Molybdenum(II), (acetylene)carbonylbis-

(diisopropylphosphinodithioato)-, 18:55
—, bis(η^3-allyl)bis(η^6-benzene)di-μ-chlorodi-, 17:57
—, bis[1,2-bis(methylthio)ethane] tetrachlorodi-, 19:131
—, dicarbonylbis(diisopropylphosphinodithioato)-, 18:53
—, tetrabromotetrakis(pyridine)di-, 19:131
—, tetrabromotetrakis(tributylphosphine)di-, 19:131
Molybdenum(III), trichlorotris(tetrahydrofuran)-, 20:121
Molybdenum(IV), bis(acetonitrite)tetrachloro-, 20:120
—, tetrachlorobis(tetrahydrofuran)-, 20:121
Molybdenum(V), (2,2'-bipyridine)trichlorooxo-:
red and green forms, 19:135, 136
Molybdenum fluoride (MoF$_5$), 19:137-139
Mossbauer spectra:
iron-57 labeled porphyrins for, 20:157

Nickel:
vaporization of, 19:64
—, bis(nitrogen trisulfide)-, 18:124
—, bis(pentafluorophenyl)(η^6-toluene)-, 19:72
—, chlorohydridobis(tricyclohexylphosphine)-:
trans-, 17:84
—, chlorohydridobis(triisopropylphosphine)-:
trans-, 17:86
—, (1-3:6-7:10-12-η-2,6,10-dodecatriene-1,2-diyl)-, 19:85
—, hydrido[(tetrahydroborato(1-)]bis(tricyclohexylphosphine)-:
trans-, 17:89
—, (mercaptosulfur diimidato)(nitrogen trisulfide)-, 18:124
Nickel(O), (axobenzene)bis(*tert*-butyl isocyanide)-, 17:122
—, (azobenzene)bis(triethylphosphine)-, 17:123
—, (azobenzene)bis(triphenylphosphine)-, 17:121

—, bis(2,2'-bipyridine)-, 17:121
—, bis(*tert*-butyl isocyanide)(diphenylacetylene)-, 17:123
—, bis(*tert*-butyl isocyanide)(ethenetetracarbonitrile)-, 17:123
—, bis[ethylenebis(dimethylphosphine)]-, 17:119
—, bis[ethylenebis(diphenylphosphine)]-, 17:121
—, bis(1,10-phenanthroline)-, 17:121
—, bis[1,2-phenylenebis(dimethylarsine)]-, 17:121
—, (butenedinitrile)bis(*tert*-butyl isocyanide)-, 17:123
—, (stilbene)bis(triphenylphosphine)-, 17:121
—, tetrakis(*tert*-butyl isocyanide)-, 17:118
—, tetrakis(cyclohexyl isocyanide)-, 17:119
—, tetrakis(diethylphenylphosphine)-, 17:119
—, tetrakis(methyldiphenylphosphine)-, 17:119
—, tetrakis(methyl diphenylphosphinite)-, 17:119
—, tetrakis(tributylphosphine)-, 17:119
—, tetrakis(triethylphosphine)-, 17:119
—, tetrakis(triethyl phosphite)-, 17:119
—, tetrakis(triisopropyl phosphite)-, 17:119
—, tetrakis(trimethylphosphine)-, 17:119
—, tetrakis(trimethyl phosphite)-, 17:119
—, tetrakis(triphenylarsine)-, 17:121
—, tetrakis(triphenylphosphine)-, 17:120
—, tetrakis(triphenyl phosphite)-, 17:119
—, tetrakis(triphenylstibine)-, 17:121
Nickel(II), bis(isothiocyanato)(tetrabenzo[*b,f,j,n*][1,5,9,13] tetraazacyclohexadecine)-, 18:31
—, bis(isothiocyanato)(2,3,9,10-tetramethyl-1,4,8,11-tetraazacyclotetradeca-1,3,8,10-tetraene)-, 18:24
—, bis(mercaptosulfur diimidato)-, 18:124
—, [6,13-diacetyl-5,14-dimethyl-1,4,8,11-tetraazacyclotetradeca-4,6,11,13-tetraenato(2-)]-, 18:39
—, dibromobis(di-*tert*-butylfluorophosphine)-:

Subject Index   267

trans-, 18:177
—, [7,16-dihydro-6,8,15,17-tetramethyldibenzo[*b,i*] [1,4,8,11] tetraazacyclotetradecinato(2-)]-, 20:115
—, (2,12-dimethyl-3,7,11,17-tetraazabicyclo[11.3.1]heptadeca-1(17),-2,11,13,15-pentaene)-: diperchlorate, 18:18
—, (2,3-dimethyl-1,4,8,11-tetraazacyclotetradeca-1,3-diene)-: tetrachlorozincate, 18:27
—, [5,14-dimethyl-1,4,8,11-tetraazacyclotetradeca-4,6,11,13-tetraenato(2-)]-, 18:42
—, [3,3'-[[ethylenebis(iminomethylidyne)]di-2,4-pentanedionato]-(2-)]-, 18:38
—, [5,6,14,15,20,21-hexamethyl-1,3,4,7,-8,10,12,13,16,17,19,22-dodecaazatetracyclo[8.8.4.1^{3,17}.1^{8,12}] tetracosa-4,6,13,15,19,21-hexaene-N^4,-N^7,N^{13},N^{16},N^{19},N^{22}]-, bis(tetrafluoroborate), 20:89
—, (5,7,7,12,14,14-hexamethyl-1,4,8,11-tetraazacyclotetradeca-4,11-diene)-: diperchlorate, *meso*- and *racemic*-, 18:5
—, (5,5,7,12,12,14-hexamethyl-1,4,8,11-tetraazacyclotetradecane)-: diperchlorate, *meso*-, 18:12
—, pentakis(trimethyl phosphite)-: bis(tetraphenylborate), 20:76
—, (1,4,8,11-tetraazacyclotetradecane)-: perchlorate, 16:221
—, (tetrabenzo[*b,f,j,n*] [1,5,9,13] tetraazacyclohexadecine)-: diperchlorate, 18:31
—, (2,3,9,10-tetramethyl-1,4,8,11-tetraazacyclotetradeca-1,3,8,10-tetraene)-: diperchlorate, 18:23
—, [5,10,15,20-tetraphenyl-21*H*,29*H*-porphinato(2-)]-, 20:143
Nickel(O) acetylene complexes, 17:117-124
Nickelate(2-), bis[2,3-dimercapto-2-butenedinitrilato(2-)]-: salt with 2,2'-bi-1,3-dithiolylidene (1:2), 19:31
Nickelate(II), tetrachloro-, dipotassium, 20:51

Nickel chromium oxide ($Cr_2 NiO_4$), 20:52
Nickel(O) diazene complexes, 17:117-124
Nickel(O) isocyanide complexes, 17:117-124
Nickel(O) olefin complexes, 17:117-124
Nickel(O) phosphine complexes, 17:117-124
Nickel(O) phosphite complexes, 17:117-124
Nickel potassium chloride ($K_2 NiCl_4$), 20:51
Nickel tetrafluorooxovanadate(IV) heptahydrate, 16:87
Niobate(1-), hexacarbonyl-: tetraphenylarsonium, 16:72
tris[bis(2-methoxyethyl)ether-, potassium, 16:69
Niobium, bis(η^5-cyclopentadienyl)-: complexes, 16:107-113
Niobium(III), bis(η^5-cyclopentadienyl)-(dimethylphenylphosphine)hydrido-, 16:110
—, bis(η^5-cyclopentadienyl)(tetrahydroborato)-, 16:109
—, bromobis(η^5-cyclopentadienyl)(dimethylphenylphosphine)dihydrido-, 16:112
Niobium(IV), bis(2,2'-bipyridine)tetrakis-(isothiocyanato)-, 16:78
—, dichlorobis (η^5-cyclopentadienyl)-, 16:107
Niobium(V), bis(η^5-cyclopentadienyl)(dimethylphenylphosphine)dihydrido-: hexafluorophosphate, 16:111
tetrafluoroborate, 16:111
Niobium chloride ($NbCl_5$), 20:42
Nioxime, *see* 1,2-Cyclohexanedione dioxime
Nitrogen: molybdenum and tungsten complexes, 20:122, 126
Nitrogen oxide, di-: as ligand, 16:75
Nitrogen sulfide (NS_3): nickel complex, 18:124
Nitrogen sulfide ($N_4 S_4$): caution in handling and storing of, 17:197
Nitrosylchromium: tetra-, 16:2

Nitrosyl complexes:
  of chromium, molybdenum, and tungsten, 18:126; 19:208
  of cobalt, 16:29, 33
  of iridium, 16:41
  of molybdenum, 16:24, 235
  of osmium, 16:11-12, 40
  of rhenium, 16:35
  of rhodium, 16:33; 17:129
  of ruthenium, 16:13, 21, 29
Nitrous oxide:
  ligand, 16:75

1,3,6,8,10,13,16,19-Octaazabicyclo-[6.6.6]eicosane:
  cobalt complex, 20:86
Olefin complexes:
  of platinum, 19:213
Olefin complexes of nickel(O), 17:117-124
Olefins, cyclo-, see Cycloolefin complexes
Organometallatranes, 16:229-234
Osmate(VI), tetrahydrooxodioxo-, dipotassium, 20:61
Osmium, pentaamminenitrosyl-:
  trihalide, monohydrates, 16:11
—, tetraamminehalonitrosyldihalides, 16:12
—, tetraamminehydroxonitrosyldihalides, 16:11
Osmium(II), carbonylbis(trifluoroacetato)bis(triphenylphosphine)-, 17:128
—, carbonylchloro(trifluoroacetato)tris(triphenylphosphine)-, 17:128
—, pentaammine(dinitrogen)-:
  diiodide, 16:9
Osmium(III), hexaammine-:
  triiodide, 16:10
—, pentaammineiodo-:
  diiodide, 16:10
Osmium(VI), bis(ethylenediamine)dioxo-, trans-, dichloride, 20:61
Osmium complexes:
  ammines, 16:9-12
Osmium fluoride (OsF_5), 19:137-139
Oxalic acid, 20:58
  chromium complex, 19:127
  platinum partially oxidized complex, 19:16

Oxalic acid chromium complexes, 17:147-151
Oxide, bis(trimethyltin), 17:181
Oxygen:
  iron complex, 20:168

Palladium:
  vaporization, 19:64
—, chlorohydridobis(tricyclohexylphosphine)-:
  trans-, 17:87
—, (η-ethylene)bis(tricyclohexylphosphine), 16:129
—, (η-ethylene)bis(triphenylphosphine)-, 16:127
—, (η-ethylene)bis(tri-o-tolyl phosphite)-, 16:129
—, hydrido[tetrahydroborato(1-)]bis(tricyclohexylphosphine)-:
  trans-, 17:90
Palladium(O), bis(dibenzylideneacetone)-, 17:135
—, bis(di-tert-butylphenylphosphine)-, 19:102
—, bis(tri-tert-butylphosphine)-, 19:103
—, bis(tricyclohexylphosphine)-, 19:103
Palladium(I), tetrakis(tert-butyl isocyanide)dichlorodi-, 17:134
Palladium(II), (η^3-allyl)(η^5-cyclopentadienyl)-, 19:221
—, bis(η^3-allyl)di-μ-chloro-di-, 19:220
—, pentakis(trimethyl phosphite)-:
  bis(tetraphenylborate), 20:77
Pentaammine type complexes of cobalt(III), rhodium(III), and iridium (III), 17:152-154
Pentaborane(9), 1-bromo-:
  preparation and detonation of, 19:247, 248
2,4-Pentanedionato complexes of molybdenum, 17:61
2,4-Pentanedione:
  indium complex, 19:261
  macrocyclic ligands from, 18:37
  platinum complex, 20:66
—, 3-(ethoxymethylene)-, 18:37
—, 3,3'-[ethylenebis(iminomethylidyne)]-di-, 18:37
—, 1,1,1,5,5,5-hexafluoro-:
  complexes, 16:118
  platinum complex, 20:67

—, 1,1,1-trifluoro-:
  complexes, 16:118
  platinum complex, 20:67
2-Pentanethione, 4,4'-(alkylenedinitrilo)-
  bis-:
  transition metal complexes, 16:227-228
—, 4,4'-(ethylenedinitrilo)bis-, 16:226
Pentathiazyl hexachloroantimonate(1-),
  17:191
Penthathiazyl salts, 17:188-192
Penthathiazyl tetrachloroaluminate(1-),
  17:190
Pentathiazyl tetrachloroferrate(1-), 17:90
1,10-Phenanthroline:
  nickel complexes, 17:119
o-Phenylenediamine:
  macrocyclic ligands from, 18:51
Phenyl isocyanide:
  platinum complex, 19:174
Phenylphosphinic acid, {ethylenebis-
  [(methylimino)methylene] }bis-
  dihydrochloride, 16:201-202
—, [ethylenebis(nitrilodimethylene)]-
  tetrakis-, 16:199
—, [nitrilotris(methylene)] tris-, 16:201-
  202
Phenyl phosphite:
  nickel complexes, 17:119
Phosphate (PnO$_{(3n+1)}$H$_2$):
  polyammonium, *catena*-poly-, 19:278
Phosphates:
  crystal growth of, in silica gel, 20:5
Phosphide, diphenyl-:
  lithium, 17:186
Phosphinates, chromium(III) complexes:
  polymers, 16:89-92
Phosphine, see Amine, tris {2-(diphenylphosphino)ethyl}-; Ethylenediamine,
  N,N,N',N'-tetrakis[(diphenylphosphino)methyl]-
—, benzyldiphenyl-, 16:159
—, bis[2-(methylthio)phenyl] phenyl-,
  16:172
—, *tert*-butyldifluoro-:
  transition metal complexes, 18:173, 174
—, butyldiphenyl-, 16:158
—, cyclohexyldiphenyl-, 16:159
—, dibenzylphenyl-, 18:122
—, di-*tert*-butylfluoro-, 18:176
—, dibutylphenyl-, 18:171

—, di-*tert*-butylphenyl-:
  palladium and platinum complexes,
  19:102, 104
—, dicyclohexylphenyl-, 18:171
—, diethylphenyl-, 18:170
  nickel(O) complex, 17:119
—, dimethyl-:
  titanium complex, 16:100
—, [(2,2-dimethyl-1,3-dioxolane-4,5-
  diylbis(methylene)] bis[diphenyl-,
  17:81
—, dimethyl(pentafluorophenyl)-, 16:181
—, dimethylphenyl-:
  molybdenum complexes, 17:58-60
  niobium complex, 16:110-112
  rhenium complexes, 17:65, 111
  titanium complex, 16:239
—, diphenyl-, 16:161
—, [2-(diphenylarsino)ethyl] diphenyl-,
  16:191
—, [2-(diphenylarsino)vinyl] diphenyl-,
  16:189
—, diphenyl[2-(phenylphosphino)ethyl]-,
  16:202
—, diphenyl(trimethylsilyl)-, 17:187
—, ethyldiphenyl-, 16:158
—, ethylenebis[dimethyl-,:
  nickel complexes, 17:119
—, ethylenebis[diphenyl-, 20:208
  cobalt complex, 16:19
  iron complexes, 17:69-71
  manganese complex, 18:191
  molybdenum and tungsten complexes,
    20:122, 125, 126
  nickel complexes, 17:121
—, [2-(isopropylphenylphosphino)ethyl]-
  diphenyl-, 16:192
—, methyl-:
  titanium complex, 16:98
—, methyldiphenyl-, 16:157
  cobalt, complex, 16:29
  nickel complexes, 17:119
—, [2-(methylthiophenyl] diphenyl-,
  16:171
—, tributyl-:
  molybdenum complex, 19:131, 133
  molybdenum complexes, 16:58-59
  nickel complexes, 17:119
  platinum complex, 19:116
—, tri-*tert*-butyl-:

Subject Index

palladium complex, 19:103
—, tricyclohexyl-:
nickel complexes, 17:84, 89
palladium complexes, 16:129; 17:87, 90
palladium and platinum complexes, 19:103, 105
platinum complex, 19:216
—, triethyl-:
nickel complexes, 17:119, 123
platinum complex, 17:132; 19:108, 110, 174-176
titanium complex, 16:101
—, triisopropyl-:
nickel complexes, 17:86
platinum complex, 19:108
—, trimethyl-, 16:153
iridium complex, 18:63, 64
iron complexes, 20:69-75
nickel complexes, 17:119
titanium complex, 16:101
—, triphenyl-:
cobalt complexes, 16:18, 33
copper complex, 19:87, 88, 93, 96
iridium complexes, 16:41; 17:129
iridium and rhodium complexes, 19:204, 206
manganese complex, 19:187
molybdenum complexes, 17:57
nickel complexes, 17:120-121
osmium complex, 17:128
palladium complexes, 16:127
platinum complex, 19:98, 115
platinum complexes, 17:130; 18:120
rhenium complexes, 17:110
rhodium complexes, 17:129
ruthenium complexes, 16:21; 17:73, 75, 77, 79, 126-127
tungsten complex, 20:124
—, tri(phenyl-d_5)-, 16:164
—, tris[2-(methylthio)phenyl-, 16:173
—, tris-*p*-tolyl-:
copper complex, 19:89
—, vinylenebis(diphenyl-):
cobalt complex, 20:207
Phosphine complexes with nickel(0), 17:117-124
Phosphine oxide, dimethylphenyl-, 17:185
—, methyldiphenyl-, 17:184

Phosphines:
(alkylamino)difluoro-, as ligand, 16:63-66
difluorohalo, as ligand, 16:66-68
—, tertiary, 18:169
Phosphine sulfide:
[(isopropylphenylphosphino)methyl] diphenyl-, 16:195
Phosphinodithioic acid, diisopropyl-, molybdenum complex, 18:53, 55
Phosphinous acid, diphenyl-:
methyl ester, nickel complex, 17:119
Phosphinous fluoride, di-*tert*-butyl-, *see* Phosphine, di-*tert*-butylfluoro-
Phosphite, difluoro-:
methyl, 16:166
Phospholanium, 1-ethyl-1-phenyl-:
perchlorate, 18:189, 191
Phosphonium, methyltriphemyl-:
heptahydrodiborate(1-), 17:24
tetrahydroborate(1-), 17:22
—, tetramethyl-:
bromide, 18:138
—, trimethyl-:
methylide, 18:137, 138
(trimethylsilyl)methanide, 18:137
Phosphonium compounds, quaternary, 17:21-26
Phosphonous difluoride, *tert*-butyl-, *see* Phosphine, *tert*-butyldifluoro-
Phosphoramidic acid $(HO)_2 PO(NH_2)$:
and salts, correction, 19:281
Phosphorane, bis(diethylamino)trifluoro-, 18:187
—, bis(dimethylamino)trifluoro-, 18:186
—, (diethylamino)tetrafluoro-, 18:185
—, (dimethylamino)tetrafluoro-, 18:181
—, trimethylmethylene-, *see* Phosphonium, trimethyl-, methylide
—, trimethyl[(trimethylsilyl)methylene-], *see* Phosphonium, trimethyl-, (trimethylsilyl)methanide
Phosphoranes:
dialkylamino fluoro, 18:179
Phosphorodifluoridous acid, methyl ester, 16:166
Phosphorous acid:
esters, nickel complexes, 17:119
tri-*o*-tolyl ester, palladium complex, 16:129
Phosphorus(1+),μ-nitrido-bis(triphenyl-,

undecacarbonylhydridotriferrate(1-), 20:218
—, µ-nitrido-bis(triphenyl-, undecacarbonyltriferrate(2-) (2:1), 20:222
Phosphorus-sulfur ligands, 16:168-172
Phosphorus ylides, 18:135
Phthatocyanine:
  iron and lithium complexes, 20:159, 160
Platinate, bis(oxalato)-:
  dipotassium, dihydrate, 19:16
  potassium (1:1.64), dihydrate, 19:16, 17
Platinate, hexacyano-:
  partially oxidized, 20:23
—, tetracyano-:
  cesium (1:1.75), dihydrate, 19:6, 7
  cesium fluoride (1:2:0.19), 20:29
  cesium (hydrogen difluoride) (1:2:0.23), 20:26
  cesium (hydrogen difluoride) (1:2:0.38), 20:28
  guanidinium bromide (1:2:0.25), monohydrate, 19:10, 12
  potassium (1:1.75), sesquihydrate, 19:8, 14
  potassium bromide (1:2:0.3), trihydrate, 19:1, 4, 15
  potassium chloride (1:2:0.3), trihydrate, 19:15
  rubidium (1:1:6), dihydrate, 19:9
  rubidium [hydrogen bis(sulfate)] (1:3:0.45), monohydrate, 20:20
  rubidium (hydrogen difluoride) (1:2:0.29), hydrate (1:1.6), 20:24
  rubidium (hydrogen difluoride) (1:2:0.38), 20:25
Platinate(1-), bis[2,3-dimercapto-2-butenedinitrilato(2-)]-:
  salt with 2,2]-bi-1,3-dithiolylidene (1:1), 19:31
Platinate(2-), bis[2,3-dimercapto-2-butenedinitrilato(2-)-]-:
  salt with 2,2'-bi-1,3-dithiolylidene (1:2), 19:31
Platinate(II), tetrabromo-:
  dipotassium, 19:2
—, tetracyano-:
  barium (1:1), tetrahydrate, 19:112
  bis(guanidinium), 19:11

dipotassium, trihydrate, 19:3
  salt with 2,2'-bi-1,3-dithiolylidene (1:2), 19:31
Platinate(IV), dibromotetracyano-:
  bis(guanidinium), hydrate, 19:11
  dipotassium, dihydrate, 19:4
—, hexabromo-:
  dihydrogen, 19:2
  dipotassium, 19:2
—, tetracyano-:
  dicesium, monohydrate, 19:6
Platinum:
  vaporization of, 19:64
—, (anilinoethoxymethylene)dichloro(triethylphosphine)-:
  cis-, 19:175
—, (aryltrimethylene)dichloro-, 16:113-116
—, (aryltrimethylene)dichlorobis(pyridine)-, 16:113-116
—, (2-benzyltrimethylene)-, 16:116
—, (2-benzyltrimethylene)dichlorobis-(pyridine)-, 16:116
—, (1-butyl-2-methyltrimethylene)dichloro-:
  trans-, 16:116
—, (dianilinomethylene)dichloro(triethylphosphine)-:
  cis-, 19:176
—, dichlorobis(pyridine)(1-p-tolyltrimethylene)-, 16:116
—, dichlorobis(pyridine)(trimethylene)-, 16:116
—, dichlorobis(tricyclohexylphosphine)-:
  trans-, 19:105
—, dichlorobis(triisopropylphosphine)-:
  trans-, 19:108
—, dichloro(1,2-diphenyltrimethylene)-:
  trans-, 16:116
—, dichloro(1,2-diphenyltrimethylene)-bis(pyridine)-, 16:116
—, dichloro(2-hexyltrimethylene)-, 16:116
—, dichloro(2-hexyltrimethylene)bis-(pyridine)-, 16:116
—, dichloro[2-(2-nitrophenyl)trimethylene]-, 16:116
—, dichloro[2-(2-nitrophenyl)trimethylene]bis(pyridine)-, 16:116
—, dichloro(phenyl isocyanide)(triethylphosphine)-:

272 Subject Index

cis-, 19:174
—, dichloro(2-phenyltrimethylene)-, 16:116
—, dichloro(2-phenyltrimethylene)bis(pyridine)-, 16:115-116
—, dichloro[1-(*p*-tolyl)trimethylmethylene]-, 16:116
—, dichloro(trimethylene)-, 16:114
—, tris(triethylphosphine)-, 19:108
—, tris(triisopropylphosphine)-, 19:108
Platinum(O), bis(1,5-cyclooctadiene)-, 19:213, 214
—, bis(di-*tert*-butylphenylphosphine)-, 19:104
—, bis(ethylene)(tricyclohexylphosphine)-, 19:216
—, bis(tricyclohexylphosphine)-, 19:105
—, (diphenylacetylene)bis(triphenylphosphine)-, 18:122
—, (ethylene)bis(triphenylphosphine)-, 18:121
—, tetrakis(triethylphosphine-, 19:110
—, tetrakis(triphenylphosphine)-:
  in preparation of mix ligand complexes, 18:120
—, tris(ethylene)-, 19:215
Platinum(1+), [anilino(ethylamino)methylene] chlorobis(triethylphosphine)-:
trans-, perchlorate, 19:176
Platinum(II), (2-ammonioethanethiolato-*S*)(2,2':6',2''-terpyridine)-, dinitrate, 20:104
—, bis(acetato)bis(triphenylphosphine)-: *cis-*, 17:130
—, bis(dimethyl phosphito-*P*)[*o*-phenylenebis(dimethylarsine)]-, 19:100
—, bis(1,1,1,5,5,5-hexafluoro-2,4-pentanedionato)-, 20:67
—, bis(2,4-pentanedionato)-, 20:66
—, bis[1,1,1-trifluoro-2,4-pentanedionato)-, 20:67
—, carbonatobis(triphenylphosphine)-, 18:120
—, chloro(dimethyl hydrogen phosphite-*P*)(dimethyl phosphito-*P*)(triphenylphosphine)-, 19:98
—, chloroethylbis(triethylphosphine)-:
trans-, 17:132

—, chloro(2,2':6',2''-terpyridine)-, chloride, dihydrate, 20:102
—, dichlorobis(tributylphosphine)-: *trans-*, 19:116
—, dichlorobis(triphenylphosphine)-: *trans-*, 19:115
—, di-μ-chloro-dichlorobis(1-dodecene)di-, 20:181, 183
—, di-μ-chloro-dichlorobis(ethylene)di-, 20:181, 182
—, di-μ-chloro-dichlorobis(styrene)di-, 20:181, 182
—, dichloro(ethylene)(pyridine)-: *trans-*, 20:181
—, (2-mercaptoethanolato-*S*)(2,2':6',2''-terpyridine)-, nitrate, 20:103
—, pentakis(trimethyl phosphite)-: bis(tetraphenylborate), 20:78
Platinum(IV), dibromodimethyl-, 20:185
—, dibromodimethylbis(pyridine)-, 20:186
—, dihydroxodimethyl-, sesquihydrate, 20:185, 186
Platinum chloride (PtCl$_2$):
  phosphine complexes, trans, 19:114
Platinum chloride (PtCl$_2$), 20:48
Platinum compounds:
  one-dimensional elec. conductors, 19:1-18
Platinum(II) β-diketonate complexes, 20:65
Platinum sulfide (PtS$_2$), 19:49
Platinum telluride (PtTe$_2$), 19:49
Polymers:
  of chromium(III), bis(phosphinates), 16:89-92
*catena-*Polyphosphate (PnO$_{(3n+1)}$H$_2$) polyammonium, 19:278
21*H*,23*H*-Porphine, *see* Porphyrin
—, 2,3,7,8,12,13,17,18-octaethyl-, iron and magnesium complexes, 20:145, 151
—, 5,10,15,20-tetraphenyl- (H$_2$tpp) nickel and vanadium complexes, 20:143, 144
21*H*,23*H*-Porphine-2,18-dipropionic acid, 7,12-diethyl-3,8,13,17-tetramethyl-, dimethyl ester, iron complex, 20:148
2,18-Porphinedipropanoic acid, 7,12-bis-

(1-hydroxyethyl)-3,8,13,17-tetramethyl-:
dimethyl ester, chloro iron(III) complexes, 16:216
Porphyrin, *see also* 21*H*,23*H*-Porphine
—, 5,10,15,20-tetrakis(2-aminophenyl)-, and *all-cis-*, 20:163, 164
—, 5,10,15,20-tetrakis[2-(2,2-dimethylpropionamido)phenyl]-, *all-cis-*, 20:165
—, 5,10,15,20-tetrakis(2-nitrophenyl)-, (H_2 TNPP), 20:162
Porphyrin(IX), hemato-:
dimethyl ester, chloro-, iron complex, 16:216
—, 5,10,15,20-tetraphenyl-:
copper complex, 16:214
Porphyrin complexes:
with copper, 16:214
with iron, 16:216
Porphyrins:
iron-57 insertion, 20:153
Potassium, tris[bis(2-methoxyethyl) ether]-:
hexacarbonylniobate(1-), 16:69
hexacarbonyltantalate(1-), 16:71
Potassium bis(oxalato)platinate:
(1.64:1), dihydrate, 19:16, 17
Potassium cobalt chloride (CoK$_2$Cl$_4$), 20:51
Potassium cyclopentadienide, 17:105, 173, 176
Potassium diaquabis(malonato)chromate(III):
cis- and *trans-*, trihydrate, 16:81
Potassium dicarbonyldichloroiridate (0.6:1):
hemihydrate, 19:20
Potassium magnesium chloride (K$_2$MgCl$_4$), 20:51
Potassium manganese chloride (K$_2$MnCl$_4$), 20:51
Potassium methylcyclopentadienide, 17:175
Potassium nickel chloride (K$_2$NiCl$_4$), 20:51
Potassium osmate (K$_2$(OsO$_2$(OH)$_4$)), 20:61
Potassium tetracyanoplatinate (1.75:1):
sesquihydrate, 19:8, 14

Potassium tetrahydridogallate(1-), 17:50
Potassium tris(malonato)chromate(III):
trihydrate, 16:80
Potassium zinc chloride (K$_2$ZnCl$_4$), 20:51
Propanedial, bromo-:
macrocyclic ligands from, 18:50
1,3-Propanediamine:
macrocyclic ligands from, 18:23
—, *N*-(3-aminopropyl)-, *see* Dipropylamine, 3,3'-diamino-
Propanedinitrile, 2,2'-(2,5-cyclohexadiene-1,4-diylidene)bis-:
radical anion, salt with 2,2'-bi-1,3-dithiolylidene (1:1), 19:32
2-Propanethiol, 2-methyl-:
iridium complex, 20:237
Propene:
palladium complexes, 19:220, 221
2-Propenyl, *see* Allyl, 19:220
Protoporphyrine, *see* 21*H*,23*H*-Porphine-2,18-dipropionic acid, 7,12-diethenyl-3,8,13,17-tetramethyl-
Pyridine:
compd. with TaS$_2$ (1:2), 19:40
molybdenum complex, 19:131
platinum complex, 20:181
—, 4-*tert*-butyl-:
cobalt complexes, 20:130-132
—, 2,6-diacetyl-:
macrocyclic ligands from, 18:18
—, 2,6-diamino-:
macrocyclic ligands from, 18:47
Pyrophosphate, *see* Diphosphate

Rare earth hexacyanotransition metalates:
crystal growth of, by double infusion, 20:12
Resolution:
of bis(ethylenediamine)(oxalato)rhodium(III) and *cis*-dichlorobis(ethylenediamine)rhodium(III) ions, 20:60
of *cis*-amminebromobis(ethylenediamine)cobalt(III) ion, 16:93
of cobalt ethylenediamine salts, 18:96
of [Co(edta)]-, 18:100
Rhenium, acetyl(1-aminoethylidene)tetracarbonyl-, *cis-*, 20:204
—, acetylpentacarbonyl-, 20:201
—, acetyltetracarbonyl(1-hydroxyethylidene)-, *cis-*, 20:200, 202

—, carbonylchloronitrosyl complexes, 16:35-38
—, octacarbonyldi-μ-chloro-di-, 16:35
—, pentacarbonyltri-μ-chloro-nitrosyldi-, 16:36
—, phosphine complexes of, 17:110-112
—, tetraacetatodichlorodi-, 20:46
—, tetracarbonyldi-μ-chloro-dichlorodinitrosyldi-, 16:37
—, tri-μ-carbonyl-nonacarbonyltetra-, 20:209
Rhenium(I), dodecacarbonyltetra-μ-hydrido-*tetrahedro*-tetra-, 18:60
—, dodecacarbonyltri-μ-hydrido-*triangulo*-tri-, 17:66
Rhenium(III), pentahydridotris(dimethylphenylphosphine)-, 17:64
—, trichlorotris(dimethylphenylphosphine)-:
mer-, 17:65, 111
Rhenium(V), trichlorooxobis(triphenylphosphine)-, 17:110
Rhenium bromide (Re$_3$Br$_9$), 20:47
Rhenium chloride (ReCl$_5$), 20:41
Rhenium chloride (Re$_3$Cl$_9$), 20:44, 47
Rhenium fluoride (ReF$_5$), 19:137-139
Rhenium halides (Re$_3$X$_9$), 20:46
Rhenium iodide (Re$_3$I$_9$), 20:47
Rhodate(2-), μ_6-carbido-nona-μ-carbonylhexacarbonylhexa-, dipotassium, 20:212
—, di-μ-carbonyl-octa-μ_3-carbonylicosacarbonyldodeca-, disodium, 20:215
Rhodium, (bromotetraphenyl-η^5-cyclopentadienyl)(η^4-1,5-cyclooctadiene)-, 20:192
—, (chloro-η^5-cyclopentadienyl)(η^4-1,5-cyclooctadiene)-, 20:192
—, (chlorotetraphenyl-η^5-cyclopentadienyl)(η^4-1,5-cyclooctadiene)-, 20:191
—, chloro(thiocarbonyl)bis(triphenylphosphine)-:
trans-, 19:204
—, (η^4-1,5-cyclooctadiene)(pentachloro-η^5-cyclopentadienyl)-, 20:194
—, dicarbonyl(chlorotetraphenyl-η^5-cyclopentadienyl)-, 20:192
—, di-μ-chloro-bis(η^4-1,5-hexadiene)di-, 19:219

—, di-μ-chloro-tetrakis(η^2-2,3-dimethyl-2-butene)di-, 19:219
—, dodecacarbonyltetra-, 17:115
—, hexadecacarbonylhexa-, 16:49
—, nitrosyltris(triphenylphosphine)-, 16:33
Rhodium(I), acetatotris(triphenylphosphine)-, 17:129
—, bis[(+)-[(2,2-dimethyl-1,3-dioxolane-4,5-diyl)bis(methylene)]bis[diphenylphosphine]]hydrido-, 17:81
—, di-μ-chloro-bis(η^4-1,5-cyclooctadiene)-di-, 19:218
—, pentakis(trimethyl phosphite)-, tetraphenylborate, 20:78
Rhodium(III), bis(acetato)nitrosylbis(triphenylphosphine)-, 17:129
—, bis(ethylenediamine)dinitro-:
cis-, nitrate, 20:59
—, bis(ethylenediamine)(oxalato)-:
perchlorate, 20:58
—, dibromobis(ethylenediamine):
cis-, bromide, 20:60
—, dichlorobis(ethylenediamine):
cis-, chloride, hydrate, 20:60
—, pentaammine(carbonato)-:
perchlorate, monohydrate, 17:153
Rhodium carbonyl:
Rh$_4$(CO)$_{12}$, 17:115
Rh$_6$(CO)$_{16}$, 16:49
Rubidium tetracyanoplatinate (1.6:1) dihydrate, 19:9
Rubidium tetracyanoplatinate [hydrogen bis(sulfate)] (3:1:0.45), monohydrate, 20:20
Rubidium tetracycloplatinate (hydrogen difluoride) (2:1:0.29), hydrate (1:1.67), 20:24
Rubidium tetracyanoplatinate (hydrogen difluoride) (2:1:0.38), 20:25
Ruthenium, acetatotetraamminenitrosyl-:
diperchlorate, 16:14
—, acidotetraamminenitrosyl-:
diperchlorate, 16:14
—, amminenitrosyl-:
complexes, 16:13-16
—, chlorodinitrosylbis(triphenylphosphine)-:
tetrafluoroborate, 16:21
—, decacarbonyldi-μ-nitrosyl-tri-, 16:39

—, dihydridotetrakis(triphenylphosphine)-, 17:75
—, dodecacarbonyltri-, 16:45, 47
—, hexacarbonyldi-μ-chloro-dichlorodi-, 16:51
—, hydrido(η^6-phenyldiphenylphosphine)bis(triphenylphosphine)-: tetrafluoroborate(1-), 17:77
—, pentakis(trimethyl phosphite)-, 20:80
—, tetraamminechloronitrosyl-: dichloride, 16:13
—, tetraamminecyanatonitrosyl-: diperchlorate, 15:15
—, tricarbonyl(1,5-cyclooctadiene)-, 16:105
—, tricarbonyl-η-diene complexes, 16:103
—, trichloronitrosyl-, 17:73
Ruthenium(I), hydridonitrosyltris(triphenylphosphine)-, 17:73
Ruthenium(II), acetatocarbonylchlorobis(triphenylphosphine)-, 17:126
—, acetatocarbonylhydridobis(triphenylphosphine)-, 17:126
—, acetatohydridotris(triphenylphosphine)-, 17:79
—, bis(acetato)dicarbonylbis(triphenylphosphine)-, 17:126
—, carbonylbis(trifluoroacetato)bis(triphenylphosphine)-, 17:127
—, nitrosyltris(trifluoroacetato)bis(triphenylphosphine)-, 17:127
—, pentaammine(dinitrogen oxide)-: dihalides, 16:75-77
—, tris(ethylenediamine)-: dichloride, 19:118
tetrachlorozincate, 19:118
Ruthenium(III), tris(ethylenediamine)-, trichloride, 19:119
Ruthenium carbonyl: Ru$_3$(CO)$_{12}$, 16:45, 47

Samarium(III) hexacyanoferrate(III), tetrahydrate:
crystal growth of, by double infusion, 20:13
Scandium(III), tris[bis(trimethylsilyl)-amido]-, 18:115
Selenate, pentafluorooxo-, hydrogen, 20:38
Selenic acid, 20:37

Selenide, bis(trimethylsilyl), see Disilaselenane, hexamethyl-
—, digermyl, see Digermaselenane
Selenides:
Mo, Ta, and Ti, 19:46
Selenocarbonyl complexes:
Mn, 19:193, 195
Selenonyl difluoride, 20:36
Selenourea:
cobalt(II) complexes, 16:83-85
mercury halide complexes, 16:85-87
Silane, cyclopentadienyl-, 17:172
—, difluorodimethyl-, 16:141
—, (dimethylamino)trimethyl-, 18:180
—, fluoro(methyl)-: derivatives, 16:139-142
—, iodo-, 19:268, 270
—, iododimethyl-, 19:271
—, iodomethyl-, 19:271
—, iodotrimethyl-, 19:272
—, (methylcyclopentadientyl)-, 17:174
—, tetramethyl-: magnesium complex, 19:262
—, trifluoro(methyl)-, 16:139
Silica gels:
crystal growth in, 20:2
Silver, bis-μ-[[dimethyl(methylene)phosphoranyl]methyl] di-, 18:142
Silver(I), η-cycloalkene(fluoro-β-diketonato)-:
complexes, 16:117-119
—, cycloheptene(1,1,1,5,5,5-hexafluoro-2,4-pentanedionato)-, 16:118
—, η-cyclohexene(1,1,1,5,5,5-hexafluoro-2,4-pentanedionato)-, 16:118
—, η-cyclooctadiene(1,1,1,5,5,5-hexafluoro-2,4-pentanedionato)-, 16:117
—, η-cyclooctadiene(1,1,1-trifluoro-2,4-pentanedionato)-, 16:118
—, η-cyclooctatetraene(1,1,1,5,5,5-hexafluoro-2,4-pentanedionato)-, 16:118
—, η-cyclooctatetraene(1,1,1-trifluoro-2,4-pentanedionato)-, 16:118
—, η-cyclooctene(1,1,1,5,5,5-hexafluoro-2,4-pentanedionato)-, 16:118
Silver chloride, 20:18
Silver periodate (Ag$_2$H$_3$IO$_6$)
crystal growth of, by double infusion, 20:15
Silver(I) sulfamate, 18:201

Silylamine, pentamethyl-, see Silane, (dimethylamino)trimethyl-
Silyl complexes of chromium, molybdenum, and tungsten, 17:104-109
Silyl sulfide, see Disilathiane
Sodium:
  compd. with TaS_2, 19:42, 44
Sodium decavanadate(V) $Na_6[V_{10}O_{28}]$: octadecahydrate, 19:140, 142
Sodium tetrahydridogallate(1-), 17:50
Sodium trihydridodimethyldizincate(1-), 17:14
Sodium trihydridozincate(1-), 17:15
Stannane, cyclopentadienyltrimethyl-, 17:178
Stannatrane, see Tin, ethyl{{2,2',2''-nitrilotris[ethanolato]}(3-)-$N,O,$-O',O''}
Stannoxane, di-, see Distannoxane
Stibine, triphenyl-:
  copper complex, 19:94
  nickel complexes, 17:121
—, tris[2-(dimethylarsino)phenyl]-, 16:187
Stilbene:
  nickel complexes, 17:121
Sulfamate:
  silver(I), 18:201
Sulfates:
  crystal growth of, in silica gel, 20:4
Sulfide, see Amine, tris[2-(methylthio)ethyl]-; Aniline, 2-(methylthio)-; Lithium, [2-(methylthio)phenyl]-; Phosphine, bis[2-(methylthio)phenyl]phenyl-; Phosphine, [2-(methylthio)phenyl]diphenyl-; Phosphine, tris[2-(methylthiophenyl]-; Thioanisole, 2-bromo-
—, 2-bromophenyl methyl, 16:169
—, 2-(diphenylphosphino)phenyl methyl, 16:171
Sulfides:
  crystal growth of, in silica gel, 20:5
  Mo, Ta, and Ti, 19:35-48
  Pt, 19:49
Sulfur diimide, mercapto-:
  nickel complex, 18:124
Sulfur nitride, see Nitrogen sulfide
Sulfur nitride ($S_2N_2H_2$), see Mercaptosulfur diimide

Sulfur tetrafluoride oxide, 20:34
Styrene:
  platinum complex, 20:181

TADA-H_2, see Dibenzo[b,i][1,4,8,11]-tetraazacyclotetradecine, 5,14-dihydro-
Tantalate(1-), hexacarbonyl-:
  tetraphenylarsonium, 16:71-72
  tris[bis(2-methoxyethyl) ether] potassium, 16:71
Tantalium chloride ($TaCl_5$), 20:42
Tantalum sulfide (TaS_2):
  compd. with B (2:1), 19:42
  compd. with NH_3 (1:1), 19:42
  compd. with pyridine (2:1), 19:40
  compd. with Sn (1:1), 19:47
  compd. with sodium, 19:42, 44
  2H(a)phase, 19:35-37
Tartrates:
  crystal growth of, in silica gel, 20:4
TCNQ, see 2,2'-(2,5-Cyclohexadiene-1,4-diylidene)bis[propanedinitrile]
Technetium(VIII) oxide, 18:155
Telluride, bis(trimethylsilyl), see Disilatellurane, hexamethyl-
—, digermyl, see Digermatellurane
Tellurides:
  Pt, 19:49
Tellurium tetrafluoride, 20:33
Terpy, see 2,2':6,2''-Terpyridine
2,2':6,2''-Terpyridine:
  platinum complexes, 20:101
3,7,11,17-Tetraazabicyclo[11.3.1]heptadeca-1(17),2,11,13,15-pentaene, 2,12-dimethyl-:
  cobalt and nickel complexes, 18:17
1,5,9,13-Tetraazacyclohexadecane, ([16]aneN_4), 20:109
1,4,10,13-Tetraazacyclooctadeca-5,8,14,17-tetraene-7,16-dione, 5,9,14,18-tetramethyl-:
  copper complex, 20:91
1,4,8,12-Tetraazacyclopentadecane, ([15]aneN_4), 20:108
1,4,8,11-Tetraazacyclotetradeca-1,3-diene:
  cobalt and nickel complexes, 18:27
1,4,8,11-Tetraazacyclotetradeca-4,11-diene, 5,7,7,12,14,14-hexamethyl-:

bis(trifluoromethanesulfonate), 18:3
diperchlorate, 18:4
iron II and nickel II complexes, 18:2
1,4,8,11-Tetraazacyclotetradecane, 16:233
—, 5,5,7,12,12,14-hexamethyl-:
cobalt(III), iron(II) and nickel(II) complexes, 18:10
meso- and *racemic*-, hydrate, 18:10
1,4,8,11-Tetraazacyclotetradeca-1,3,8,10-tetraene:
cobalt and nickel complexes, 18:22
1,4,8,11-Tetraazacyclotetradeca-4,6,11,13-tetraene:
nickel complex, 18:42
—, 6,13-diacetyl-5,14-dimethylnickel complex, 18:39
—, 5,14-dimethyl-:
bis(hexafluorophosphate), 18:40
1,4,7,10-Tetraazacyclotridecane, ([13]aneN$_4$), 20:106
29H,31H-Tetrabenzo[b,g,l,q]porphine, 1,4,8,11,15,18,22,25-octamethyl-, 20:158
cobalt complex, 20:156
Tetrabenzo[b,f,j,n][1,5,9,13]tetraazacyclohexadecine:
transition metal complexes, 18:30
Tetrabutylammonium tetrathionitrate, 18:203, 205
Tetracyanoquinodimethanide, *see* 2,2'-(2,5-cyclohexadiene-1,4-diylidene)bis[propanedinitrile]
Tetraethylenepentamine:
cobalt(III) complex, 17:153-154
Tetrahydrofuran:
precautions in drying, 17:3
Tetrathiafulvalene (TTF), *see* 2,2'-Bi-1,3-ditholylidene
Tetrathionitrate, tetrabutylammonium, 18:203, 205
Thallium, tris(pentacarbonylmanganese)-, 16:61
Thioacetylacetonimine, N,N'-ethylenebis-, *see* 2-Pentanethione, 4,4'-(ethylenedinitrilo)bis-
Thioacetylacetonimine:
cobalt(II) complex, 16:227
Thioanisole, 2-bromo-, 16:169
Thiocarbamic acid, diethyl-:
molybdenum complex, 16:235

—, dimethyl-:
cobalt complex, 16:7
iron complex, 16:5
Thiocarbonyl complexes, 16:53; 17:100
Cr, Mn, Ir, Rh, 19:188, 207
tungsten, 19:181-187
Thiocyanate:
nickel macrocyclic complex, 18:24
(Thiothenoyl)acetone, 16:206
Thorium(IV), chlorotris(η^5-cyclopentadienyl)-, 16:149
Tin:
compd. with TaS$_2$ (1:1), 19:47
—, ethyl{{2,2',2''-nitrilotris[ethanolato]}-(3-)-N,O,O',O''}-, 16:230
—, {{2,2'-iminobis[ethanolato]}(2-)-N,O,O'}-, 16:232
—, {{2,2'-(methylimino)bis[ethanolato]}-(2-)-N,O,O'}-, 16:234
—, oxobis(trimethyl-, *see* Oxide, bis(trimethyltin)
Titanium, dibromobis(2,4-pentanedionato)-:
cis-, 19:146
—, methyl-:
trihalides, 16:120-126
—, tribromomethyl-, 16:124
—, trichlorobis(dimethylphosphine)-, 16:100
—, trichlorobis(methylphosphine)-, 16:98
—, trichlorobis(triethylphosphine)-, 16:101
—, trichlorobis(trimethylphosphine)-, 16:100
—, trichloromethyl-, 16:122
Titanium(III), bis(alkylphosphine)trichloro-, 16:97
—, dichloro(η^5-cyclopentadienyl)-polymer, 16:238
—, dichloro(η^5-cyclopentadienyl)bis(dimethylphenylphosphine)-, 16:239
—, tris[bis(trimethylsilyl)amido]-, 18:116
Titanium(IV), dichlorobis(2,4-pentadionato)-:
cis-, 19:146
—, difluorobis(2,4-pentanonato)-:
cis-, 19:145
Titanium bis(η^5-cyclopentadienyl)[tetrahydroborato(1-)]-, 17:91

Subject Index

Toluene:
  cadmium slurry, 19:78
  cobalt complex, 20:226, 228
  nickel complex, 19:72
  tungsten complex, 19:172
o-Tolyl phosphite:
  palladium complex, 16:129
Transition metalate, hexacyano-:
  rare earth metal, crystal growth:
    of, by double infusion, 20:12
1,3,5,2,4,6-Triazatriphosphorine, 2,4-bis(dimethylamino)-2,4,6,6-tetrafluoro-2,2,4,4,6,6-hexahydro-, 18:197
—, 2,4-dibromo-2,4,6,6-tetrafluoro-2,2,-4,4,6,6-hexahydro-, 18:197
—, 2,2-dichloro-4,6-bis(dimethylamino)-4,6-difluoro-2,2,4,4,6,6-hexahydro-, 18:195
—, 2,4,6,6-tetrachloro-2,4-bis(dimethylamino)-2,2,4,4,6,6-hexahydro-, 18:194
—, 2,4,6-tribromo-2,4,6-trifluoro-2,2,4,-4,6,6-hexahydro-, 18:197
—, 2,4,6-trichloro-2,4,6-tris(dimethylamino)-2,2,4,4,6,6-hexahydro-, 18:194
—, 2,4,6-tris(dimethylamino)-2,4,6-trifluoro-2,2,4,4,6,6-hexahydro-, 18:195
Triborate(1-), octahydro-:
  manganese complex, 19:227, 228
Triethylamine:
  aluminum complex, 17:37
  gallium complex, 17:42
—, 2-(diphenylphosphino)-, 16:160
—, 2,2',2''-tris(diphenylarsino)-, 16:177
—, 2,2',2''-tris(diphenylphosphino)-, 16:176
Triethyl orthoformate:
  macrocyclic ligands from, 19:37
Trimethylamine:
  boron complex, 19:233, 234
Trimethyl phosphite, 20:76-82
Triphenyl phosphite:
  chromium complex, 19:202
Tripod ligands, 16:174
1,3,5,2,4,6-Trithiatriazine, 17:188
TTF, see 2,2'-Bi-1,3-dithiolyidene
Tungstate(1-), tetracarbonyliodo(thiocarbonyl)-:
  tetrabutylammonium, trans-, 19:186
Tungsten:
  vaporization of, 19:64
—, bis(dinitrogen)bis[ethylenebis(diphenylphosphine)]-:
  trans-, 20:126
—, bromotetracarbonyl(phenylmethylidyne)-:
  trans-, 19:172, 173
—, chloro(η^5-cyclopentadienyl)dinitrosyl-, 18:129
—, (η^5-cyclopentadienyl)methyldinitrosyl-, 19:210
—, dicarbonyl(η^5-cyclopentadienyl)nitrosyl-, 18:127
—, pentacarbonyl[(dimethylamino)phenylmethylene]-, 19:169
—, pentacarbonyl(methoxyphenylmethylene)-, 19:165
—, pentacarbonyl(thiocarbonyl)-, 19:181, 187
—, pentacarbonyl[p-tolyl(trimethylsiloxy)methylene]-, 19:167
—, tricarbonyl(η^5-cyclopentadienyl)silyl-, 17:104
Tungsten(O), pentacarbonyl(diphenylmethylene)-, 19:182
—, pentacarbonyl(methoxymethylcarbene)-, 17:97
—, pentacarbonyl[1-(phenylthio)ethylidene]-, 17:98
Tungsten(IV), tetrachlorobis(triphenylphosphine)-, 20:124
—, tetrachloro[ethylenebis(diphenylphosphine)]-, 20:125
Tungsten bromide, 20:42

Uranium, bis(η^8-cyclooctatetraene)-, 19:149, 150
Uranium(IV), chlorotris(η^5-cyclopentadienyl)-, 16:148
—, [[7,12:21,26-diimino-19,14:28,33:-35,5-trinitrilo-5H-pentabenzo[c,h,-m,r,w][1,6,11,16,21]pentaazacyclopentacosinato] (2-)] dioxo-, 20:97
Uranium fluoride (UF$_5$), 19:137-139
Uranium hexachloride, 16:143

Uranocene, *see* Uranium, bis(η^6-cyclo-
  octatetraene)
Uranyl superphthalocyanine (UO$_2$ spc),
  20:97

Vanadate(2-), tetrafluorooxo-:
  nickel, heptahydrate, 16:87
Vanadate(V) [V$_{10}$O$_{28}$]$^{6-}$:
  hexaammonium, hexahydrate, 19:140,
    143
  hexasodium, octadecahydrate, 19:140,
    142
Vanadium:
  vaporization of, 19:64
Vanadium(III), tris[bis(trimethylsilyl)-
  amido]-, 18:117
Vanadium(IV), aqua-μ-[[6,6'-(ethylenedinitrilo)bis(2,4-heptanedionato)]-
  (4-)-N,N',O^4,$O^{4'}$:O^2,$O^{2'}$,O^4,$O^{4'}$]-
  oxocopper(II), 20:95
—, [5,26:13,18-diimino-7,11:20,24-
  dinitrilodibenzo[c,n][1,6,12,17]-
  tetraazacyclodocosinato(2-)]oxo-,
  18:48
—, oxo[5,10,15,20-tetraphenyl-21H,23H-
  porphinato(2-)]-, 20:144
Vanadium chloride (VCl$_4$), 20:42

o-Xylene, *see* Benzene, o-dimethyl-
p-Xylene, α-(trimethylsiloxy)-:
  tungsten complex, 19:167

Zinc:
  vaporization of, 19:64
—, dimethyl-, 17:7, 10; 19:253
Zinc(II), (tetrabenzo[b,f,j,n][1,5,9,13]-
    tetraazacyclohexadecine)-:
  tetrachlorozincate, 18:33
Zincate, tetrachloro-:
  dipotassium, 20:51
  tris(ethylenediamine)ruthenium(II),
    19:118
Zincate(1-), trihydrido-:
  lithium, 17:10
  sodium, 17:15
—, trihydridodimethyldi-:
  sodium, 17:14
Zincate(2-), tetrahydrido-:
  lithium, 17:12
—, tetramethyl-:
  lithium, 17:12
Zinc chromium oxide (Cr$_2$ZnO$_4$),
  20:52
Zinc dihydride, 17:7
Zinc potassium chloride (K$_2$ZnCl$_4$),
  20:51
Zirconium, bis(η^5-cyclopentadienyl)di-
  hydrido-, 19:224, 225
—, chlorobis(η^5-cyclopentadienyl)-
  hydrixo-, 19:226
—, μ-oxo-bis[chlorobis(η^5-cyclopentadi-
  enyl)-, 19:224

FORMULA INDEX

The Formula Index, as well as the Subject Index, is a cumulative index for Volumes XVI, XVII, XVIII, XIX, and XX. The chief aim of this index, like that of other formula indexes, is to help in locating specific compounds or ions, or even groups of compounds, that might not be easily found in the Subject Index, or in the case of many coordination complexes, are to be found only as general entries in the Subject Index. *All* specific compounds, or in some cases ions, with definite formulas (or even a few less definite) are entered in this index or noted under a related compound, whether entered specifically in the Subject Index or not.

Wherever it seemed best, formulas have been entered in their usual form (i.e., as used in the text) for easy recognition: Si_2H_6, XeO_3, NOBr. However, for the less simple compounds, including coordination complexes, the significant or central atom has been placed first in the formula in order to throw together as many related compounds as possible. This procedure often involves placing the cation last as being of relatively minor interest (e.g., $Co(C_5H_7O_2)_3$, $NaB_{12}H_{12}O$. Where they may be of almost equal interest in two or more parts of a formula, two or more entries have been made: Fe_2O_4Ni and $NiFe_2O_4$; $NH(SO_2F)_2^{2-}$ $(SO_2F)_2NH$, and $(FSO_2)_2NH$ (halogens other than fluorine are entered only under the other elements or groups in most cases); $(B_{10}C_{11})_2Ni^{2-}$ and $Ni(B_{10}CH_{11})^{2-}$.

Formulas for organic compounds are structural or semistructural so far as feasible: $CH_3COCH(NHCH_3)CH_3$. Consideration has been given to probable interest for inorganic chemists, i.e., any element other than carbon, hydrogen, or oxygen in an organic molecule is given priority in the formula if only one entry is made, or equal rating if more than one entry: only $Co(C_5H_7O_2)_2$, but $AsO(+)-C_4H_4O_6Na$ and $(+)-C_4H_4O_6AsONa$. Names are given only where the formula for an organic compound, ligand, or radical may not be self-evident, but not for frequently occurring relatively simple ones like C_5H_5 (cyclopentadienyl), $C_5H_7O_2$ (2,4-pentanedionato), C_6H_{11} (cyclohexyl), C_5H_5N (pyridine). A few abbreviations for ligands used in the text, including macrocyclic ligands, are retained here for simplicity and are alphabetized as such, "bipy" for bipyridine, "en" for ethylenediamine or 1,2-ethanediamine, "diphos" for ethylenebis(diphenylphosphine) or 1,2-bis(diphenylphosphino)ethane or 1,2-ethanediylbis(diphenylphosphine), and "tmeda" for N,N,N',N'-tetramethylethylenediamine or N,N,N',N'-tetramethyl-1,2-ethanediamine.

Footnotes are indicated by *n*, following the page number.

Ag(C_5H$F_8$$O_2$)($C_6H_{10}$), 16:118
Ag(C_5H$F_6$$O_2$)($C_7H_{12}$), 16:118
Ag(C_5H$F_6$$O_2$)($C_8H_8$), 16:118
Ag(C_5H$F_6$$O_2$)($C_8H_{12}$), 16:117
Ag(C_5H$F_6$$O_2$)($C_8H_{14}$), 16:118
Ag(C_5H$_4$$F_3$$O_2$)($C_8H_8$), 16:118
Ag(C_5H$_4$$F_3$$O_2$)($C_8H_{12}$), 16:118
AgCl, 20:18
Ag(O_3SNH_2), 18:201
Ag$_2$H$_3$IO_6, 20:15
Ag$_2$ μ-[P(CH$_3$)$_2$(CH$_2$)]$_2$, 18:142
Al[(CH$_3$CO)$_2$(CO)$_4$Mn]$_3$, 18:56, 58
AlH[N(C$_2$H$_5$)$_2$]$_2$, 17:41
AlH$_2$[N(C$_2$H$_5$)$_2$], 17:40
[AlH$_2$[OC$_2$H$_4$(2-CH$_3$O)]$_2$]Na, 18:149
AlH$_3$·N(CH$_3$)$_3$, 17:37
AlI$_3$(C$_5$H$_5$N)$_3$, 20:83
Al$_{11}$GaO$_{17}$, 19:56
Al$_{11}$KO$_{17}$, 19:55
Al$_{11}$LiO$_{17}$, 19:54
Al$_{11}$(NH$_4$)O$_{17}$, 19:56
Al$_{11}$(NO)O$_{17}$, 19:56
Al$_{11}$RbO$_{17}$, 19:55
Al$_{11}$TlO$_{17}$, 19:53
trans-[([13]aneN$_4$)CoCl$_2$]Cl, 20:111
trans-[([15]aneN$_4$)CoCl$_2$]Cl (isomers I and II), 20:112
trans-[([16]aneN$_4$CoCl$_2$]ClO$_4$ (isomers I and II), 20:113
As(CH$_3$)$_2$(C$_6$F$_5$), 16:183
As(CH$_3$)$_2$(C$_6$H$_4$Br), 16:185
[As(CH$_3$)$_2$C$_6$H$_4$]$_3$As, 16:186
[As(CH$_3$)$_2$C$_6$H$_4$]$_3$Sb, 16:187
[[As(CH$_3$)$_2$]$_2$-1,2-C$_6$H$_4$]Pt[P(O)(OCH$_3$)$_2$]$_2$, 19:100
As[C$_6$H$_4$As(CH$_3$)$_2$]$_3$, 16:186
[As(C$_6$H$_5$)$_2$CH=CH]P(C$_6$H$_5$)$_2$, 16:189
[As(C$_6$H$_5$)$_2$C$_2$H$_4$]P(C$_6$H$_5$)$_2$, 16:191
[As(C$_6$H$_5$)$_2$C$_2$H$_4$]$_3$N, 16:177
[As(C$_6$H$_5$)$_3$]$_3$Cu(NO$_3$), 19:95
[As(C$_6$H$_5$)$_4$](CN), 16:135
[As(C$_6$H$_5$)$_4$](OCN), 16:134
[As(C$_6$H$_5$)$_4$][Nb(CO)$_6$], 16:72
[As(C$_6$H$_5$)$_4$][Ta(CO)$_6$], 16:71
[AsF$_6$]Hg$_{2.86}$, 19:25
[AsF$_6$]$_2$Hg$_3$, 19:24
Au(CH$_3$)[P(CH$_3$)$_3$(CH$_2$)], 18:141

[{(BF)$_2$(dmg)$_3$}Co], 17:140
[{(BF)$_2$(dmg)$_3$}Fe], 17:142

[{(BF)$_2$(nox)$_3$}Fe], 17:143
(BF$_4$)[Nb(C$_6$H$_5$)$_2$H$_2${P(CH$_3$)$_2$(C$_6$H$_5$)}], 16:111
[BF$_4$][RuH(P(C$_6$H$_5$)$_3$)$_2$(η^6-C$_6$H$_5$-P(C$_6$H$_5$)$_2$)], 17:77
BH[(-CH$_2$-NCH$_3$-)(CH$_2$)$_2$], 17:165
BH[(-CH$_2$-NCH$_3$-)$_2$(CH$_2$)$_3$], 16:166
[BH(sec-C$_4$H$_9$)$_3$]K, 17:26
BH[N(CH$_3$)$_2$]$_2$, 17:30
[(BH$_2$)N(CH$_3$)$_2$]$_2$, 17:32
BH$_2$(NC)[N(CH$_3$)$_3$], 19:233, 234
[BH$_4$]$_2$Ca, 17:17
(BH$_4$)Cu[P(C$_6$H$_5$)$_3$]$_2$, 19:96
[BH$_4$][N(n-C$_4$H$_9$)$_4$], 17:23
(BH$_4$)Nb(C$_5$H$_5$)$_2$, 16:109
[BH$_4$][P(C$_6$H$_5$)$_3$CH$_3$], 17:22
(BH$_4$)[P(C$_6$H$_{11}$)$_3$]$_2$NiH, 17:89
(BH$_4$)Ti(η^5-C$_5$H$_5$)$_2$, 17:91
B[N(C$_2$H$_5$)$_2$]$_3$, 17:159
B[N(CH$_3$)C$_6$H$_5$]$_3$, 17:162
B[(NHCH)(CH$_3$)C$_2$H$_5$]$_3$, 17:160
[{(BOC$_4$H$_9$)$_2$(dmg)$_3$}Fe], 17:144
[{(BOC$_4$H$_9$)$_2$(dpg)$_3$}Fe], 17:145
[{(BOH)$_2$(nox)$_3$}Fe], 17:144
B(TaS$_2$)$_2$, 19:42
B$_2$Cl$_4$, 19:74
B$_2$D$_5$Br, 18:146
B$_2$D$_5$I, 18:147
B$_2$H$_2$(C$_4$H$_8$)$_2$ (1,6-diboracyclodecane), 19:239, 241
B$_2$H$_3$(C$_4$H$_8$)$_2$·N(n-C$_4$H$_9$)$_4$ (tetrabutyl-ammonium μ-hydro-bis(μ-tetramethylene)-diborate(1-)), 19:243
B$_2$H$_5$Br, 18:146
B$_2$H$_5$CH$_3$, 19:237
B$_2$H$_5$ μ-[(CH$_3$)$_2$N], 17:34
B$_2$H$_5$I, 18:147
(μ-B$_2$H$_6$)Mn$_3$(CO)$_{10}$ μ-H, 20:240
[B$_2$H$_7$][N(n-C$_4$H$_9$)$_4$], 17:25
[B$_2$H$_7$][P(C$_6$H$_5$)$_3$CH$_3$], 17:24
(B$_3$H$_8$)Mn(CO)$_4$, 19:227, 228
B$_5$H$_8$Br (1-bromopentaborane(9)), 19:247, 248
B$_6$H$_{10}$, 19:247, 248
[BpyH$_2$][MoOCl$_5$], 19:135
(bpy)MoCl$_3$O (red and green forms), 19:135, 136
BrCH(CHO)$_2$, 18:50
BrC$_6$D$_5$, 16:164
BrC$_6$H$_4$(SCH$_3$), 16:169

Formula Index   283

$Br_{0.25}[Pt(CN)_4][C(NH_2)_3]_2 \cdot H_2O$,
   19:10, 12; 19:19, 2
$Br_{0.3}[Pt(CN)_4]K_2 \cdot 3H_2O$, 19:14, 15
Br_3Cr, anhydrous, 19:123, 124

trans-$[(CC_6H_5)WBr(CO)_4]$, 19:172, 173
$[C(C_6H_5)_2]W(CO)_5$, 19:180
$(CF_3COCHCOCH_3)_2Pt$, 20:67
$[(CF_3CO)_2CH]_2Pt$, 20:67
$(\mu_3\text{-CH})Co_3(CO)_9$, 20:226, 227
$[CH_2CH(OEt)_2]Co(C_{62}H_{88}N_{13}O_{14}P)$
   (cobalamin, (2,2-diethoxyethyl)-),
   20:138
$(CH_2OEt)Co[(C=NO)_2H(CH_3)_2]_2[4-(t-C_4H_9)C_5H_4N]$, 20:131
$[(CH_2)_2OEt]Co[(C=NO)_2H(CH_3)_2]_2$
   $[4-(t-C_4H_9)C_5H_4N]$, 20:131
$[CH_2[Si(CH_3)_3]_2Mg$, 19:262
$[CH_3(CH_2)_2CO_2]_4Mo_2$, 19:133
CH_3CN, 18:6
$[(CH_3CN)_4Cu][PF_6]$, 19:90
$[CH_3C(NH_2)]Re[CH_3C(O)](CO)_4$,
   20:204
$[CH_3C(O)]Re[CH_3C(NH_2)](CO)_4$,
   20:204
$(CH_3CO)_2C[CH(OC_2H_5)]$, 18:37
$[(CH_3CO)_2C](CH_2NHCH)_2$, 18:37
cis-$[CH_3C(O)]_2ReH(CO)_4$, 20:200, 202
$[\eta^6\text{-}o\text{-}(CH_3)_2C_6H_4]Cr(CO)_2(CS)$, 19:197, 198
$[\eta^6\text{-}m\text{-}(CH_3)C_6H_4CO_2CH_3]Cr(CO)(CS)$
   $[P(OC_6H_5)_3]$, 19:202
$[\eta^6\text{-}m\text{-}(CH_3)C_6H_4CO_2CH_3]Cr(CO)_2(CS)$,
   19:201
$[(CH_3)Co(H_2O)(C_{52}H_{73}N_4O_{14})](ClO_4)$,
   (cobyrinic acid, aquamethyl-, heptamethyl
   ester, perchlorate), 20:141
$[\mu_3\text{-}[(CH_3NH)CO]C]Co_3(CO)_9$, 20:230, 232
$(CH_3)_2PtBr_2$, 20:185
$(CH_3)_2PtBr_2(C_5H_5N)_2$, 20:186
$(CH_3)_2Pt(OH)_2 \cdot 1.5H_2O$, 20:185, 186
$[\mu_3\text{-}[[[(CH_3)_3C]O]CO]C]Co_3(CO)_9$,
   20:234, 235
$[CNC(CH_3)_3]_2(C_6H_5N=NC_6H_5)Ni$, 17:122
$[CNC(CH_3)_3]_4Ni$, 17:118
$[C[N(CH_3)_2]C_6H_5]W(CO)_5$, 19:169
cis-$[(CNC_6H_5)PtCl_2[P(C_2H_5)_3]]$, 19:174
cis-$[[C(NHC_6H_5)(C_2H_5O)]PtCl_2[P-(C_2H_5)_3]]^-$, 19:175

$[(CN)Co(H_2O)(C_{52}H_{73}N_4O_{14})](ClO_4)$,
   (cobyrinic acid, aquacyano-, heptamethyl
   ester, perchlorate), 20:141
trans-$[[C(NHC_6H_5)(NHC_2H_5)]PtCl[P-(C_2H_5)_3]]ClO_4$, 19:176
cis-$[[C(NHC_6H_5)_2]PtCl_2[P(C_2H_5)_3]]$,
   19:176
$[C(NH_2)]_2Br_{0.25}[Pt(CN)_4]\cdot H_2O$, 19:10, 12
$[C(NH_2)_3][PtBr_2(CN)_4]\cdot XH_2O$, 19:11
$[C(NH_2)_3][Pt(CN)_4]$, 19:11
$(CN)[N(C_2H_5)_4]$, 16:133
$(CN)_2Co(C_{52}H_{73}N_4O_{14})$(cobyrinic acid,
   dicyano-, heptamethyl ester), 20:139
$[(CN)_6Fe]Sm \cdot 4H_2O$, 20:13
$(C=NO)_2H(CH_3)_2]_2CoBr[(CH_3)_2S]$,
   20:128
$[(C=NO)_2H(CH_3)_2]_2CoBr[4-(t-C_4H_9)C_5H_4N]$-, 20:130
$[(C=NO)_2H(CH_3)_2]_2Co[4-(t-C_4H_9)C_5H_4N]$
   $[EtO(CH_2)_2]$, 20:131
$(COCH_3)_2$, 18:23
$[C(OCH_3)(C_6H_5)]W(CO)_5$, 19:165
$[C(O)CH_3]Re(CO)_5$, 20:201
$[C(OC_2H_5)[N(C_2H_5)_2]]Cr(CO)_5$, 19:168
$(CO)Cr[\eta^6\text{-}m\text{-}(CH_3)C_6H_4CO_2CH_3](CS)^-$
   $[P(OC_6H_5)_3]$, 19:202
$(CO)IrCl[P(CH_3)_3]_2$, 18:64
$[(CO)Ir[P(CH_3)_3]_4]Cl$, 18:63
$(CO)Mn(\eta^5\text{-}C_6H_5)(CS)[P(C_6H_5)_3]$, 19:189
$(CO)Mn(NO)_3$, 16:4
$(CO)Mo(C_2H_2)[S_2P(i\text{-}C_3H_7)_2]_2$, 18:55
$(CO)(OCOCH_3)[P(C_6H_5)_3]_2RuCl$, 17:126
$(CO)(OCOCH_3)[P(C_6H_5)_3]_2RuH$, 17:126
$(CO)(OCOCF_3)[P(C_6H_5)_3]_3OsCl$, 17:128
$(CO)(OCOCF_3)_2[P(C_6H_5)_3]_2Os$, 17:128
$(CO)(OCOCF_3)_2[P(C_6H_5)_3]_2Ru$, 17:127
$[C[OSi(CH_3)_3]C_6H_4\text{-}p\text{-}(CH_3)]W(CO)_5$,
   19:167
$(CO)_2Cr[\eta^6\text{-}m\text{-}(CH_3)C_6H_4CO_2CH_3](CS)$,
   19:201
$(CO)_2Cr[\eta^6\text{-}o\text{-}(CH_3)_2C_6H_4](CS)$, 19:197, 198
$(CO)_2Cr(\eta^6\text{-}C_6H_5CO_2CH_3)(CS)$, 19:200
$(CO)_2Cr(\eta^5\text{-}C_5H_5)(NO)$, 18:127
$[(CO)_2Fe(C_5H_5)]_2Mg(C_4H_8O)_2$, 16:56
$(CO)_2Mn(C_5H_5)(CS)$, 16:53
$(CO)_2Mn(\eta^5\text{-}C_5H_5)(CSe)$, 19:193, 195
$(CO)_2Mo(\eta^5\text{-}C_5H_5)(NO)$, 18:127
$(CO)_2Mo(NO)(C_5H_5)$, 16:24

Formula Index

[(CO)$_2$Mo P(C$_4$H$_9$)$_3$(C$_5$H$_5$)]$_2$Mg(C$_4$H$_8$O)$_4$, 16:59
(CO)$_2$Mo[S$_2$P(i-C$_3$H$_7$)$_2$]$_2$, 18:53
(CO)$_2$(OCOCH$_3$)$_2$[P(C$_6$H$_5$)$_3$]$_2$Ru, 17:126
[(CO)$_2$ReCl$_2$(NO)]$_2$, 16:37
(CO)$_2$Rh(η^5-C$_5$Ph$_4$Cl), 20:192
(CO)$_2$W(η^5-C$_5$H$_5$)(NO), 18:127
(CO)$_3$(η^5-C$_5$H$_5$)CrSiH$_3$, 17:104
(CO)$_3$(η^5-C$_5$H$_5$)MoSiH$_3$, 17:104
(CO)$_3$(η^5-C$_5$H$_5$)WSiH$_3$, 17:104
[(CO)$_3$Co{P(CH$_3$)(C$_6$H$_5$)$_2$}]$_2$Mg(tmeda)$_2$, 16:59
[(CO)$_3$Co[P(C$_4$H$_9$)$_3$]]$_2$Mg(C$_4$H$_8$O)$_4$, 16:58
(CO)$_3$Cr[η^6-C$_6$H$_5$(CH$_3$O)], 19:155
(CO)$_3$Cr(η^6-C$_6$H$_5$Cl), 19:157
(CO)$_3$Cr(η^6-C$_6$H$_5$CO$_2$CH$_3$), 19:157
(CO)$_3$Cr(η^6-C$_6$H$_5$F), 19:157
(CO)$_3$Cr[η^6-C$_6$H$_5$[N(CH$_3$)$_2$]], 19:157
(CO)$_3$Cr(η^6-C$_6$H$_6$), 19:157
(CO)$_3$Fe[C$_6$H$_5$CH=CHC(O)CH$_3$], 16:104
(CO)$_3$IrCl$_{1\cdot10}$, 19:19
(CO)$_3$Mn(η^5-C$_5$Cl$_4$Br), 20:194, 195
(CO)$_3$Mn(η^5-C$_5$Cl$_5$), 20:194
(CO)$_3$Mn(η^5-C$_5$H$_4$Br), 20:193
(CO)$_3$Mn(η^5-C$_5$H$_4$Cl), 20:192
(CO)$_3$Mn(η^5-C$_5$H$_4$I), 20:193
(CO)$_3$Ru(C$_8$H$_{12}$), 16:105
[(CO)$_3$RuCl$_2$]$_2$, 16:51
[(CO)$_4$Co]$_2$Mg(C$_5$H$_5$N)$_4$, 16:58
(CO)$_4$Fe(PClF$_2$), 16:66
(CO)$_4$Fe[PF$_2$N(C$_2$H$_5$)$_2$], 16:64
(CO)$_4$Fe(PF$_3$), 16:67
(CO)$_4$Ir$_2$[μ-SC(CH$_3$)$_3$]$_2$, 20:237
(CO)$_4$Ir$_2$(μ-SC$_6$H$_5$)$_2$, 20:238
(CO)$_4$Mn(B$_3$H$_8$), 19:227, 228
[(CO)$_4$Mn(CH$_3$CO)$_2$]$_3$Al, 18:56, 58
cis-(CO)$_4$Mo[P(t-C$_4$H$_9$)F$_2$]$_2$, 18:175
(CO)$_4$Re[CH$_3$C(O)][CH$_3$C(NH$_2$)], 20:204
[(CO)$_4$ReCl]$_2$, 16:35
cis-(CO)$_4$ReH[CH$_3$C(O)]$_2$, 20:200, 202
trans-[(CO)$_4$WBr(CC$_6$H$_5$)], 19:172, 173
trans-[(CO)$_4$WI(CS)][N(n-(C$_4$H$_9$)$_4$], 19:186
[(CO)$_5${C(CH$_3$)(OCH$_3$)}Cr], 17:96
[(CO)$_5${C(CH$_3$)(SC$_6$H$_5$)}W], 17:99
(CO)$_5$Cr[C(OC$_2$H$_5$)[N(C$_2$H$_5$)$_2$]], 19:168
(CO)$_5$Cr(C$_4$H$_6$O)(pentacarbonyl(dihydro-2(3H)-furanylidene)chromium), 19:178, 179
(CO)$_5$MnBr, 19:160
(CO)$_5$Mn(CH$_3$CO), 18:57
(CO)$_5$Mn(η^1-C$_5$BrCl$_4$), 20:194
(CO)$_5$Mn(η^1-C$_5$Cl$_5$), 20:193
(CO)$_5$MnCl, 19:159
(CO)$_5$MnI, 19:161, 162
[(CO)$_5$Mn]$_3$Tl, 16:61
(CO)$_5$Re[CH$_3$C(O)], 20:201
(CO)$_5$ReCl$_3$(NO), 16:36
(CO)$_5$W[C(C$_6$H$_5$)$_2$], 19:180
(CO)$_5$W[C[N(CH$_3$)$_2$]C$_6$H$_5$], 19:169
(CO)$_5$W[C(OCH$_3$)(C$_6$H$_5$)], 19:165
(CO)$_5$W[C[OSi(CH$_3$)$_3$]C$_6$H$_4$-p-(CH$_3$)], 19:167
(CO)$_5$W(CS), 19:183, 187
[(CO)$_6$Nb][As(C$_6$H$_5$)$_4$], 16:72
[(CO)$_6$Nb][K{(CH$_3$OCH$_2$CH$_2$)$_2$O}$_3$], 16:69
(CO)$_6$Ru$_2$Cl$_4$, 16:51
[(CO)$_6$Ta][As(C$_6$H$_5$)$_4$], 16:71
[(CO)$_6$Ta][K{(CH$_3$OCH$_2$CH$_2$)$_2$O}$_3$], 16:71
(CO)$_9$Co$_3$(μ_3-CH), 20:226, 227
(CO)$_9$Co$_3$[μ_3-[[(CH$_3$)C]O]C]], 20:234, 235
(CO)$_9$Co$_3$[μ_3-[(CH$_3$NH)CO]C], 20:230, 232
(CO)$_9$Co$_3$[μ_3-[(C$_2$H$_5$O)CO]C], 20:230
(CO)$_9$Co$_3$(μ_3-C$_6$H$_5$C), 20:226, 228
(CO)$_9$Co$_3$(μ_3-ClC), 20:234
(CO)$_{10}$Mn$_3$(μ-B$_2$H$_6$)μ-H, 20:240
(CO)$_{10}$Os$_3$(NO)$_2$, 16:40
(CO)$_{10}$Ru$_3$(NO)$_2$, 16:39
[(CO)$_{11}$Fe$_3$H][(PPh$_3$)$_2$N], 20:218
[(CO)$_{11}$Fe$_3$][(PPh$_3$)$_2$N]$_2$, 20:222
(CO)$_{12}$Re$_3$H$_3$, 17:66
(CO)$_{12}$Re$_4$H$_4$, 18:60
(CO)$_{12}$Rh$_4$, 17:115; 20:209
(CO)$_{12}$Ru$_3$, 16:45, 47
[(CO)$_{15}$Rh$_6$C]K$_2$, 20:212
(CO)$_{16}$Rh$_6$, 16:49
[(CO)$_{30}$Rh$_{12}$]Na$_2$, 20:215
(CO$_2$)Fe[P(CH$_3$)$_3$]$_4$, 20:73
(CS)Cr(CO)[η^6-m-(CH$_3$)C$_6$H$_4$CO$_2$CH$_3$][P(OC$_6$H$_5$)$_3$], 19:202
(CS)Cr(CO)$_2$[η^6-m-(CH$_3$)C$_6$H$_4$CO$_2$CH$_3$], 19:201
(CS)Cr(CO)$_2$[η^6-o-(CH$_3$)$_2$C$_6$H$_4$], 19:197, 198
(CS)Cr(CO)$_2$(η^6-C$_6$H$_5$CO$_2$CH$_3$), 19:200
trans-[(CS)IrCl[P(C$_6$H$_5$)$_3$]$_2$], 19:206
(CS)Mn(CO)(η^5-C$_5$H$_5$)[P(C$_6$H$_5$)$_3$], 19:189

Formula Index   285

(CS)Mn(C_5H_5)(CO)$_2$, 16:53
(CS)Mn(η^5-C_5H_5)(diphos), 19:191
trans-[(CS)RhCl[P(C_6H_5)$_3$]$_2$], 19:204
(CS)W(CO)$_5$, 19:183, 187
trans-[(CS)WI(CO)$_4$][N(n-C_4H_9)$_4$], 19:186
(CSe)Mn(CO)$_2$(η^5-C_5H_5), 19:193, 195
[(CSH)$_2$(CN)$_2$] (2,3-dimercapto-2-butenedinitrile), 19:31
(C_2H_2)Mo(CO)[S_2P(i-C_3H_7)$_2$]$_2$, 18:55
$C_2H_2O_4$ (oxalic acid), 19:16
($C_2H_3O_2$)$_3$[Co(NH$_3$)$_6$], 18:68
C_2H_4OH(2-CH$_3$O), 18:145
trans-[(C_2H_4)PtCl$_2$(C_5H_5N)], 20:181
(C_2H_4)$_2$Pt$_2$Cl$_4$, 20:181, 182
(C_2H_4)$_2$Pt[P(cyclo-C_6H_{11})$_3$], 19:216
(C_2H_4)$_3$Pt, 19:215
(C_2H_5)CdI, 19:78
[[(C_2H_5)$_2$COCO$_2$]$_2$CrO]Na, 20:63
(C_2H_5O)CH, 18:37
[μ_3-[(C_2H_5O)CO]C]Co$_3$(CO)$_9$, 20:230
$C_2H_8N_2$ (en), 18:37
($C_2H_7O_2$)$_3$In, 19:261
[(C_2O_4)Co(en)$_2$]$^+$, 18:96
[(C_2O_4)Rh(en)$_2$]ClO$_4$, 20:58
[(C_2O_4)$_3$Cr]K$_3$·3H$_2$O, 19:127
C_3H_3BrO$_2$ (bromomalonaldehyde), 18:50
($C_3H_3S_2$)[BF$_4$] (1,3-dithiolylium tetrafluoroborate), 19:28
(η^3-C_3H_5)Pd(η^5-C_5H_5), 19:221
(η^3-C_3H_5)$_2$Pd$_2$Cl$_2$, 19:220
(C_4H_3S)C(SH)=CHC(O)CF$_3$, 16:206
(C_4H_4O)Cr(CO)$_5$ (pentacarbonyl(dihydro-2(3H)-furanylidene)chromium), 19:178, 179
$C_4H_{10}N_4$ (2,4-butanedione, dihydrazone), 20:88
[($C_4H_{10}N_4$)$_3$Fe][BF$_4$]$_2$, 20:88
(η^1-C_5BrCl$_4$)Mn(CO)$_5$, 20:194
(η^5-C_5Cl$_4$Br)Mn(CO)$_3$, 20:194, 195
C_5Cl$_4$(N$_2$) (1,3-cyclopentadiene, tetrachloro-5-diazo-), 20:189, 190
C_5Cl$_4$(N$_2$H$_2$) (2,4-cyclopentadien-1-one, hydrazone), 20:190
(η^1-C_5Cl$_5$)Mn(CO)$_5$, 20:193
(η^5-C_5Cl$_5$)Mn(CO)$_3$, 20:194
(η^5-C_5Cl$_5$)Rh(η^4-C_8H_{12}), 20:194
C_5H_3N-2,6-(CH$_3$CO)$_2$, 18:18
C_5H_3N-2,6-(NH$_2$)$_2$, 18:47
(η^5-C_5H_4Br)Mn(CO)$_3$, 20:193
C_5H_4(CH$_3$)(SiH$_3$), 17:174

(η^5-C_5H_4Cl)Mn(CO)$_3$, 20:192
(η^5-C_5H_4Cl)Rh(η^4-C_8H_{12}), 20:192
(η^5-C_5H_4I)Mn(CO)$_3$, 20:193
$C_5H_4N_2$ (1,3-cyclopentadiene, 5-diazo-), 20:191
C_5H_5(GeH$_3$), 17:176
(η^5-C_5H_5)Mo(η^6-C_6H_6), 20:196, 197
(η^5-C_5H_5)Mo(η^6-C_6H_6)Cl, 20:198
(η^5-C_5H_5)Mo(η^6-C_6H_6)I, 20:199
(η^5-C_5H_5)MoBr(η^6-C_6H_6), 20:199
(C_5H_5N)(TaS$_2$)$_2$, 19:40
C_5H_5(SiH$_3$), 17:172
(C_5H_5)Sn(CH$_3$)$_3$, 17:178
($C_5H_7O_2$)$_2$Pt, 20:66
$C_5H_8O_2$ (acac), 18:37
(η^5-C_5Ph$_4$Br)Rh(η^4-C_8H_{12}), 20:192
(η^5-C_5Ph$_4$Cl)Rh(CO)$_2$, 20:192
(η^5-C_5Ph$_4$Cl)Rh(η^4-C_8H_{12}), 20:191
(C_6F_5)$_2$Ni[η^6-C_6H_5(CH$_3$)], 19:72
C_6H_4(Br)(SCH$_3$), 16:169
[C_6H_4(CF$_3$)$_2$]Cr, 19:70
C_6H_4(CHO)(NH$_2$), 18:31
C_6H_4(NH$_2$), 18:51
$C_6H_4S_4$ (2,2'-bi-1,3-dithiolylidene)(TTF), 19:28
($C_6H_4S_4$)($C_{10}H_4N_4$) (2,2'-bi-1,3-dithiolylidene salt with 2,2'-(2,5-cyclohexadiene-1,4-diylidene)bis[propanedinitrile] (1:1)), 19:32
($C_6H_4S_4$[Pt[(CS)$_2$(CN)$_2$]$_2$] (2,2'-bi-1,3-dithiolylidene radical cation bis[2,3-dimercapto-2-butenedinitrilato(2-)] platinate(1-) (1:1)), 19:31
($C_6H_4S_4$)$_2$[Cu[(CS)$_2$(CN)$_2$]$_2$] (2,2'-bi-1,3-dithiolylidene radical cation bis[2,3-dimercapto-2-butenedinitrilato(2-)] cuprate(2-) (2:1)), 19:31
($C_6H_4S_4$)$_2$[Ni[(CS)$_2$(CN)$_2$]$_2$] (2,2'-bi-1,3-dithiolidene radical cation bis[2,3-dimercapto-2-butenedinitrilato(2-)] nickelate(2-) (2:1)), 19:31
($C_6H_4S_4$)$_2$[Pt(CN)$_4$] (2,2'-bi-1,3-dithiolylidene radical cation tetracyanoplatinate (II) (2:1)), 19:31
($C_6H_4S_4$)$_2$[Pt[(CS)$_2$(CN)$_2$]$_2$] (2,2'-bi-1,3-dithiolylidene radical cation bis[2,3-dimercapto-2-butenedinitrilato(2-)] platinate(2-) (2:1)), 19:31
($C_6H_4S_4$)$_3$[BF$_4$]$_2$ (2,2'-bi-1,3-dithiolylidene radical cation tetrafluoroborate (1-)(3:2)), 19:31

$(C_6H_4S_4)_8I_{15}$ (2,2'-bi-1,3-dithiolylidene radical cation iodide (8:15)), 19:31

$(C_6H_4S_4)_{11}I_8$ (2,2'-bi-1,3-dithiolylidene radical cation iodide (11:8)), 19:31

$(C_6H_4S_4)_{14}(NCS)_8$ (2,2'-bi-1,3-dithiolylidene radical cation thiocyanate (14:8)), 19:31

$(C_6H_4S_4)_{14}(NCSe)_8$ (2,2'-bi-1,3-dithiolylidene radical cation selenocyanate (14:8)), 19:31

$(C_6H_4S_4)_{24}I_{63}$ (2,2'-bi-1,3-dithiolylidene radical cation iodide (24:63)), 19:31

$(\mu_3\text{-}C_6H_5C)Co_3(CO)_9$, 20:226, 228

$(C_6H_5CH=CH_2)_2PtCl_4$, 20:181, 182

$[\eta^6\text{-}C_6H_5(CH_3)]Ni(C_6F_5)_2$, 19:72

$[\eta^6\text{-}C_6H_5(CH_3O)]Cr(CO)_3$, 19:155

$(C_6H_5)(CH_3)_2PO$, 17:185

$(C_6H_5)_2(CH_3)PO$, 17:184

$(\eta^6\text{-}C_6H_5CO_2CH_3)Cr(CO)_2(CS)$, 19:200

$(\eta^6\text{-}C_6H_5CO_2CH_3)Cr(CO)_3$, 19:157

$(C_6H_5CO_2)Mo_2Br_2[P(n\text{-}C_4H_9)_3]_2$, 19:133

$(\eta^6\text{-}C_6H_5Cl)Cr(CO)_3$, 19:157

$(\eta^6\text{-}C_6H_5Cl)_2Mo$, 19:81, 82

$(\eta^6\text{-}C_6H_5F)Cr(CO)_3$, 19:157

$[\eta^6\text{-}C_6H_5[N(CH_3)_2]]Cr(CO)_3$, 19:157

$[\eta^6\text{-}C_6H_5[N(CH_3)_2]]_2Mo$, 19:81

$(C_6H_5)_2PLi$, 17:186

$(C_6H_5)_2PSi(CH_3)_3$, 17:187

$(\eta^6\text{-}C_6H_6)Cr(CO)_3$, 19:157

$(\eta^6\text{-}C_6H_6)Mo(\eta^5\text{-}C_5H_5)$, 20:196, 197

$(\eta^6\text{-}C_6H_6)Mo(\eta^5\text{-}C_5H_5)I$, 20:199

$(\eta^6\text{-}C_6H_6)MoBr(\eta^5\text{-}C_5H_5)$, 20:199

$(\eta^6\text{-}C_6H_6)MoCl(\eta^5\text{-}C_5H_5)$, 20:198

$(\eta^4\text{-}C_6H_{10})_2Rh_2Cl_2$ (di-μ-chloro-bis(η^4-1,5-hexadiene)dirhodium), 19:219

$(\eta^2\text{-}C_6H_{12})_4Rh_2Cl_2$ (di-μ-chloro-tetrakis(η^2-2,3-dimethyl-2-butene)dirhodium), 19:219

$C_8H_4N_2$ (benzenedicarbonitrile), 18:47

(cyclo-C_8H_8)Li$_2$, 19:214

$(\eta^8\text{-}cyclo\text{-}C_8H_8)_2U$, 19:149, 150

$(\eta^4\text{-}C_8H_{12})Rh(\eta^5\text{-}C_5Cl_5)$, 20:194

$(\eta^4\text{-}C_8H_{12})Rh(\eta^5\text{-}C_5H_4Cl)$, 20:192

$(\eta^4\text{-}C_8H_{12})Rh(\eta^5\text{-}C_5Ph_4Br)$, 20:192

$(\eta^4\text{-}C_8H_{12})Rh(\eta^5\text{-}C_5Ph_4Cl)$, 20:191

$(C_8H_{12})_2Pt$ (bis(1,5-cyclooctadiene)-platinum), 19:213, 214

$(\eta^4\text{-}cyclo\text{-}C_8H_{12})_2Rh_2Cl_2$, 19:218

$C_9H_{22}N_4$ (1,4,7,10-tetraazacyclotridecane([13]aneN$_4$)), 20:106

$(C_{10}H_4N_4)(C_6H_4S_4)$ (2,2'-(2,5-cyclohexadiene-1,4-diylidene)bis[propanedinitrile] salt with 2,2'-bi-1,3-dithiolylidene (1:1)), 19:32

$C_{10}H_{24}N_4$, 16:223

$C_{11}H_{26}N_4$ (1,4,8,12-tetraazacyclopentadecane([15]aneN$_4$)), 20:108

(1-3:6-7:10-12-η^8-$C_{12}H_{16}$)Ni ((η^8-dodecatriene-1,12-diyl)nickel), 19:85

$(C_{12}H_{18}P)(ClO_4)$ (1-ethyl-1-phenphospholanium perchlorate), 18:189, 191

$C_{12}H_{20}N_4$ (5,14-Me$_2$[14]-4,6,11,13-tetraeneN$_4$), 18:42

$C_{12}H_{22}N_4$ (H$_2$ 5,14-Me$_2$[14]-4,6,11,13-tetraene-1,4,8,11-N_4)(PF$_6$)$_2$, 18:40

$C_{12}H_{24}N_4$ (2,3-Me$_2$[14]-1,3-diene-1,4,8,11-N$_4$), 18:27

$(C_{12}H_{24})_2Pt_2Cl_4$, 20:181, 183

$C_{12}H_{28}N_4$ (1,5,9,13-tetraazacyclohexadecane([16]aneN$_4$)), 20:109

$C_{12}H_{30}N_8$ (bicyclo[6.6.6]ane-1,3,6,8,10,13,16,19-N$_8$), 20:86

$C_{14}H_{14}Br_2N_4$ (6,13-Br$_2$-2,3-Bzo[14]-2,4,6,11,13-pentene-1,4,8,11-N$_4$), 18:50

$C_{14}H_{24}N_4$ (Me$_4$[14]-1,3,8,10-tetraene-1,4,8,11-N$_4$), 18:22

$C_{15}H_{22}N_4$ (Me$_2$Pyro[14]trieneN$_4$), 18:17

$C_{16}H_{26}N_4O_2$ (6,13-Ac$_2$-5,14-Me$_2$[14]tetraeneN$_4$), 18:39

$C_{16}H_{32}N_4$ (Me$_6$[14]-4,11-diene-1,4,8,11-N$_4$), 18:2

$C_{16}H_{36}N_4 \cdot XH_2O$ (Me$_6$[14]ane-1,4,8,11-N$_4 \cdot XH_2O$), 18:10

$C_{18}H_{16}N_4$ (2,3,9,10-Bzo$_2$[14]-2,4,6,9,11,13-hexaene-1,4,8,11-N_4)(TADA-H$_2$), 18:45

$C_{18}H_{28}N_4O_2$ (1,4,10,13-tetraazacyclooctadeca-5,8,14,17-tetraene-7,16-dione, 5,9,14,18-tetramethyl-, H$_4$daen), 20:91

$C_{18}H_{30}N_{12}$ (1,3,4,7,8,10,12,13,16,17,19,22-dodecaazatetracyclo[8.8.4.1^{3,17}.1^{8,12}]tetracosa-4,6,13,15,19,21-hexene, 5,6,14,15,20,21-hexamethyl-), 20:88

$[(C_{18}H_{30}N_{12})Co][BF_4]_2$, 20:89

$[(C_{18}H_{30}N_{12})Co][BF_4]_3$, 20:89

$[(C_{18}H_{30}N_{12})Fe][BF_4]_2$, 20:88

$[(C_{18}H_{30}N_{12})Ni][BF_4]_2$, 20:89

$C_{22}H_{22}N_4$ (dibenzo[b,i][1,4,8,11]tetraazacyclotetradecine, 7,16-dihydro-

6,8,15,17-tetramethyl-(5,7,12,14-Me$_4$-2,3:9,10-Bzo$_2$[14]hexaeneN$_4$)), 20:117
C$_{26}$H$_{16}$N$_8$ (5,26:13,18-diimino-7,11:20,24-dinitrilodibenzo[c,n][1,6,12,17]tetraazacyclodocosine)(Hp-H$_2$), 18:47
C$_{28}$H$_{20}$N$_4$ (2,3;6,7;10,11;14,15-Bzo$_4$[16]octaene-1,5,9,13-N$_4$), 18:30
(C$_{32}$H$_{16}$N$_4$)Fe (iron(II), [phthalocyaninato(2-)]-), 20:159
(C$_{32}$H$_{16}$N$_4$)Li$_2$ (lithium, -[phthalocyaninato(2-)]di-), 20:159
(C$_{34}$H$_{36}$N$_4$O$_4$)FeCl (iron(III), chloro[7,12-diethyl-3,8,13,17-tetramethyl-21H,23H-porphine-2,18-dipropionato(2-)]-), 20:152
(C$_{36}$H$_{36}$N$_4$O$_4$)FeCl (iron(III), chloro-[dimethyl 7,12-diethenyl-3,8,13,17-tetramethyl-21H,23H-porphine-2,18-dipropionato(2-)]-), 20:148
(C$_{36}$H$_{44}$N$_4$)FeCl (iron(III), [2,3,7,8,12,13,17,18-octaethyl-21H,23H-porphinato(2-)]-), 20:151
(C$_{36}$H$_{44}$N$_4$)Mg (magnesium(II), [2,3,7,8,12,13,17,18-octaethyl-21H,23H-porphinato(2-)]-), 20:145
C$_{44}$H$_{26}$N$_8$O$_8$ (porphyrin, 5,10,15,20-tetrakis(2-nitrophenyl)-, (H$_2$TNPP)), 20:162
(C$_{44}$H$_{28}$N$_4$)Ni (nickel(II), [5,10,15,20-tetraphenyl-21H,23H-porphinato(2-)]-), 20:143
(C$_{44}$H$_{28}$N$_4$)VO (vanadium(IV), oxo[5,10,15,20-tetraphenyl-21H,23H-prophinato(2-)]-), 20:144
C$_{44}$H$_{34}$N$_8$ (porphyrin, 5,10,15,20-tetrakis(2-aminophenyl)-, and all-cis-), 20:163, 164
(C$_{44}$H$_{36}$N$_4$)Co (cobalt(II), [1,4,8,11,15,18,22,25-octamethyl-29H,31H-tetrabenzo[b,g,l,q]porphinato(2-)]-), 20:156
(C$_{44}$H$_{36}$N$_4$)Mg(C$_6$H$_5$N)$_2$ (magnesium(II), [1,4,8,11,15,18,22,25-octamethyl-29H,31H-tetrabenzo[b,g,l,q]porphinato(2-)]bis(pyridine)-), 20:158
C$_{44}$H$_{38}$N$_4$ (29H,31H-tetrabenzo[b,g,l,q]porphine, 1,4,8,11,15,18,22,25-octamethyl-), 20:158
(C$_{52}$H$_{73}$N$_4$O$_{14}$)Co(CN)$_2$ (cobyrinic acid, dicyano-, heptamethyl ester), 20:139
[(C$_{52}$H$_{73}$N$_4$O$_{14}$)Co(H$_2$O)(CH$_3$)](ClO$_4$) (cobyrinic acid, aquamethyl-, heptamethyl ester, perchlorate), 20:141
[(C$_{52}$H$_{73}$N$_4$O$_{14}$)Co(H$_2$O)(CN)](ClO$_4$) (cobyrinic acid, aquacyano-, heptamethyl ester, perchlorate), 20:141
(C$_{62}$H$_{88}$N$_{13}$O$_{14}$P)Co[CH$_2$CH(OEt)$_2$] (cobalamin, (2,2-diethoxyethyl)-), 20:138
(C$_{62}$H$_{88}$N$_{13}$O$_{14}$P)Co(CH$_3$) (cobalamin, methyl-), 20:136
(C$_{62}$H$_{88}$N$_{13}$O$_{14}$P)Co(OH) (cobalamin, hydroxo-), 20:138
C$_{64}$H$_{66}$N$_8$O$_4$ (porphyrin, 5,10,15,20-tetrakis[2-(2,2-dimethylpropionamido)phenyl]-, all-cis-), 20:165
(C$_{66}$H$_{62}$N$_8$O$_4$)FeBr (iron(III), bromo[(all-cis)-5,10,15,20-tetrakis[2(2,2-dimethylpropionamido)phenyl]porphyrinato(2-)]-), 20:166
(C$_{66}$H$_{62}$N$_8$O$_4$)Fe(N-MeC$_3$H$_3$N$_2$) (iron(II), bis(N-methylimidazole)[(all-cis-)-5,10,15,20-tetrakis[2-(2,2-dimethylpropion-amido)phenyl]porphyrinato(2-)]-), 20:167
(C$_{66}$H$_{62}$N$_8$O$_4$)Fe(O$_2$)(N-MeC$_3$H$_3$N$_2$) (iron(II), (dioxygen) (N-methylimidazole)[(all-cis)-5,10,15,20-tetrakis[2-(2,2-dimethylpropionamido)phenyl]porphyrinato(2-)]-), 20:168
Ca[BH$_4$]$_2$, 17:17
Ca((CHOH)$_2$(CO$_2$)$_2$), 20:9
Cd(C$_2$H$_5$)I, 19:78
(μ_3-ClC)Co$_3$(CO)$_9$, 20:234
Cl$_{0.3}$[Pt(CN)$_4$]K$_2$·3H$_2$O, 19:15
Cl$_{1.10}$Ir(CO)$_3$, 19:19
trans-[Cl$_2$Co(en)$_2$](NO$_3$), 18:73
Cl$_3$[Co(NH$_3$)$_6$], 18:68
Cl$_3$MoO$_3$C$_{12}$H$_{24}$ (molybdenum(III), trichlorotris(tetrahydrofuran)-), 20:121
Cl$_4$C$_5$(N$_2$) (1,3-cyclopentadiene, tetrachloro-5-diazo-), 20:189, 190
Cl$_4$C$_5$(N$_2$H$_2$) (2,4-cyclopentadien-1-one, hydrazone), 20:190
Cl$_4$K$_2$Mg, 20:51
[Co(bicyclo[6.6.6]ane-1,3,6,8,10,13,16,19-N$_8$)]Cl$_3$, 20:85
CoBr[(C=NO)$_2$H(CH$_3$)$_2$]$_2$[(CH$_3$)$_2$S], 20:128
CoBr[(C=NO)$_2$H(CH$_3$)$_2$]$_2$[4-(t-C$_4$H$_9$)C$_5$H$_4$N], 20:130
[CoBr(C$_{15}$H$_{22}$N$_4$)]Br·H$_2$O, 18:19
[CoBr(en)$_2$(NH$_3$)]Br$_2$, 16:93

[CoBr(en)$_2$(NH$_3$)]Cl$_2$, 16:93, 95, 96
[CoBr(en)$_2$(NH$_3$)](NO$_3$)$_2$, 16:93
[CoBr(en)$_2$(NH$_3$)](O$_3$SOC$_{10}$H$_{14}$Br)$_2$, 16:93
[CoBr(en)$_2$(NH$_3$)](S$_2$O$_6$), 16:94
[CoBr$_2$(C$_{12}$H$_{24}$N$_4$)]ClO$_4$, 18:28
[CoBr$_2$(C$_{14}$H$_{24}$N$_4$)]Br, 18:25
[CoBr$_2$(C$_{15}$H$_{22}$N$_4$)]Br·H$_2$O, 18:21
meso-trans-[CoBr$_2$(C$_{16}$H$_{36}$N$_4$)]ClO$_4$, 18:14
[CoBr$_2$(C$_{28}$H$_{20}$N$_4$)]Br, 18:34
meso-trans-[CoBr$_2$(Me$_6$[14]aneN$_4$)]ClO$_4$, 18:73
Co(2,3:9,10-Bzo$_2$[14]-2,4,6,9,11,13-hexaeneato(2-)N$_4$) (Co(TADA)), 18:46
[Co(Bzo$_4$[16]octeneN$_4$)Br$_2$]Br, 18:34
Co[(C=NO)$_2$H(CH$_3$)$_2$]$_2$[4-(t-C$_4$H$_9$)C$_5$H$_4$N][EtO(CH$_2$)$_2$], 20:131
[Co(CO)$_3$ {P(CH$_3$)(C$_6$H$_5$)$_2$}]$_2$Mg(tmeda), 16:59
[Co(CO)$_3$ {P(C$_4$H$_9$)$_3$}]$_2$Mg(C$_4$H$_8$O)$_4$, 16:58
[Co(CO)$_4$]$_2$Mg(C$_5$H$_5$N)$_4$, 16:58
[Co(CO$_3$)(edda)]Na, 18:104
Co(C$_2$H$_2$N$_2$)[C(CH$_3$)CH$_2$C(S)CH$_3$]$_2$, 16:227
Co(C$_8$H$_{12}$)(C$_8$H$_{13}$), 17:112
[Co(C$_{12}$H$_{30}$N$_8$]Cl$_3$ ([Co(bicyclo[6.6.6]ane-1,3,6,8,10,13,16,19-N$_8$)] trichloride), 20:86
Co(C$_{18}$H$_{14}$N$_4$), 18:46
[Co(C$_{18}$H$_{30}$N$_{12}$)][BF$_4$]$_2$, 20:89
[Co(C$_{18}$H$_{30}$N$_{12}$)][BF$_4$]$_3$, 20:89
Co(C$_{44}$H$_{36}$N$_4$) (cobalt(II), [1,4,8,11,15,18,22,25-octamethyl-29H,31H-tetrabenzo[b,g,l,q]porphinato(2-)]-), 20:156
Co(C$_{52}$H$_{73}$N$_4$O$_{14}$)(CN)$_2$ (cobyrinic acid, dicyano-, heptamethyl ester), 20:139
[Co(C$_{52}$H$_{73}$N$_4$O$_{14}$)(H$_2$O)(CH$_3$)](ClO$_4$) (cobyrinic acid, aquamethyl-, heptamethyl ester, perchlorate), 20:141
[Co(C$_{52}$H$_{73}$N$_4$O$_{14}$)(H$_2$O)(CN)](ClO$_4$) (cobyrinic acid, aquacyano-, heptamethyl ester, perchlorate), 20:141
Co(C$_{62}$H$_{88}$N$_{13}$O$_{14}$P)[CH$_2$CH(OEt)$_2$] (cobalamin, (2,2-diethoxyethyl)-), 20:138
Co(C$_{62}$H$_{88}$N$_{13}$O$_{14}$P)(CH$_3$) (cobalamin, methyl-), 20:136
Co(C$_{62}$H$_{88}$N$_{13}$O$_{14}$P)(OH) (cobalamin, hydroxo-), 20:138

trans-[CoCl$_2$([13]aneN$_4$)]Cl, 20:111
trans-[CoCl$_2$([15]aneN$_4$)]Cl (isomers I and II), 20:112
trans-[CoCl$_2$([16]aneN$_4$)]ClO$_4$ (isomers I and II), 20:113
trans-[CoCl$_2$(en)$_2$](NO$_3$), 18:73
CoCl$_2$(NO)[P(CH$_3$)(C$_6$H$_5$)$_2$]$_2$, 16:29
CoCr$_2$O$_4$, 20:52
[Co{(dmg)$_3$(BF)$_2$}], 17:140
(+)-asym-cis-[Co(edda)(en)]Cl·3H$_2$O, 18:106
(−)-sym-cis-[Co(edda)(NO$_2$)]K, 18:101
(−)-[Co(edda)(NO$_2$)$_2$] (−)-[Co(en)$_2$(C$_2$O$_4$)], 18:101
sym-cis-[Co(edda)(NO$_2$)$_2$]K, 18:100
(−)-[Co(edta)] (−)-[Co(en)$_2$(C$_2$O$_4$)]·3H$_2$O, 18:100
[Co(edta)]K·2H$_2$O, 18:100
(−)-[Co(edta)]K·2H$_2$O, 18:100
(+)-[Co(edta)]K·2H$_2$O, 18:100
(−)-asym-cis-[Co(en)(edda)] (+)-brcamsul, 18:106
(+)-asym-cis-[Co(en)(edda)] (+)-brcamsul, 18:106
asym-cis-[Co(en)(edda)]Cl, 18:105
(−)-asym-cis-[Co(en)(edda)]Cl·3H$_2$O, 18:106
(−)-sym-cis-[Co(en)(edda)](HC$_4$H$_4$O$_6$), 18:109
(+)-sym-cis-[Co(en)(edda)](HC$_4$H$_4$O$_6$), 18:109
(−)-sym-cis-[Co(en)(edda)](NO$_3$), 18:109
(+)-sym-cis-[Co(en)(edda)](NO$_3$), 18:109
[Co(en)$_2$(C$_2$O$_4$)]$^+$, 18:96
(−)-[Co(en)$_2$(C$_2$O$_4$)]Br·H$_2$O, 18:99
[Co(en)$_2$(C$_2$O$_4$)]Cl·H$_2$O, 18:97
(−)-[Co(en)$_2$(C$_2$O$_4$)] (−)-[Co(edda)(NO$_2$)$_2$], 18:101
(−)-[Co(en)$_2$(C$_2$O$_4$)] (−)-[Co(edta)]·3H$_2$O, 18:100
(−)-[Co(en)$_2$(C$_2$O$_4$)][(+)-HC$_4$H$_4$O$_6$], 18:98
(+)-[Co(en)$_2$(C$_2$O$_4$)][(+)-HC$_4$H$_4$O$_6$], 18:98
(+)-[Co(en)$_2$(C$_2$O$_4$)]I, 18:99
[Co(en)$_2$(NH$_3$)(CO$_3$)]Br·0.5 5H$_2$O, 17:152
[Co(en)$_2$(NH$_3$)(CO$_3$)]ClO$_4$, 17:152
CoH(diphos)$_2$, 20:208
CoH(dp)$_2$, 20:207
CoH[(PPh$_2$)$_2$(CH$_2$)]$_2$, 20:207
CoH[(PPh$_2$)$_2$(CH$_2$)$_2$]$_2$, 20:208
Co(H$_2$O)$_2$(edda), 18:100
CoK$_2$Cl$_4$, 20:51

Formula Index 289

[Co(Me$_2$[14]-1,3-diene)Br$_2$] ClO$_4$, 18:28
[Co(Me$_2$-Pyo[14] trieneN$_4$)Br] Br·H$_2$O, 18:19
[Co(Me$_2$-Pyo[14] trieneN$_4$)Br$_2$] Br·H$_2$O, 18:21
[Co(Me$_4$[14]-1,3,8,10-tetraeneN$_4$)Br$_2$] Br, 18:25
[Co{(NH$_2$)$_2$CSe}$_3$](ClO$_4$)$_2$, 16:48
[Co{(NH$_2$)$_2$CSe}$_3$](SO$_4$), 16:85
cis-[Co(NH$_3$)$_4$(H$_2$O)(OH)](S$_2$O$_6$), 18:81
cis-[Co(NH$_3$)$_4$(H$_2$O)$_2$](ClO$_4$)$_3$, 18:83
cis- and, trans-[Co(NH$_3$)$_4$(NO$_2$)$_2$](NO$_3$), 18:70, 71
[Co(NH$_3$)$_5$(CO$_3$)]ClO$_4$·H$_2$O, 17:152
[Co(NH$_3$)$_6$](C$_2$H$_3$O$_2$)$_3$, 18:68
[Co(NH$_3$)$_6$]Cl$_3$, 18:68
Co(NO)[P(C$_6$H$_5$)$_3$]$_3$, 16:33
Co(NO)[S$_2$CN(CH$_3$)$_2$]$_2$, 16:7
[Co(NO)$_2$(diphos)][B(C$_6$H$_5$)$_4$], 16:19
[Co(NO)$_2${P(C$_6$H$_5$)$_3$}$_2$][B(C$_6$H$_5$)$_4$], 16:18
[Co(NO)$_2$(tmeda)][B(C$_6$H$_5$)$_4$], 16:17
Co(N$_2$C$_2$H$_2$)[C(CH$_3$)CH$_2$C(S)CH$_3$]$_2$, 16:227
[Co[P(OCH$_3$)$_3$]$_5$][BPh$_4$], 20:81
[Co(tetren)(CO$_3$)]ClO$_4$, 17:152
[Co$_2$(en)$_4$-μ-(OH)$_2$] Br$_4$·2H$_2$O, 18:92
[Co$_2$(en)$_4$-μ-(OH)$_2$] Cl$_4$·5H$_2$O, 18:93
[Co$_2$(en)$_4$-μ-(OH)$_2$] (ClO$_4$)$_4$, 18:94
[Co$_2$(en)$_4$-μ-(OH)$_2$] (S$_2$O$_6$)$_2$, 18:92
[Co$_2$(NH$_3$)$_8$-μ-(OH)$_2$] Br$_4$·4H$_2$O, 18:88
[Co$_2$(NH$_3$)$_8$-μ-(OH)$_2$] (ClO$_4$)$_4$·2H$_2$O, 18:88
Co$_3$(μ$_3$-CH)(CO)$_9$, 20:226, 227
Co$_3$[μ$_3$-[[[(CH$_3$)$_3$C]O]CO]C](CO)$_9$, 20:234, 235
Co$_3$[μ$_3$-[(CH$_3$NH)CO]C](CO)$_9$, 20:230, 232
Co$_3$[μ$_3$-[(C$_2$H$_5$O)CO]C](CO)$_9$, 20:230
Co$_3$(μ$_3$-C$_6$H$_5$C)(CO)$_9$, 20:226, 228
Co$_3$(μ$_3$-ClC)(CO)$_9$, 20:234
CrBr$_3$, anhydrous, 19:123, 124
Cr[[(CH$_3$)$_3$Si]$_2$N]$_3$, 18:118
Cr(CO)[η^6-m-(CH$_3$)C$_6$H$_4$CO$_2$CH$_3$](CS)[P(OC$_6$H$_5$)$_3$], 19:202
Cr(CO)$_2$[η^6-m-(CH$_3$)C$_6$H$_4$CO$_2$CH$_3$](CS), 19:201
Cr(CO)$_2$[η^6-o-(CH$_3$)$_2$C$_6$H$_4$](CS), 19:197, 198
Cr(CO)$_2$(η^6-C$_6$H$_5$CO$_2$CH$_3$)(CS), 19:200

Cr(CO)$_2$(η^5-C$_5$H$_5$)(NO), 18:127
Cr(CO)$_3$(η^5-C$_5$H$_5$)SiH$_3$, 17:104
Cr(CO)$_3$(η^6-C$_6$H$_5$CO$_2$CH$_3$), 19:157
Cr(CO)$_3$(η^6-C$_6$H$_5$Cl), 19:157
Cr(CO)$_3$(η^6-C$_6$H$_5$F), 19:157
Cr(CO)$_3$[η^6-C$_6$H$_5$[N(CH$_3$)$_2$]], 19:157
Cr(CO)$_5${C[(CH$_3$)(OCH$_3$)]}, 17:96
Cr(CO)$_5${C[(CH$_3$)(SC$_6$H$_5$)]}, 17:98
Cr(CO)$_5$[C(OC$_2$H$_5$)[N(C$_2$H$_5$)$_2$]], 19:168
Cr(CO)$_5$(C$_4$H$_6$O) (pentacarbonyl(dihydro-2(3H)-furanylidene)chromium), 19:178, 179
[Cr[(C$_2$H$_5$)$_2$COCO$_2$]$_2$O]Na, 20:63
[Cr(C$_2$O$_4$)$_3$]K$_3$·3H$_2$O, 19:127
[Cr(C$_3$H$_2$O$_4$)$_2$(H$_2$O)$_2$]K·3H$_2$O, 16:81
[Cr(C$_3$H$_2$O$_4$)$_3$]K$_3$·3H$_2$O, 16:80
Cr(η^5-C$_5$H$_5$)(i-C$_4$H$_9$)(NO)$_2$, 19:209
[Cr(η^5-C$_5$H$_5$)(NO)$_2$]$_2$, 19:211
Cr[C$_6$H$_4$(CF$_3$)$_2$]$_2$, 19:70
Cr[η^6-C$_6$H$_5$(CH$_3$O)](CO)$_3$, 19:155
Cr(η^6-C$_6$H$_5$)(CO)$_3$, 19:157
CrCl(η^5-C$_5$H$_5$)(NO)$_2$, 18:129
cis-[Cr(en)$_2$(H$_2$O)(OH)](S$_2$O$_6$), 18:84
[Cr(en)$_3$]Br$_3$, 19:125
[Cr(H$_2$O)(OH){OP(CH$_3$)(C$_6$H$_5$)(O)}$_2$]$_x$, 16:90
[Cr(H$_2$O)(OH){OP(C$_6$H$_5$)$_2$(O)}$_2$]$_x$, 16:90
[Cr(H$_2$O)(OH){OP(C$_8$H$_{17}$)$_2$(O)}$_2$]$_x$, 16:90
[Cr(H$_2$O)$_2$(C$_2$O$_4$)]K·2H$_2$O, 17:148
[Cr(H$_2$O)$_2$(C$_2$O$_4$)$_2$]K·3H$_2$O, 17:149
cis-[Cr(H$_2$O)$_2$(en)$_2$]Br$_3$, 18:85
CrLiO$_2$, 20:50
cis-[Cr(NH$_3$)$_4$(H$_2$O)Cl](SO$_4$), 18:78
cis-[Cr(NH$_3$)$_4$(H$_2$O)(OH)](S$_2$O$_6$), 18:80
cis-[Cr(NH$_3$)$_4$(H$_2$O)$_2$](ClO$_4$), 18:82
Cr(NO)$_4$, 16:2
[Cr(OH){OP(CH$_3$)(C$_6$H$_5$)(O)}$_2$]$_x$, 16:91
[Cr(OH){OP(C$_6$H$_5$)$_2$(O)}$_2$]$_x$, 16:91
[Cr(OH){OP(C$_8$H$_{17}$)$_2$(O)}$_2$]$_x$, 16:91
[Cr(OS(CH$_3$)$_2$]Br$_3$, 19:126
Cr$_2$CoO$_4$, 20:52
[Cr$_2$(en)$_4$-μ-(OH)$_2$]Br$_4$·2H$_2$O, 18:90
[Cr$_2$(en)$_4$-μ-(OH)$_2$]Cl$_4$·2H$_2$O, 18:91
[Cr$_2$(en)$_4$-μ-(OH)$_2$](ClO$_4$)$_4$, 18:91
[Cr$_2$(en)$_4$-μ-(OH)$_2$](S$_2$O$_6$)$_2$, 18:90
Cr$_2$MgO$_4$, 20:52
Cr$_2$MnO$_4$, 20:52
[Cr$_2$(NH$_3$)$_8$-μ-(OH)$_2$]Br$_4$·4H$_2$O, 18:86
[Cr$_2$(NH$_3$)$_8$-μ-(OH)$_2$](ClO$_4$)$_4$·2H$_2$O, 18:87

Cr$_2$NiO$_4$, 20:52
Cr$_2$ZnO$_4$, 20:52
Cs$_{1.75}$[Pt(CN)$_4$]·2H$_2$O, 19:6, 7
Cs$_2$[Pt(CN)$_4$]F$_{0.19}$, 20:29
Cs$_2$[Pt(CN)$_4$](HF$_2$)$_{0.23}$, 20:26
Cs$_2$[Pt(CN)$_4$](HF$_2$)$_{0.38}$, 20:28
Cu(BH$_4$)[P(C$_6$H$_5$)$_3$]$_2$, 19:96
Cu[6,13-Br$_2$-2,3-Bzo[14]-2,4,6,11,13-pentaenato(2-)-1,4,8,11-N$_4$], 18:50
[Cu(Bzo$_4$[16]octaeneN$_4$](NO$_3$)$_2$, 18:32
[Cu(CH$_3$CN)$_4$][PF$_6$], 19:90
CuCl, 20:10
[Cu[(CS)$_2$(CN)$_2$]$_2$](C$_6$H$_4$S$_4$)$_2$ (bis[2,3-dimercapto-2-butenedinitrilato(2-)]cuprate(2-) salt with 2,2'-bi-1,3-dithiolylidene (1:2)), 19:31
Cu(C$_{14}$H$_{12}$Br$_2$N$_4$), 18:50
[Cu(C$_{28}$H$_{20}$N$_4$)](NO$_3$)$_2$, 18:32
CuC$_{44}$H$_{28}$N$_4$, 16:214
CuCl[(C$_6$H$_5$)$_3$P]$_3$, 19:88
Cu(daenH$_2$), 20:92
CuH[(p-CH$_3$C$_6$H$_4$)$_3$P], 19:89
CuH[(C$_6$H$_5$)$_3$P], 19:87, 88
Cu(H$_2$daaen-N$_2$O$_2$), 20:93
Cu(NO$_3$)[As(C$_6$H$_5$)$_3$]$_3$, 19:95
Cu(NO$_3$)[P(C$_6$H$_5$)$_3$]$_2$, 19:93
Cu(NO$_3$)[Sb(C$_6$H$_5$)$_3$]$_3$, 19:94
CuVO(daaen)(H$_2$O), 20:95
Cu$_2$(C$_2$H$_3$O$_2$)$_2$, 20:53
Cu$_2$(daaen), 20:94

(daaen)CuVO(H$_2$O), 20:95
(daaen)Cu$_2$, 20:94
(daaenH$_2$)N$_2$O$_2$)Cu, 20:93
(daenH$_2$)Cu, 20:92
[diop–(+)]$_2$RhH, 17:81
(diphos)Mn(η^5-C$_5$H$_5$)(CS), 19:191

[(edda)Co(CO$_3$)]Na, 18:104
(–))asym-cis-[(edda)Co(en)] (+)-brcamsul, 18:106
(+)-asym-cis-[(edda)Co(en)] (+)-brcamsul, 18:106
asym-cis-[(edda)Co(en)]Cl, 18:105
(–)-asym-cis-[(edda)Co(en)]Cl·3H$_2$O, 18:106
(+)-asym-cis-[(edda)Co(en)]Cl·3H$_2$O, 18:106
(–)-sym-cis-[(edda)Co(en)](HC$_4$H$_4$O$_6$), 18:109

(+)-sym-cis-[(edda)Co(en)](HC$_4$H$_4$O$_6$), 18:109
(–)-sym-cis-[(edda)Co(en)](NO$_3$), 18:109
(+)-sym-cis-[(edda)Co(en)](NO$_3$), 18:109
(edda)Co(H$_2$O)$_2$, 18:100
(–)-sym-cis-[(edda)Co(NO$_2$)], 18:101
(–)-[(edda)Co(NO$_2$)$_2$] (–))[Co(en)$_2$(C$_2$O$_4$)], 18:101
sym-cis-[(edda)Co(NO$_2$)$_2$]K, 18:100
[(en)$_2$Co(C$_2$O$_4$)]$^+$, 18:96
trans-[(en)$_2$CoCl$_2$](NO$_3$), 18:73
cis-[(en)$_2$Cr(H$_2$O)(OH)](S$_2$O$_6$), 18:84
cis-[(en)$_2$Cr(H$_2$O)$_2$]Br$_3$, 18:85
cis-[(en)$_2$RhBr$_2$]Br, 20:60
[(en)$_2$Rh(C$_2$O$_4$)]ClO$_4$, 20:58
cis-[(en)$_2$RhCl$_2$]Cl, 20:60
cis-[(en)$_2$Rh(NO$_2$)$_2$]NO$_3$, 20:59
[(en)$_4$Co$_2$-µ-(OH)$_2$]Br$_4$·2H$_2$O, 18:92
[(en)$_4$Co$_2$-µ-(OH)$_2$]Cl$_4$·4H$_2$O, 18:93
[(en)$_4$Co$_2$-µ-(OH)$_2$](ClO$_4$)$_4$, 18:94
[(en)$_4$Co$_2$-µ-(OH)$_2$](S$_2$O$_6$)$_2$, 18:92
[(en)$_4$Cr$_2$-µ-(OH)$_2$]Br$_4$·2H$_2$O, 18:90
[(en)$_4$Cr$_2$-µ-(OH)$_2$]Cl$_4$·2H$_2$O, 18:91
[(en)$_4$Cr$_2$-µ-(OH)$_2$](ClO$_4$)$_4$, 18:91
[(en)$_4$Cr$_2$-µ-(OH)$_2$](S$_2$O$_6$)$_2$, 18:90

F$_{0.19}$Cs$_2$[Pt(CN)$_4$], 20:29
F$_2$PO(CH$_3$), 16:166
FeBr(C$_{64}$H$_{62}$N$_8$O$_4$) (iron(III), bromo [(all-cis)-5,10,15,20-tetrakis[2-(2,2-dimethylpropionamido)phenyl]porphyrinato(2-)]-), 20:166
FeBr(TpivPP), 20:166
Fe[[(CH$_3$)$_3$Si]$_2$N]$_3$, 18:18
[Fe(CN)$_6$]Sm·4H$_2$O, 20:13
[Fe(CO)$_2$(η^5-C$_5$H$_5$)(CS)][PF$_6$], 17:100
Fe(CO)$_3$(C$_6$H$_5$CH=CHCOCH$_3$), 16:104
Fe(CO)$_4$(PClF$_2$), 16:66
Fe(CO)$_4$[PF$_2$N(C$_2$H$_5$)$_2$], 16:64
Fe(CO)$_4$(PF$_3$), 16:67
[Fe$_3$(CO)$_{11}$H][(PPh$_3$)$_2$N], 20:218
Fe(CO$_2$)[P(CH$_3$)$_3$]$_4$, 20:73
meso-[Fe(C$_2$H$_4$N)$_2$(C$_{16}$H$_{36}$N$_4$)](CF$_3$O$_3$S)$_2$, 18:15
[Fe(C$_5$H$_5$)(CO)$_2$]$_2$Mg(C$_4$H$_8$O)$_2$, 16:56
[Fe(C$_4$H$_{10}$N$_4$)$_3$][BF$_4$]$_2$, 20:88
[Fe(C$_{16}$H$_{30}$N$_4$)(C$_2$H$_3$N)](CF$_3$O$_3$S)$_2$, 18:6
[Fe(C$_{18}$H$_{30}$N$_{12}$)][BF$_4$]$_2$, 20:88
Fe(C$_{32}$H$_{16}$N$_4$) (iron(II), [phthatocyaninato(2-)]-), 20:159

FeCl($C_{34}H_{36}N_4O_4$) (iron(III), chloro[7,12-diethyl-3,8,13,17-tetramethyl-21H,23H-porphine-2,18-dipropionato(2-)]-), 20:152

FeCl($C_{36}H_{36}N_4O_4$) (iron(III), chloro [dimethyl 7,12-diethenyl-3,8,13,17-tetramethyl-21H,23H-porphine-2,18-dipropionato(2-)]-), 20:148

FeCl($C_{36}H_{40}N_4O_6$), 16:216

FeCl($C_{36}H_{44}N_4$) (iron(III), chloro[2,3,7,8,12,13,17,18-octaethyl-21H,23H-porphinato(2-)]-), 20:151

FeCl(oep), 20:151

FeCl$_2$[(CH$_3$)$_3$P]$_2$, 20:70

[Fe{(dmg)$_3$(BF)$_2$}], 17:142

[Fe{(dmg)$_3$(BOC$_4$H$_9$)$_2$}], 17:144

[Fe{(dpg)$_3$(BOC$_4$H$_9$)$_2$}], 17:145

FeH[(C$_6$H$_5$)$_2$PCH$_2$CH$_2$P(C$_6$H$_5$)$_2$]$_2$, 17:71

[FeH(C$_6$H$_5$)$_2$PCH$_2$CH$_2$P(C$_6$H$_5$)$_2$]$_2$ [B(C$_6$H$_5$)$_4$], 17:70

FeHCl[(C$_6$H$_5$)$_2$PCH$_2$CH$_2$P(C$_6$H$_5$)$_2$]$_2$, 17:69

Fe(N-MeC$_3$H$_3$N$_2$)$_2$(C$_{66}$H$_{62}$N$_8$O$_4$) (iron(II), bis(N-methylimidazole)[(all-cis)-5,10,15,20-tetrakis[2-(2,2-dimethyl-propionamido)phenyl] porphyrinato(2-)]-), 20:167

meso-[Fe(Me$_6$[14]aneN$_4$)(C$_2$H$_3$N)$_2$](CF$_3$O$_3$S)$_2$, 18:15

[Fe(Me$_6$[14]-4,11-dieneN$_4$)(CH$_3$CN)$_2$](CF$_3$O$_3$S)$_2$, 18:6

Fe(NO)[S$_2$CN(C$_2$H$_5$)$_2$]$_2$, 16:5

[Fe{(nox)$_3$(BF)$_2$}], 17:143

[Fe{(nox)$_3$(BOH)$_2$}], 17:144

Fe(N$_3$H$_3$C$_3$-N-Me)$_2$ (all-cis-TpivPP), 20:167

Fe(O$_2$)(N-MeC$_3$H$_3$N$_2$)(C$_{66}$H$_{62}$N$_8$O$_4$) (iron(II), (dioxygen) (N-methylimidazole) [(all-cis)-5,10,15,20-tetrakis[2-(2,2-dimethylpropionamido)phenyl] por–phyrinato(2-)]-), 20:168

Fe(O$_2$)(N-MeC$_3$H$_3$N$_2$)(all-cis-TpivPP), 20:168

Fe[P(CH$_2$)(CH$_3$)$_2$]H[P(CH$_3$)$_3$]$_3$, 20:71

Fe[P(CH$_3$)$_3$]$_4$, 20:71

Fe[P(OCH$_3$)$_3$]$_5$, 20:79

[Fe$_3$(CO)$_{11}$][(PPh$_3$)$_2$N]$_2$, 20:222

GaAl$_{11}$O$_{17}$, 19:56

GaCl$_3$, 17:167

GaH$_3$N(CH$_3$)$_3$, 17:42

[GaH$_4$]K, 17:50n

[GaH$_4$]Li, 17:45

[GaH$_4$]Na, 17:50

GeBr(CH$_3$)$_3$, 18:153

Ge(CH$_3$)$_4$, 18:153

[Ge(CH$_3$)$_3$]$_2$O, 20:176, 179

Ge(C$_2$H$_5$)(OCH$_2$CH$_2$)$_3$N, 16:229

GeC$_8$H$_{17}$O$_3$N, 16:229

GeHBr(CH$_3$)$_2$, 18:157

[GeH(CH$_3$)$_2$]$_2$O, 20:176, 179

GeHCl(CH$_3$)$_2$, 18:157

GeHF(CH$_3$)$_2$, 18:159

GeHI(CH$_3$)$_2$, 18:158

GeH$_2$(CH$_3$)$_2$, 18:154, 156

(GeH$_2$CH$_3$)$_2$O, 20:176, 179

(GeH$_3$)C$_5$H$_5$, 17:176

GeH$_3$I, 18:162

[(GeH$_3$)N]$_2$C, 18:163

GeH$_3$(SCH$_3$), 18:165

GeH$_3$(SC$_6$H$_5$), 18:165

(GeH$_3$)$_2$O, 20:176, 178

(GeH$_3$)$_2$S, 18:164

(GeH$_3$)$_2$Se, 20:175

(GeH$_3$)$_2$Te, 20:175

[(+)-HC$_4$H$_4$O$_6$] (−)-[Co(en)$_2$(C$_2$O$_4$)], 18:98

[(+)-HC$_4$H$_4$O$_6$] (+)-[Co(en)$_2$(C$_2$O$_4$)], 18:98

HCo[(PPh$_2$)$_2$(CH)$_2$]$_2$, 20:207

HCo[(PPh$_2$)$_2$(CH$_2$)$_2$]$_2$, 20:208

(HF$_2$)$_{0.23}$Cs$_2$[Pt(CN)$_4$], 20:26

(HF$_2$)$_{0.29}$Rb$_2$[Pt(CN)$_4$] 1.67H$_2$O, 20:24

(HF$_2$)$_{0.38}$Cs$_2$[Pt(CN)$_4$], 20:28

(HF$_2$)$_{0.38}$Rb$_2$[Pt(CN)$_4$], 20:25

[HFe$_3$(CO)$_{11}$][(PPh$_3$)$_2$N], 20:218

μ-HMn$_3$(CO)$_{10}$(μ-B$_2$H$_6$), 20:240

cis-HRe[CH$_3$C(O)]$_2$(CO)$_4$, 20:200, 202

HSeOF$_5$, 20:38

HZrCl(η^5-C$_5$H$_5$)$_2$, 19:226

H$_2$SeO$_4$, 20:37

H$_2$Zr(η^5-C$_5$H$_5$)$_2$, 19:224, 225

H$_4$Re(CO)$_{12}$, 18:60

Hg[(NH$_2$)$_2$CSe]$_2$Br$_2$, 18:86

Hg[(NH$_2$)$_2$CSe]$_2$Cl$_2$, 16:85

[Hg(NH$_2$)$_2$CSeCl$_2$]$_2$, 16:86

[Hg[P(CH$_3$)$_3$(CH$_2$)]$_2$]Cl$_2$, 18:140

Hg$_{2.86}$[AsF$_6$], 19:25

Hg$_{2.91}$[SbF$_6$], 19:26

Hg$_3$[AsF$_6$]$_2$, 19:24

Hg$_3$[Sb$_2$F$_{11}$]$_2$, 19:23
Hp-H$_2$ (see 5,26:13,18-diimino-7,11:
   20,24-dinitrilobenzo[c,n] [1,6,12,17]
   tetraazacyclodocosine)

ICd(C$_2$H$_5$), 19:78
IO$_6$H$_3$Ag$_2$, 20:15
IO$_6$H$_3$(NH$_4$)$_2$, 20:15
ISi(CH$_3$)$_3$, 19:272
ISiH(CH$_3$)$_2$, 19:271
ISiH$_2$(CH$_3$), 19:271
ISiH$_3$, 19:268, 270
I$_3$Al(C$_5$H$_5$N)$_3$, 20:83
InBr$_3$, 19:259
InBr$_3$[OS(CH$_3$)$_2$]$_3$, 19:260
In(C$_5$H$_7$O$_2$)$_3$, 19:261
InCl$_3$, 19:258
InCl$_3$[OS(CH$_3$)$_2$]$_3$, 19:259
[InCl$_5$][(C$_2$H$_5$)$_4$N]$_2$, 19:260
Ir, 18:131
[IrBr(NO){P(C$_6$H$_5$)$_3$}$_2$](BF$_4$), 16:42
IrBr(N$_2$)[((C$_6$H$_5$)$_3$]$_2$, 16:42
[Ir(CO)[P(CH$_3$)$_3$]$_4$]Cl, 18:63
[Ir(CO)$_2$Cl$_2$]K$_{0.6}$·½H$_2$O, 19:20
Ir(CO)$_3$Cl$_{1.10}$, 19:19
IrCl(CO)[P(CH$_3$)$_3$]$_2$, 18:64
trans-[IrCl(CS)[P(C$_6$H$_5$)$_3$]$_2$], 19:206
[IrCl(NO){P(C$_6$H$_5$)$_3$}$_2$](BF$_4$), 16:41
IrCl(N$_2$)[P(C$_6$H$_5$)$_3$]$_2$, 16:42
[IrCl$_6$](NH$_4$)$_2$, 18:132
IrH$_2$(OCOCH$_3$)[P(C$_6$H$_5$)$_3$]$_3$, 17:129
[Ir(NH$_3$)$_5$(CO$_3$)]ClO$_4$, 17:152
[Ir[P(OCH$_3$)$_3$]$_5$][BPh$_4$], 20:79
Ir$_2$(CO)$_4$[μ-SC(CH$_3$)$_3$]$_2$, 20:237
Ir$_2$(μ-SC$_6$H$_5$)$_2$(CO)$_4$, 20:238

KAl$_{11}$O$_{17}$, 19:55
[K{(CH$_3$OCH$_2$CH$_2$)$_2$O}$_3$]Nb(CO)$_6$, 16:69
[K{(CH$_3$OCH$_2$CH$_2$)$_2$O}$_3$]Ta(CO)$_6$, 16:71
K[sec-C$_4$H$_9$)$_3$BH], 17:26
K[Cr(C$_3$H$_2$O$_4$)$_2$(H$_2$O)$_2$]·3H$_2$O, 16:81
K[Cr(H$_2$O)(C$_2$O$_4$)$_2$]·3H$_2$O, 17:149
K[Cr(H$_2$O)$_2$(C$_2$O$_4$)$_2$]·2H$_2$O, 17:148
K[GaH$_4$], 17:50n
K$_{0.6}$[Ir(CO)$_2$Cl$_2$]·½H$_2$O, 19:20
K$_{1.64}$[Pt(C$_2$O$_4$)]·2H$_2$O, 19:16, 17
K$_{1.75}$[Pt(CN)$_4$]·1.5H$_2$O, 19:8, 14
K$_2$[CoCl$_4$], 20:51
K$_2$[MgCl$_4$], 20:51
K$_2$[MnCl$_4$], 20:51
K$_2$[NiCl$_4$], 20:51
K$_2$[ZnCl$_4$], 20:51
K$_3$[Cr(C$_3$H$_2$O$_4$)$_3$]·3H$_2$O, 16:80

LiAl$_{11}$O$_{17}$, 19:54
Li[[(CH$_3$)$_3$Si]$_2$N], 18:115
LiC$_6$H$_4$S(CH$_3$), 16:170
LiCrO$_2$, 20:50
Li[GaH$_4$], 17:45
LiP(C$_6$H$_5$)$_2$, 17:186
Li[ZnH$_3$], 17:10
Li$_2$(cyclo-C$_8$H$_8$), 19:214
Li$_2$(C$_{32}$H$_{16}$N$_4$) (lithium, μ-[phthalo-
   cyaninato(2-)]di-), 20:159
Li$_2$[ZnH$_4$], 17:12

[5,7,12,14-Me$_4$-2,3:9,10-Bzo$_2$[14]hexa-
   enato(2-)N$_4$]Ni, 20:115
Me$_6$[14]-4,11-diene-1,4,8,11-N$_4$, 18:2
Mg[CH$_2$[Si(CH$_3$)$_3$]]$_2$, 19:262
Mg(C$_4$H$_8$O)$_2$[Fe(C$_5$H$_5$)(CO)$_2$]$_2$, 16:56
Mg(C$_4$H$_8$O)$_4$[Co(CO)$_3$P(C$_4$H$_9$)$_3$]$_2$, 16:58
Mg(C$_4$H$_8$O)$_4$[Mo(CO)$_2$P(C$_4$H$_9$)$_3$(C$_5$H$_5$)]$_2$, 16:59
Mg(C$_5$H$_5$N)$_4$[Co(CO)$_4$]$_2$, 16:58
Mg(C$_9$H$_7$)$_2$, 16:137
Mg(C$_{36}$H$_{44}$N$_4$) (magnesium(II), [2,3,7,8,
   12,13,17,18-octaethyl-21H,23H-
   porphinato(2-)]-), 20:145
Mg(C$_{44}$H$_{36}$N$_4$)(C$_6$H$_5$N)$_2$ (magnesium(II),
   [1,4,8,11,15,18,22,25-octamethyl-
   29H,31H-tetrabenzo[b,g,l,q]porphinato
   (2-)]bis(pyridine)-), 20:158
Mg[Co(CO)$_3$ {P(CH$_3$)(C$_6$H$_5$)$_2$}]$_2$(tmeda)$_2$, 16:59
Mg[Co(CO)$_3$ {P(C$_4$H$_9$)$_3$}]$_2$(C$_4$H$_8$O)$_4$, 16:58
Mg[Co(CO)$_4$]$_2$(C$_5$H$_5$N)$_4$, 16:58
MgCr$_2$O$_4$, 20:52
Mg[Fe(C$_5$H$_5$)(CO)$_2$]$_2$(C$_4$H$_8$O)$_2$, 16:56
MgH$_2$, 17:2
MgK$_2$Cl$_4$, 20:51
Mg(oep), 20:145
MnBr(CO)$_5$, 19:160
Mn(CH$_3$CO)(CO)$_5$, 18:57
[Mn(CH$_3$CO)$_2$(CO)$_4$]$_3$Al, 18:56, 58
Mn(CO)(η^5-C$_5$H$_5$)(CS)[P(C$_6$H$_5$)$_3$], 19:189
Mn(CO)(NO)$_3$, 16:64
Mn(CO)$_2$(CS)(C$_5$H$_5$), 16:53
Mn(CO)$_2$(η^5-C$_5$H$_5$)(CSe), 19:193, 195

Formula Index   293

Mn(CO)$_3$(η^5-C$_5$Cl$_5$), 20:194
Mn(CO)$_3$(η^5-C$_5$H$_4$Cl), 20:192
Mn(CO)$_3$(η^5-C$_5$H$_4$I), 20:193
Mn(CO)$_4$(B$_3$H$_8$), 19:227, 228
Mn(CO)$_5$(η^1-C$_5$Cl$_5$), 20:193
Mn(CO)$_5$Cl, 19:159
Mn(CO)$_5$I, 19:161, 162
[Mn(CO)$_5$]$_3$Tl, 16:61
Mn(η^1-C$_5$BrCl$_4$)(CO)$_5$, 20:194
Mn(η^5-C$_5$Cl$_4$Br)(CO)$_3$, 20:194, 195
Mn(η^5-C$_5$H$_4$Br)(CO)$_3$, 20:193
Mn(C$_5$H$_5$)(CO)$_2$(CS), 16:53
Mn(C$_{26}$H$_{14}$N$_8$)(MnHp), 18:48
MnCr$_2$O$_4$, 20:52
Mn(diphos)(η^5-C$_5$H$_5$)(CS), 19:191
MnHp, 18:48
MnK$_2$Cl$_4$, 20:51
Mn$_2$P$_2$O$_7$, 19:121
Mn$_3$(CO)$_{10}$(μ-B$_2$H$_6$)-μ-H, 20:240
[MoBr$_2$(NO)(C$_5$H$_5$)]$_2$, 16:27
Mo(CH$_3$CN)$_2$Cl$_4$, 20:120
Mo(CO)$_2$(η^5-C$_5$H$_5$)(NO), 18:127
Mo(CO)$_2$(NO)(C$_5$H$_5$), 16:24
[Mo(CO)$_2$ {P(C$_4$H$_9$)$_3$}(C$_5$H$_5$)]$_2$Mg
    (C$_4$H$_8$O)$_4$, 16:59
Mo(CO)$_2$[S$_2$P(i-C$_3$H$_7$)$_2$]$_2$, 18:53
Mo(CO)$_3$(η^5-C$_5$H$_5$)SiH$_3$, 17:104
Mo(C$_2$H$_2$)(CO)[S$_2$P(i-C$_3$H$_7$)$_2$]$_2$, 18:55
[Mo(C$_5$H$_5$)Br$_2$(NO)]$_2$, 16:27
Mo(C$_5$H$_5$)(CO)$_2$(NO), 16:24
Mo(η^5-C$_5$H$_5$)(C$_2$H$_5$)(NO)$_2$, 19:210
[Mo(C$_5$H$_5$)Cl$_2$(NO)]$_2$, 16:26
[Mo(C$_5$H$_5$)I$_2$(NO)]$_2$, 16:28
Mo(η^5-C$_5$H$_5$)(NO)$_2$(C$_6$H$_5$), 19:209
Mo(η^6-C$_6$H$_5$Cl)$_2$, 19:81, 82
Mo[η^6-C$_6$H$_5$[N(CH$_3$)$_2$]]$_2$, 19:81
Mo(η^6-C$_6$H$_6$)(η^5-C$_5$H$_5$), 20:196, 197
Mo(η^6-C$_6$H$_6$)(η^5-C$_5$H$_5$)Br, 20:199
Mo(η^6-C$_6$H$_6$)(η^5-C$_5$H$_5$)Cl, 20:198
Mo(η^6-C$_6$H$_6$)(η^5-C$_5$H$_5$)I, 20:199
Mo(η^6-C$_6$H$_6$)$_2$, 17:54
MoCl(η^5-C$_5$H$_5$)(NO)$_2$, 18:129
[MoCl$_2$(NO)(C$_5$H$_5$)]$_2$, 16:26
MoCl$_3$(C$_4$H$_8$O)$_3$, 20:121
MoCl$_4$(C$_4$H$_8$O)$_2$, 20:121
MoF$_5$, 19:137-139
MoH(acac)[(C$_6$H$_5$)$_2$PCH$_2$CH$_2$P(C$_6$H$_5$)$_2$]$_2$,
    17:61
[MoH{P(C$_6$H$_5$)(CH$_3$)$_2$}(η^6-C$_6$H$_6$)][PF$_6$],
    17:58

[MoH$_2$ {P(C$_6$H$_5$)(CH$_3$)$_2$}(η^6-C$_6$H$_6$)][PF$_6$]$_2$,
    17:60
MoH$_2$[P(C$_6$H$_5$)$_3$]$_2$(η^6-C$_6$H$_6$), 17:57
[MoI$_2$(NO)(C$_5$H$_5$)]$_2$, 16:28
Mo(NO)$_2$[S$_2$CN(C$_2$H$_5$)$_2$]$_2$, 16:235
$trans$-Mo(N$_2$)$_2$[(PPh$_2$)$_2$(CH$_2$)$_2$]$_2$, 20:122
MoOCl$_3$(bpy)(red and green forms),
    19:135, 136
[MoOCl$_5$][BpyH$_2$], 19:135
cis-Mo[P(t-C$_4$H$_9$)F$_2$]$_2$(CO)$_4$, 18:125
Mo$_2$Br$_4$(C$_5$H$_5$N)$_4$, 19:131
Mo$_2$Br$_4$[P(n-C$_4$H$_9$)$_3$]$_4$, 19:131
[Mo$_2$Br$_8$H]Cs$_3$, 19:130
[Mo$_2$Cl$_4$[C$_2$H$_4$(SCH$_3$)$_2$]$_2$], 19:131
[Mo$_2$Cl$_8$]$^{4-}$ (see [Mo$_2$Cl$_8$H]$^{3-}$)
[Mo$_2$Cl$_8$H]Cs$_3$, 19:129
[Mo$_2$Cl$_9$](NH$_4$)$_5$·H$_2$O, 19:129
Mo$_2$[O$_2$C(CH$_2$)$_2$CH$_3$]$_4$, 19:133
Mo$_2$(O$_2$CC$_6$H$_5$)$_2$Br$_2$[P(n-C$_4$H$_9$)$_3$]$_2$, 19:133

NAlH$_3$(CH$_3$)$_3$, 17:37
(NC)BH$_2$[N(CH$_3$)$_3$], 19:233, 234
(NCCH$_3$)$_2$MoCl$_4$, 20:120
[N{CH$_2$P(C$_6$H$_5$)$_2$}(CH$_3$)CH$_2$]$_2$, 16:199
N[CH$_2$P(C$_6$H$_5$)$_2$](CH$_3$)C$_2$H$_4$N(CH$_3$)$_2$,
    16:199
[N{CH$_2$P(C$_6$H$_5$)$_2$}$_2$CH$_2$]$_2$, 16:198
N[CH$_2$P(C$_6$H$_5$)$_2$]$_2$C$_2$H$_4$N[CH$_2$P(C$_6$H$_5$)$_2$]
    (CH$_3$), 16:199
N[CH$_2$P(C$_6$H$_5$)$_2$]$_2$C$_2$H$_4$N(CH$_3$)$_2$, 16:199
[N{CH$_2$P(O)(OH)(C$_6$H$_5$)}$_2$CH$_2$]$_2$, 16:199
N[CH$_2$P(O)(OH)(C$_6$H$_5$)]$_3$, 16:202
[N(CH$_3$)C$_6$H$_5$]$_3$B, 17:162
[N(CH$_3$)$_2$]$_2$BH, 17:30
[N(CH$_3$)$_2$BH$_2$]$_2$, 17:32
μ-[N(CH$_3$)$_2$]B$_2$H$_5$, 17:34
[N(CH$_3$)$_2$]PF$_4$, 18:181
[N(CH$_3$)$_2$]Si(CH$_3$)$_3$, 18:180
[N(CH$_3$)$_2$]$_2$PF$_3$, 18:186
[N(CH$_3$)$_3$]BH$_2$(NC), 19:233, 234
(NCN)(GeH$_3$)$_2$, 17:163
(NCS)$_2$Ni(Bzo$_4$[16]octaeneN$_4$), 18:31
(NCS)$_2$Ni(Me$_4$[14]-1,3,8,10-tetraeneN$_4$),
    18:24
N[C$_2$H$_4$As(C$_6$H$_5$)$_2$]$_3$, 16:177
N[C$_2$H$_4$P(C$_6$H$_5$)$_2$]$_3$, 16:176
N(C$_2$H$_5$)$_2$[C$_2$H$_4$P(C$_6$H$_5$)$_2$], 16:160
[N(C$_2$H$_5$)$_2$]PF$_4$, 18:85
[N(C$_2$H$_5$)$_2$]AlH$_2$, 17:40
[N(C$_2$H$_5$)$_2$]$_2$AlH, 17:41

[N(C$_2$H$_5$)$_2$]$_2$PF$_3$, 18:187
[N(C$_2$H$_5$)$_2$]$_3$B, 17:159
[N(C$_2$H$_5$)$_4$](CN), 16:133
[N(C$_2$H$_5$)$_4$](OCN), 16:131
[N(n-C$_4$H$_9$)$_4$][BH$_4$], 17:23
[N(n-C$_4$H$_9$)$_4$][B$_2$H$_7$], 17:25
NC$_5$H$_3$-2,6-(CH$_3$CO)$_2$, 18:18
[NC$_5$H$_4$-4-(t-C$_4$H$_9$)]CoBr[(C=NO)$_2$H(CH$_3$)$_2$]$_2$, 20:130
[NC$_5$H$_4$-4-(t-C$_4$H$_9$)]Co[(C=NO)$_2$H(CH$_3$)$_2$]$_2$(CH$_2$OEt), 20:131
[NC$_5$H$_4$-4-(t-C$_4$H$_9$)]Co[(C=NO)$_2$H(CH$_3$)$_2$]$_2$[EtO(CH$_2$)$_2$], 20:131
trans-[(NC$_5$H$_5$)PtCl$_2$(C$_2$H$_4$)], 20:181
(NC$_5$H$_5$)$_2$PtBr$_2$(CH$_3$)$_2$, 20:186
(NC$_5$H$_5$)$_3$AlI$_3$, 20:83
(NC$_5$H$_5$)$_4$MoBr$_4$, 19:131
NGaH$_3$(CH$_3$)$_3$, 17:42
NGe(C$_2$H$_5$)(OCH$_2$CH$_2$)$_3$, 16:229
[(NHCH)(CH$_3$)C$_2$H$_5$]$_3$B, 17:160
NH{(CH$_2$)$_3$NH}$_2$, 18:18
NH[(CH$_3$)$_3$Si]$_2$, 18:12
(NH)S[N(SH)], 18:124
[(NH)S(NS)]Ni(NS$_3$), 18:124
[(NH)S(NS)]$_2$Ni, 18:124
NHS$_7$, 18:203, 204
NH$_2$(C$_6$H$_4$)(CHO), 18:31
(NH$_2$)C$_6$H$_4$(SCH$_3$), 16:169
1,3-(NH$_2$)$_2$(CH$_2$)$_3$, 18:23
(NH$_2$)PO(OH)$_2$ (correction), 19:281
[[NH$_3$(CH$_2$)$_2$S]Pt(terpy)](NO$_3$)$_2$, 20:104
(NH$_3$)(TaS$_2$), 19:42
(NH$_4$)Al$_{11}$O$_{17}$, 19:56
(NH$_4$)(OCN), 16:136
(NH$_4$)$_2$H$_3$IO$_6$, 20:15
(NH$_4$)$_6$[V$_{10}$O$_{28}$]·6H$_2$O, 19:140, 143
(NH$_4$)$_n$(P$_n$O$_{(3n+1)}$H$_2$), 19:278
(NO)Al$_{11}$O$_{17}$, 19:56
(NO)CoCl$_2$[P(CH$_3$)(C$_6$H$_5$)$_2$]$_2$, 16:29
(NO)Co[P(C$_6$H$_5$)$_3$]$_3$, 16:33
(NO)Co[S$_2$CN(CH$_3$)$_2$]$_2$, 16:7
(NO)Cr(CO)$_2$(η^5-C$_5$H$_5$), 18:127
Fe[S$_2$CN(C$_2$H$_5$)$_2$]$_2$, 16:5
[(NO)IrBr{P(C$_6$H$_5$)$_3$}$_2$](BF$_4$), 16:42
[(NO)IrCl{P(C$_6$H$_5$)$_3$}$_2$](BF$_4$), 16:41
[(NO)MoBr$_2$(C$_5$H$_5$)]$_2$, 16:27
(NO)Mo(CO)$_2$(η^5-C$_5$H$_5$), 15:24; 18:127
[(NO)MoCl$_2$(C$_5$H$_5$)]$_2$, 16:26
[(NO)MoI$_2$(C$_5$H$_5$)]$_2$, 16:28
(NO)(OCOCH$_3$)$_2$[P(C$_6$H$_5$)$_3$]$_2$Rh, 17:129

(NO)(OCOCF$_3$)$_3$[P(C$_6$H$_5$)$_3$]$_2$Ru, 17:127
[(NO)OsBr(NH$_3$)$_4$]Br$_2$, 16:12
[(NO)OsCl(NH$_3$)$_4$]Cl$_2$, 16:12
[(NO)OsI(NH$_3$)$_4$]I$_2$, 16:12
[(NO)Os(NH$_3$)$_4$]Br$_2$, 16:11
[(NO)Os(NH$_3$)$_4$]Cl$_2$, 16:11
[(NO)Os(NH$_3$)$_4$]I$_2$, 16:11
[(NO)Os(NH$_3$)$_5$]Br$_3$·H$_2$O, 16:11
[(NO)Os(NH$_3$)$_5$]Cl$_3$·H$_2$O, 16:11
[(NO)Os(NH$_3$)$_5$]I$_3$·H$_2$O, 16:11
(NO)[P(C$_6$H$_5$)$_3$]$_3$RuH, 17:73
[(NO)Re(CO)$_2$Cl$_2$]$_2$, 16:37
(NO)Re$_2$(CO)$_5$Cl$_3$, 16:36
(NO)Rh[P(C$_6$H$_5$)$_3$]$_3$, 16:33
[(NO)Ru(C$_2$H$_3$O$_2$)(NH$_3$)$_4$](ClO$_4$), 16:14
[(NO)RuCl(NH$_3$)$_4$]Cl$_2$, 16:13
[(NO)Ru(NCO)(NH$_3$)$_4$](ClO$_4$), 16:15
(NO)W(CO)$_2$(η^5-C$_5$H$_5$), 18:127
[(NO)$_2$Co(diphos)][B(C$_6$H$_5$)$_4$], 16:19
[(NO)$_2$Co{P(C$_6$H$_5$)$_3$}$_2$][B(C$_6$H$_5$)$_4$], 16:18
[(NO)$_2$Co(tmeda)][B(C$_6$H$_5$)$_4$], 16:17
(NO)$_2$Cr(η^5-C$_5$H$_5$)(i-C$_4$H$_9$), 19:209
[(NO)$_2$Cr(η^5-C$_5$H$_5$)]$_2$, 19:211
(NO)$_2$CrCl(η^5-C$_5$H$_5$), 18:129
(NO)$_2$Mo(η^5-C$_5$H$_5$)(C$_2$H$_5$), 19:210
(NO)$_2$Mo(η^5-C$_5$H$_5$)(C$_6$H$_5$), 19:209
(NO)$_2$MoCl(η^5-C$_5$H$_5$), 18:129
(NO)$_2$Mo[S$_2$CN(C$_2$H$_5$)$_2$]$_2$, 16:235
(NO)$_2$Os$_3$(CO)$_{10}$, 16:40
[(NO)$_2$RuCl{P(C$_6$H$_5$)$_3$}$_2$](BF$_4$), 16:21
(NO)$_2$Ru$_2$(CO)$_{10}$, 16:39
(NO)$_2$W(η^5-C$_5$H$_5$)(CH$_3$), 19:210
(NO)$_2$WCl(η^5-C$_5$H$_5$), 18:129
(NO)$_3$Mn(CO), 16:4
(NO)$_4$Cr, 16:2
cis-, and trans-[(NO$_2$)$_2$Co(NH$_3$)$_3$](NO$_3$), 18:70, 71
cis-[(NO$_2$)$_2$Rh(en)$_2$]NO$_3$, 20:59
[N(PPh$_3$)$_2$][Fe(CO)$_{11}$H], 20:218
[N(PPh$_3$)$_2$]$_2$[Fe$_3$(CO)$_{11}$], 20:222
NS$_3$, 18:124
(NS$_3$)Ni[(NH)S(NS)], 18:124
(NS$_3$)$_2$Ni, 18:124
(NS$_4$)[(n-C$_4$H$_9$)$_4$N], 18:203, 205
[N[Si(CH$_3$)$_3$]$_2$]$_3$Cr, 18:118
[N[Si(CH$_3$)$_3$]$_2$]$_3$Fe, 18:18
[N[Si(CH$_3$)$_3$]$_2$]$_3$Sc, 18:115
[N[Si(CH$_3$)$_3$]$_2$]$_3$Ti, 18:116
[N[Si(CH$_3$)$_3$]$_2$]$_3$V, 18:117
NSn(C$_2$H$_5$)(OCH$_2$CH$_2$)$_3$, 16:230

$(N_2 C_2H_4)[C(CH_3)CH_2C(S)CH_3]_2$, 16:226
$(N_2 C_3 H_3 \text{-}N\text{-Me})Fe(O_2)(C_{66}H_{62}N_8O_4)$ (iron(II), (dioxygen) (N-methylimidazole) [(all-cis)-5,10,15,20-tetrakis[2-(2,2-dimethylpropionamido)phenyl] porphyrinato(2-)]-), 20:168
$(N_2 C_3 H_3 \text{-}N\text{-Me})_2 Fe(C_{66}H_{62}N_8O_4)$ (iron(II), bis(N-methylimidazole) [(all-cis)-5,10,15,20-tetrakis[2-(2,2-dimethylpropionamido)phenyl] porphyrinato(2-)]-), 20:167
$(N_2)C_5Cl_4$ (1,3-cyclopentadiene, tetrachloro-5-diazo-), 20:189, 190
$(N_2)C_5H_4$ (1,3-cyclopentadiene, 5-diazo-), 20:191
$(N_2 H_2)C_5Cl_4$ (2,4-cyclopentadien-1-one, hydrazone), 20:190
trans-$(N_2)_2 Mo[(PPh_2)_2(CH_2)_2]_2$, 20:122
trans-$(N_2)_2 W[(PPh_2)_2(CH_2)_2]_2$, 20:126
$N_3P_3Br_2F_4$, 18:197
$N_3P_3Br_3F_3$, 18:197
$N_3P_3Cl_2F_2[N(CH_3)_2]_2$, 18:195
$N_3P_3Cl_3[N(CH_3)_2]_3$, 18:194
$N_3P_3Cl_4[N(CH_3)_2]_2$, 18:194
$N_3P_3F_3[N(CH_3)_2]_3$, 18:195
$N_3P_3F_4[N(CH_3)_2]_2$, 18:197
$N_4C_9H_{22}$ (1,4,7,10-tetraazacyclotridecane([13]aneN$_4$)), 20:106
$N_4C_{10}H_{24}$, 16:223
$N_4C_{11}H_{26}$ (1,4,8,12-tetraazacyclopentadecane([15]aneN$_4$)), 20:108
$N_4C_{12}H_{20}$ (5,14-Me$_2$[14]-4,6,11,13-tetraene), 18:42
$N_4C_{12}H_{22}$ (H$_2$ 5,14-Me$_2$[14]-4,6,11,13-tetraene-1,4,8,11-N$_4$) (PF$_6$)$_2$, 18:40
$N_4C_{12}H_{24}$ (2,3-Me$_2$[14]-1,3-diene-1,4,8,11-N$_4$), 18:27
$N_4C_{12}H_{28}$ (1,5,9,13-tetraazacyclohexadecane([16]aneN$_4$)), 20:109
$N_4C_{14}H_{14}Br_2$ (6,13-Br$_2$-2,3-Bzo[14]-2,4,6,11,13-pentene-1,4,8,11-N$_4$), 18:50
$N_4C_{14}H_{24}$ (Me$_4$[14]-1,3,8,10-tetraene-1,4,8,11-N$_4$), 18:22
$N_4C_{15}H_{22}$ (Me$_2$Pyro[14]trieneN$_4$), 18:17
$N_4C_{16}H_{32}$ (Me$_6$[14]-4,11-diene-1,4,8,11-N$_4$), 18:2
$N_4C_{16}H_{32} \cdot 2CHF_3O_3S$ (Me$_6$[14]-4,11-diene-1,4,8,11-N$_4 \cdot 2CF_3SO_3H$), 18:3
$N_4C_{16}H_{32} \cdot 2HClO_4$ (Me$_6$[14]-4,11-diene-1,4,8,11-N$_4 \cdot 2HClO_4$), 18:4

$N_4C_{16}H_{36} \cdot XH_2O$ (Me$_6$[14]ane-1,4,8,11-N$_4 \cdot XH_2O$), 18:10
$N_4C_{18}H_{16}$ (2,3;9,10-Bzo$_2$[14]-2,4,6,9,11,13-hexaene-1,4,8,11-N$_4$) (TADA-H$_2$), 18:45
$N_4C_{18}H_{28}O_2$ (1,4,10,13-tetraazacyclooctadeca-5,8,14,17-tetraene-7,16-dione, 5,9,14,18-tetramethyl-, H$_4$daen), 20:91
$N_4C_{22}H_{22}$ (dibenzo[b,i][1,4,8,11]tetracyclotetradecine, 7,16-dihydro-6,8,15,17-tetramethyl-(5,7,12,14-Me$_4$-2,3:9,10-Bzo$_2$[14]hexaeneN$_4$]), 20:117
$N_4C_{28}H_{20}$ (2,3;6,7;10,11;14,15-Bzo$_4$-[16]octaene-1,5,9,13-N$_4$), 18:30
$N_4C_{44}H_{38}$ (29H,31H-tetrabenzo[b,g,l,q]-porphine, 1,4,8,11,15,18,22,25-octamethyl-), 20:158
$N_4O_2C_{16}H_{26}$ (6,13-Ac$_2$-5,14-Me[14]tetraeneN$_4$), 18:39
N_4S_4, 17:197
$N_8C_{26}H_{16}$ (5,26:13,18-diimino-7,11:20,24-dinitrilodibenzo[c,n][1,6,12,17]tetraazacyclodocosine) (Hp-H$_2$), 18:47
$N_8C_{44}H_{34}$ (porphyrin, 5,10,15,20-tetrakis(2-aminophenyl)-, and all-cis-), 20:163, 164
$N_8O_4C_{64}H_{66}$ (porphyrin, 5,10,15,20-tetrakis[2-(2,2-dimethylpropionamido)phenyl]), all-cis-), 20:165
$N_8O_8C_{44}H_{26}$ (porphyrin, 5,10,15,20-tetrakis(2-nitrophenyl)-, (H$_2$TNPP), 20:162
Na[GaH$_4$], 17:50
Na$_6$[V$_{10}O_{28}$]·18H$_2$O, 19:140, 142
Na$_X$(TaS$_2$), 19:42, 44
Na[ZnH$_3$], 17:15
Na[Zn$_2$(CH$_3$)$_2$H$_3$], 17:13
Nb(BH$_4$)(C$_5$H$_5$)$_2$, 16:109
Nb(bipy)$_2$(NCS)$_4$, 16:78
NbBr(C$_5$H$_5$)$_2$[P(CH$_3$)$_2$(C$_6$H$_5$)], 16:112
[Nb(CO)$_6$][As(C$_6$H$_5$)$_4$], 16:72
[Nb(CO)$_6$][K{(CH$_3$OCH$_2$CH$_2$)$_2$O}$_3$], 16:69
Nb(C$_5$H$_5$)$_2$(BH$_4$), 16:109
Nb(C$_5$H$_5$)$_2$Cl$_2$, 16:107
Nb(C$_5$H$_5$)$_2$H[P(CH$_3$)$_2$(C$_6$H$_5$)], 16:110
[Nb(C$_5$H$_5$)$_2$H$_2${P(CH$_3$)$_2$(C$_6$H$_5$)}](BF$_4$), 16:111
NbCl$_2$(C$_5$H$_5$)$_2$, 16:107
NbCl$_5$, 20:42

Nb(NCS)$_4$(bipy)$_2$, 16:78
[Ni[6,13-Ac$_2$-5,14-Me$_2$[14]tetraenato-(2-)-1,4,8,11-N$_4$], 18:39
Ni[As(C$_6$H$_5$)$_3$]$_4$, 17:121
Ni[2,2'-bipyridine]$_2$, 17:121
trans-NiBr$_2$[P(t-C$_4$H$_9$)$_2$F]$_2$, 18:177
Ni[t-BuNC]$_2$[C$_6$H$_5$C≡CC$_6$H$_5$], 17:122
Ni[t-BuNC]$_2$[C$_6$H$_5$N=NC$_6$H$_5$], 17:122
Ni[t-BuNC]$_2$[(NC)$_2$C=C(CN)$_2$], 17:122
Ni[t-BuNC]$_2$[(NC)HC=CH(CN)], 17:122
[Ni(Bzo$_4$[16]octaeneN$_4$)](ClO$_4$)$_2$, 18:31
Ni(Bzo$_4$[16]octaeneN$_4$)(NCS)$_2$, 18:31
Ni[[(CH$_3$CO)$_2$C][(CH$_2$NCH)$_2$]], 18:38
Ni[(CH$_3$)$_2$AsC$_6$H$_4$As(CH$_3$)$_2$]$_2$, 17:121
Ni[(CH$_3$)$_2$PC$_2$H$_4$P(CH$_3$)$_2$]$_2$, 17:119
Ni[(CH$_3$)$_3$CNC]$_2$(C$_6$H$_5$N=NC$_6$H$_5$), 17:122
Ni[(CH$_3$)$_3$CNC]$_4$, 17:118
[Ni[(CS)$_2$(CN)$_2$]$_2$] (C$_6$H$_4$S$_4$)$_2$ (bis[2,3-dimercapto-2-butenedinitrilato(2-)]-nickelate(2-) salt with 2,2'-bi-1,3-dithiolylidene (1:2)), 19:31
Ni(C$_6$F$_5$)$_2$[η^6-C$_6$H$_5$(CH$_3$)], 19:72
Ni[(C$_6$H$_5$)$_2$PC$_2$H$_4$P(C$_5$H$_5$)$_2$]$_2$, 17:121
Ni(C$_6$H$_{11}$NC)$_4$, 17:119
Ni(C$_{10}$N$_4$H$_{24}$)(ClO$_4$)$_2$, 16:221
Ni(1-3:6-7:10-12-η-C$_{12}$H$_{16}$)-((η^8-dodecatriene-1,12-diyl)nickel), 19:85
Ni(C$_{12}$H$_{18}$N$_4$), 18:42
[Ni(C$_{12}$H$_{24}$N$_4$)][ZnCl$_4$], 18:27
[Ni(C$_{14}$H$_{24}$N$_4$)](ClO$_4$)$_2$, 18:23
Ni(C$_{14}$H$_{24}$N$_4$)(NCS)$_2$, 18:24
[Ni(C$_{15}$H$_{22}$N$_4$)](ClO$_4$)$_2$, 18:18
Ni(C$_{16}$H$_{26}$N$_4$O$_2$), 18:39
[Ni(C$_{16}$H$_{30}$N$_4$)](ClO$_4$)$_2$, 18:5
meso-[Ni(C$_{16}$H$_{36}$N$_4$)](ClO$_4$)$_2$, 18:12
[Ni(C$_{18}$H$_{30}$N$_{12}$)][BF$_4$]$_2$, 20:89
[Ni(C$_{28}$H$_{20}$N$_4$)](ClO$_4$)$_2$, 18:31
[Ni(C$_{28}$H$_{20}$N$_4$)(NCS)$_2$], 18:31
Ni(C$_{44}$H$_{28}$N$_4$) (nickel(II), [5,10,15,20-tetraphenyl-21H,23H-porphinato(2-)]-), 20:143
NiCr$_2$O$_4$, 20:52
NiH(BH$_4$)[P(C$_6$H$_{11}$)$_3$]$_2$, 17:89
NiHCl[P(i-C$_3$H$_7$)$_3$]$_2$, 17:86
NiHCl[P(C$_6$H$_{11}$)$_3$]$_2$, 17:84
NiK$_2$Cl, 20:51
[Ni(Me$_2$[14]-1,3-dieneN$_4$)][ZnCl$_4$], 18:27
[Ni(Me$_2$Pyro[14]trieneN$_4$)](ClO$_4$)$_2$, 18:18
Ni(Me$_2$[14]-4,6,11,13-tetraenato(2-)-N$_4$), 18:42

Ni[5,7,12,14-Me$_4$-2,3:9,10-Bzo$_2$[14]-hexaenato(2-)N$_4$], 20:115
[Ni(Me$_4$[14]-1,3,8,10-tetraeneN$_4$)](ClO$_4$), 18:23
Ni(Me$_4$[14]-1,3,8,10-tetraeneN$_4$)(NCS)$_2$, 18:24
meso-[Ni(Me$_6$[14]aneN$_4$)](ClO$_4$)$_2$, 18:12
meso-, and racemic-[Ni(Me$_6$[14]-4,11-dieneN$_4$)](ClO$_4$)$_2$, 18:5
Ni[(NH)S(NS)](NS$_3$), 18:124
Ni[(NH)S(NS)]$_2$, 18:124
Ni(NS$_3$)$_2$, 18:124
Ni[P(CH$_3$)(C$_6$H$_5$)$_2$]$_4$, 17:119
Ni[P(CH$_3$)$_3$]$_4$, 17:119
Ni[P(C$_2$H$_5$)$_2$(C$_6$H$_5$)]$_4$, 17:119
Ni[P(C$_2$H$_5$)$_3$]$_4$, 17:119
Ni[P(C$_2$H$_5$)$_3$]$_2$[C$_6$H$_5$N=NC$_6$H$_5$], 17:123
Ni[P(n-C$_4$H$_9$)$_3$]$_4$, 17:119
Ni[P(C$_6$H$_5$)$_3$]$_2$[(C$_6$H$_5$)HC=CH(C$_6$H$_5$)], 17:121
Ni[P(C$_6$H$_5$)$_3$]$_2$[(C$_6$H$_5$)N=N(C$_6$H$_5$)], 17:121
Ni[P(C$_6$H$_5$)$_3$]$_4$, 17:120
Ni[1,10-phenanthroline]$_2$, 17:121
Ni[P(OCH$_3$)(C$_6$H$_5$)$_2$]$_4$, 17:119
[Ni[P(OCH$_3$)$_3$]$_5$][BPh$_4$]$_2$, 20:76
Ni[P(OCH$_3$)$_3$]$_4$, 17:119
Ni[P(OC$_2$H$_5$)$_3$]$_4$, 17:119
Ni[P(O-i-C$_3$H$_7$)$_3$]$_4$, 17:119
Ni[P(OC$_6$H$_5$)$_3$]$_4$, 17:119
Ni(tpp), 20:143
Ni[VOF$_4$]·7H$_2$O, 16:87

(OCN)[N(C$_2$H$_5$)$_4$], 16:131
(OCN)(NH$_4$), 16:136
(OC$_4$H$_8$)MoCl$_4$, 20:21
(OC$_4$H$_8$)$_3$MoCl$_3$, 20:121
O[Ge(CH$_3$)$_3$]$_2$, 20:176, 179
O[GeH(CH$_3$)$_2$]$_2$, 20:176, 179
O(GeH$_2$CH$_3$)$_2$, 20:176, 179
O(GeH$_3$)$_2$, 20:176, 178
(1-OH-n-C$_4$H$_9$)-4-[P(C$_2$H$_5$)(C$_6$H$_5$)], 18:189, 190
[(OH$_2$)Co(CH$_3$)(C$_{52}$H$_{73}$N$_3$O$_{14}$)](ClO$_4$) (cobyrinic acid, aquamethyl-, heptamethyl ester, perchlorate), 20:141
[(OH$_2$)Co(CN)(C$_{52}$H$_{73}$N$_4$O$_{14}$)](ClO$_4$), (cobyrinic acid, aquacyano-, heptamethyl ester, perchlorate), 20:141
(OH)$_2$Pt(CH$_3$)$_2$1.5H$_2$O, 20:185, 186

Formula Index   297

[OS(CH$_3$)$_2$]$_3$InBr$_3$, 19:260
[OS(CH$_3$)$_2$]$_3$InCl$_3$, 19:259
O$_2$C$_{18}$H$_{28}$N$_4$ (1,4,10,13-tetraazacyclo–
  octadeca-5,8,14,17-tetraene-7,16-dione,
  5,9,14,18-tetramethyl-, H$_4$daen), 20:91
(O$_2$)Fe(N-MeC$_3$H$_3$N$_2$)(C$_{66}$H$_{62}$N$_8$O$_4$)
  (iron(II), (dioxygen)(N-methylimidazole)
  [(all-cis)-5,10,15,20-tetrakis[2-(2,2-
  dimethylpropionamido)phenyl]por–
  phyrinato(2-)]-), 20:168
[OsBr(NH$_3$)$_4$(NO)]Br$_2$, 16:12
[OsCl(NH$_3$)$_4$(NO)]Cl$_2$, 16:12
OsCl(OCOCF$_3$)(CO)[P(C$_6$H$_5$)$_3$]$_3$, 17:128
OsF$_5$, 19:137-139
[OsI(NH$_3$)$_4$(NO)]I$_2$, 16:12
[OsI(NH$_3$)$_5$]I$_2$, 16:10
[Os(NH$_3$)$_4$(NO)(OH)]Br$_2$, 16:11
[Os(NH$_3$)$_4$(NO)(OH)]Cl$_2$, 16:11
[Os(NH$_3$)$_4$(NO)(OH)]I$_2$, 16:11
[Os(NH$_3$)$_5$(NO)]Br$_2$·H$_2$O, 16:11
[Os(NH$_3$)$_5$(NO)]Cl$_2$·H$_2$O, 16:11
[Os(NH$_3$)$_5$(NO)]I$_2$·H$_2$O, 16:11
[Os(NH$_3$)$_5$(N$_2$)]I$_2$, 16:9
[Os(NH$_3$)$_6$]I$_3$, 16:10
Os(OCOCF$_3$)$_2$(CO)[P(C$_6$H$_5$)$_3$]$_2$, 17:128
[Os(OH)$_4$O$_2$]K$_2$, 20:61
trans-[OsO$_2$(en)$_2$]Cl$_2$, 20:62
Os$_3$(CO)$_{10}$(NO)$_2$, 16:40

[P(t-Bu)$_3$]$_2$Pd, 19:103
P[CH=CHAs(C$_6$H$_5$)$_2$](C$_6$H$_5$)$_2$, 16:189
[P(CH$_2$)(CH$_3$)$_2$]FeH[P(CH$_3$)$_3$]$_3$, 20:71
[(P(CH$_3$)(CH$_2$)]$_1$Hg]Cl$_2$, 18:140
[P(p-CH$_3$C$_6$H$_4$)$_3$]CuH, 19:89
[{P(CH$_3$)(C$_6$H$_5$)(O)O}Cr(H$_2$O)(OH)]$_x$, 16:90
[{P(CH$_3$)(C$_6$H$_5$)(O)O}Cr(OH)]$_x$, 16:91
P(CH$_3$)(C$_6$H$_5$)$_2$, 16:157
[{P(CH$_3$)(C$_6$H$_5$)$_2$}Co(CO)$_3$]Mg(tmeda)$_2$, 16:59
[P(CH$_3$)(C$_6$H$_5$)$_2$]CoCl$_2$(NO), 16:29
[P(CH$_3$)H$_2$]$_2$TiCl$_3$, 16:98
µ-[P(CH$_3$)$_2$(CH$_2$)$_2$]$_2$Ag$_2$, 18:142
P(CH$_3$)$_2$(C$_6$F$_5$), 16:181
[P(CH$_3$)$_2$(C$_6$H$_5$)]NbBr(C$_5$H$_5$)$_2$, 16:112
[P(CH$_3$)$_2$(C$_6$H$_5$)]Nb(C$_5$H$_5$)$_2$H, 16:110
[{P(CH$_3$)$_2$(C$_6$H$_5$)}Nb(C$_5$H$_5$)$_2$H$_2$](BF$_4$), 16:111
[{P(CH$_3$)$_2$(C$_6$H$_5$)}Nb(C$_5$H$_5$)$_2$H$_2$](PF$_6$), 16:111

[P(CH$_3$)$_2$(C$_6$H$_5$)]$_2$Ti(C$_5$H$_5$)Cl$_2$, 16:239
[P(CH$_3$)$_2$C$_6$H$_5$]$_3$ReH$_5$, 17:64
[P(CH$_3$)$_2$H]$_2$TiCl$_3$, 16:100
[P(CH$_3$)$_2$N]F$_4$, 18:181
[P(CH$_3$)$_2$N]$_2$F$_3$, 18:186
P(CH$_3$)$_3$, 16:153
P(CH$_3$)$_3$[CH[Si(CH$_3$)$_3$]], 18:137
P(CH$_3$)$_3$(CH$_2$), 18:137
P(CH$_3$)$_3$(CH$_2$)]Au(CH$_3$), 18:141
[P(CH$_3$)$_3$]$_2$FeCl$_2$, 20:70
[P(CH$_3$)$_3$]$_2$IrCl(CO), 18:64
[P(CH$_3$)$_3$]$_2$TiCl$_3$, 16:100
[P(CH$_3$)$_3$]$_4$Fe, 20:71
[P(CH$_3$)$_3$]$_4$Fe(CO$_2$), 20:73
[P(CH$_3$)$_3$]$_4$Ir(CO)]Cl, 18:63
P(CH$_3$)$_4$Br, 18:138
P[C$_2$H$_4$As(C$_6$H$_5$)$_2$](C$_6$H$_5$)$_2$, 16:191
P[C$_2$H$_4$N(C$_2$H$_5$)$_2$](C$_6$H$_5$)$_2$, 16:160
P[C$_2$H$_4$P(C$_3$H$_7$)(C$_6$H$_5$)](C$_6$H$_5$)$_2$, 16:192
P[C$_2$H$_4$P(C$_6$H$_5$)H](C$_6$H$_5$)$_2$, 16:202
P(C$_2$H$_5$)(C$_6$H$_5$)$_2$, 16:158
P(C$_2$H$_5$)$_2$(C$_6$H$_5$), 18:170
P[(C$_2$H$_5$)$_2$N]F$_4$, 18:185
P[(C$_2$H$_5$)$_2$N]$_2$F$_3$, 18:187
trans-[[P(C$_2$H$_5$)$_3$]PtCl[C(NHC$_6$H$_5$)(NH= C$_2$H$_5$)]]ClO$_4$, 19:176
cis-[[P(C$_2$H$_5$)$_3$]PtCl$_2$(CNC$_6$H$_5$)], 19:174
cis-[[P(C$_2$H$_5$)$_3$]PtCl$_2$[C(NHC$_6$H$_5$)= (C$_2$H$_5$O)]], 19:175
cis-[[P(C$_2$H$_5$)$_3$]PtCl$_2$[C(NHC$_6$H$_5$)$_2$]], 19:176
[{P(C$_2$H$_5$)$_3$}$_2$Cl(C$_2$H$_5$)Pt], 17:132
[P(C$_2$H$_5$)$_3$]$_2$TiCl$_3$, 16:101
[P(C$_2$H$_5$)$_3$]$_3$Pt, 19:108
[P(C$_2$H$_5$)$_3$]$_4$Pt, 19:110
[P(C$_3$H$_7$)(C$_6$H$_5$)CH$_2$]P(S)(C$_6$H$_5$)$_2$, 16:195
[P(C$_3$H$_7$)(C$_6$H$_5$)C$_2$H$_4$]P(C$_6$H$_5$)$_2$, 16:192
P(C$_3$H$_7$)(C$_6$H$_5$)$_2$, 16:158
trans-[[P(i-C$_3$H$_7$)$_3$]$_2$PtCl$_2$], 19:108
[P(i-C$_3$H$_7$)$_3$]$_3$Pt, 19:108
P(t-C$_4$H$_9$)F$_2$, 18:174
cis-[P(t-C$_4$H$_9$)F$_2$]$_2$Mo(CO)$_4$, 18:175
P(n-C$_4$H$_9$)$_2$(C$_6$H$_5$), 18:171
P(t-C$_4$H$_9$)$_2$F, 18:176
trans-[P(t-C$_4$H$_9$)$_2$F]$_2$NiBr$_2$, 18:177
[P(C$_4$H$_9$)$_3$Co(CO)$_3$]$_2$Mg(C$_4$H$_8$O)$_4$, 16:58
[{P(C$_4$H$_9$)$_3$}Mo(CO)$_2$(C$_5$H$_5$)]Mg(C$_4$H$_8$O)$_4$, 16:59
[P(n-C$_4$H$_9$)$_3$]$_2$Mo(C$_6$H$_5$CO$_2$)$_2$Br$_2$, 19:133
trans-[[P(n-C$_4$H$_9$)$_3$]$_2$PtCl$_2$], 19:116

[P(n-C$_4$H$_9$)$_3$]$_4$Mo$_2$Br$_4$, 19:131
P(C$_6$D$_5$)$_3$, 16:163
P(C$_6$H$_4$CH$_3$)(C$_6$H$_5$)$_2$, 16:159
P(C$_6$H$_4$SCH$_3$)(C$_6$H$_5$)$_2$, 16:171
P(C$_6$H$_4$SCH$_3$)$_2$(C$_6$H$_5$), 16:172
P(C$_6$H$_4$SCH$_3$)$_3$, 16:173
[P(C$_6$H$_5$)(t-Bu)$_2$]$_2$Pd, 19:102
[P(C$_6$H$_5$)(t-Bu)$_2$]$_2$Pt, 19:104
P(C$_6$H$_5$CH$_2$)$_2$(C$_6$H$_5$), 18:172
[{P(C$_6$H$_5$)(CH$_3$)$_2$}(η^6-C$_6$H$_6$)MoH][PF$_6$], 17:58
[{P(C$_6$H$_5$)(CH$_3$)$_2$}(η^6-C$_6$H$_6$)MoH$_2$][PF$_6$]$_2$, 17:60
[P(C$_6$H$_5$)(H)C$_2$H$_4$]P(C$_6$H$_5$)$_2$, 16:202
[{P(C$_6$H$_5$)$_2$CH$_2$}(CH$_3$)NCH$_2$]$_2$, 16:199
[P(C$_6$H$_5$)$_2$CH$_2$](CH$_3$)NC$_2$H$_4$N(CH$_3$)$_2$, 16:199
{[P(C$_6$H$_5$)$_2$CH]$_2$}MoH(acac), 17:61
[{P(C$_6$H$_5$)$_2$CH$_2$}$_2$NCH$_2$]$_2$, 16:198
[P(C$_6$H$_5$)$_2$CH$_2$]$_2$NC$_2$H$_4$N[CH$_2$P(C$_6$H$_5$)$_2$](CH$_3$), 16:199
[P(C$_6$H$_5$)$_2$CH$_2$]$_2$NC$_2$H$_4$N(CH$_3$)$_2$, 16:199
{[P(C$_6$H$_5$)$_2$CH$_2$]$_2$}$_2$FeHCl, 17:69
[P(C$_6$H$_5$)$_2$C$_2$H$_4$]$_3$N, 17:176
P(C$_6$H$_5$)$_2$(C$_6$H$_{11}$, 16:159
P(C$_6$H$_5$)$_2$(C$_7$H$_7$), 16:159
P(C$_6$H$_5$)$_2$H, 16:161
[{P(C$_6$H$_5$)$_2$(O)O}Cr(H$_2$O)(OH)]$_x$, 16:90
[{P(C$_6$H$_5$)$_2$(O)O}Cr(OH)]$_x$, 16:91
P(C$_6$H$_5$)$_3$, 18:120
[P(C$_6$H$_5$)$_3$CH$_3$][B$_2$H$_7$], 17:24
[P(C$_6$H$_5$)$_3$CH$_3$][BH$_4$], 17:22
[P(C$_6$H$_5$)$_3$]CuH, 19:87, 88
[P(C$_6$H$_5$)$_3$]Mn(CO)(η^5-C$_5$H$_5$)(CS), 19:189
[P(C$_6$H$_5$)$_3$]PtCl[P(O)(OCH$_3$)$_2$][P(OH)=(OCH$_3$)$_2$], 19:98
[P(C$_6$H$_5$)$_3$]$_2$(CO)(OCOCF$_3$)$_2$Ru, 17:127
[P(C$_6$H$_5$)$_3$]$_2$(CO)(OCOCH$_3$)RuCl, 17:126
[P(C$_6$H$_5$)$_3$]$_2$(C$_2$H$_4$)Pd, 16:127
[{P(C$_6$H$_5$)$_3$}$_2${η^6-C$_6$H$_5$)P(C$_6$H$_5$)$_2$}RuH][BF$_4$], 17:77
[P(C$_6$H$_5$)$_3$]$_2$(η^6-C$_6$H$_6$)MoH$_2$, 17:57
[P(C$_6$H$_5$)$_3$]$_2$Cu(BH$_4$), 19:96
[P(C$_6$H$_5$)$_3$]$_2$Cu(NO$_3$), 19:93
[{P(C$_6$H$_5$)$_3$}$_2$IrBr(NO)](BF$_4$), 16:42
[P(C$_6$H$_5$)$_3$]$_2$IrBr(N$_2$), 16:42
trans-[[P(C$_6$H$_5$)$_3$]$_2$IrCl(CS)], 19:206
[{P(C$_6$H$_5$)$_3$}$_2$IrCl(NO)](BF$_4$), 16:41
[P(C$_6$H$_5$)$_3$]$_2$(NO)(OCOCH$_3$)$_2$Rh, 17:129
[P(C$_6$H$_5$)$_3$]$_2$(OCOCH$_3$)(CO)RuH, 17:126

[P(C$_6$H$_5$)$_3$]$_2$(OCOCH$_3$)$_2$(CO)$_2$Ru, 17:126
[P(C$_6$H$_5$)$_3$]$_2$(OCOCH$_3$)$_2$Pt, 17:130
[P(C$_6$H$_5$)$_3$]$_2$(OCOCF$_3$)$_2$(CO)Os, 17:128
[P(C$_6$H$_5$)$_3$]$_2$(OCOCF$_3$)$_2$(NO)Ru, 17:127
[P(C$_6$H$_5$)$_3$]$_2$Pt(CO$_3$), 18:120
[P(C$_6$H$_5$)$_3$]$_2$Pt[C$_2$(C$_6$H$_5$)$_2$], 18:122
[P(C$_6$H$_5$)$_3$]$_2$Pt(C$_2$H$_4$), 18:121
trans-[[P(C$_6$H$_5$)$_3$]$_2$PtCl$_2$], 19:115
[P(C$_6$H$_5$)$_3$]$_2$ReOCl$_3$, 17:110
trans-[[P(C$_6$H$_5$)$_3$]$_2$RhCl(CS)], 19:204
[{P(C$_6$H$_5$)$_3$}$_2$RuCl(NO)$_2$](BF$_4$), 16:21
[P(C$_6$H$_5$)$_3$]$_3$Co(NO), 16:33
[P(C$_6$H$_5$)$_3$]$_3$CuCl, 19:88
P(C$_6$H$_5$)$_3$]$_3$(NO)RuH, 17:73
[P(C$_6$H$_5$)$_3$]$_3$(OCOCH$_3$)H$_2$Ir, 17:129
[P(C$_6$H$_5$)$_3$]$_3$(OCOCH$_3$)Rh, 17:129
[P(C$_6$H$_5$)$_3$]$_3$(OCOCF$_3$)(CO)OsCl, 17:128
[P(C$_6$H$_5$)$_3$]$_3$Rh(NO), 16:33
[P(C$_6$H$_5$)$_3$]$_4$Ni, 17:120
[P(C$_6$H$_5$)$_3$]$_4$Pt, 18:120
[P(C$_6$H$_5$)$_3$]$_4$RuH$_2$, 17:75
P(C$_6$H$_{11}$)$_2$(C$_6$H$_5$), 18:171
[P(cyclo-C$_6$H$_{11}$)$_3$]Pt(C$_2$H$_4$)$_2$, 19:216
[P(C$_6$H$_{11}$)$_3$]$_2$(BH$_4$)NiH, 17:89
[P(C$_6$H$_{11}$)$_3$]$_2$(C$_2$H$_4$)Pd, 16:129
[P(C$_6$H$_{11}$)$_3$]$_2$NiHCl, 17:84
[P(cyclo-C$_6$H$_{11}$)$_3$]$_2$Pd, 19:103
[P(C$_6$H$_{11}$)$_3$]$_2$PdHCl, 17:87
[P(cyclo-C$_6$H$_{11}$)$_3$]$_2$Pt, 19:105
trans-[[P(cyclo-C$_6$H$_{11}$)$_3$]$_2$PtCl$_2$], 19:105
[{P(C$_8$H$_{17}$)$_2$(O)O}Cr(H$_2$O)(OH)]$_x$, 16:90
[{P(C$_8$H$_{17}$)$_2$(O)O}Cr(OH)]$_x$, 16:91
(PC$_{12}$H$_{18}$)(ClO$_4$) (1-ethyl-1-phenylphospholanium perchlorate), 18:189, 191
(PClF$_2$)Fe(CO)$_4$, 16:66
[PF$_2$N(C$_2$H$_5$)$_2$]Fe(CO)$_4$, 16:64
PF$_2$(OCH$_3$), 16:166
(PF$_3$)Fe(CO)$_4$, 16:67
[PF$_6$][MoH{P(C$_6$H$_5$)(CH$_3$)$_2$}(η^6-C$_6$H$_6$)], 17:58
(PF$_6$)Nb(C$_5$H$_5$)$_2$(H)$_2$P(CH$_3$)$_2$(C$_6$H$_5$), 16:111
[PF$_6$]$_2$[MoH$_2${P(C$_6$H$_5$)(CH$_3$)$_2$}(η^6-C$_6$H$_6$)], 17:60
[P(isopropyl)$_3$]$_2$NiHCl, 17:86
PLi(C$_6$H$_5$)$_2$, 17:186
P(NH$_2$)O(OH)$_2$ (correction), 19:281
P(O)(CH=CH$_2$)(O-i-C$_3$H$_7$)(C$_6$H$_5$), 16:203
POCH$_3$(C$_6$H$_5$)$_2$, 17:184
PO(CH$_3$)$_2$(C$_6$H$_5$), 17:185

Formula Index 299

[[P(OCH$_3$)$_3$]$_5$Co][BPh$_4$], 20:81
[P(OCH$_3$)$_3$]$_5$Fe, 20:79
[[P(OCH$_3$)$_3$]$_5$Ir][BPh$_4$], 20:79
[[P(OCH$_3$)$_3$]$_5$Ni][BPh$_4$]$_2$, 20:76
[[P(OCH$_3$)$_3$]$_5$Pd][BPh$_4$]$_2$, 20:77
[[P(OCH$_3$)$_3$]$_5$Pt][BPh$_4$]$_2$, 20:78
[[P(OCH$_3$)$_3$]$_5$Rh][BPh$_4$], 20:78
[P(OCH$_3$)$_3$]$_5$Ru, 20:80
[P(OC$_6$H$_4$CH$_3$)$_3$]$_2$(C$_2$H$_4$)Pd, 16:129
[P(OC$_6$H$_5$)$_3$]Cr(CO)[η^6-m-(CH$_3$)C$_6$H$_4$-CO$_2$=CH$_3$](CS), 19:202
P(4-OH-n-C$_4$H$_9$)(C$_2$H$_5$)(C$_6$H$_5$), 18:189, 190
[P(O)(OCH$_3$)$_2$]PtCl[P(OH)(OCH$_3$)$_2$][P=(C$_6$H$_5$)$_3$], 19:98
[P(O)(OCH$_3$)$_2$]$_2$Pt[[As(CH$_3$)$_2$]$_2$-o-C$_6$H$_4$], 19:100
[{P(O)(OH)(C$_6$H$_5$)CH$_2$}$_2$NCH$_2$]$_2$, 16:199
[P(O)(OH)(C$_6$H$_5$)CH$_2$]$_3$N, 16:201
[P(O)(OH)(C$_6$H$_5$)CH$_2$N(CH$_3$)CH$_2$]$_2$·2HCl, 16:202
POF$_2$(CH$_3$), 16:166
PO$_3$SNa$_3$, 17:193
[(PPh$_2$)$_2$(CH)$_2$]$_2$CoH, 20:207
[(PPh$_2$)$_2$(CH$_2$)$_2$]WCl$_4$, 20:125
[(PPh$_2$)$_2$(CH$_2$)$_2$]$_2$CoH, 20:208
trans-[(PPh$_2$)$_2$(CH$_2$)$_2$]$_2$Mo(N$_2$)$_2$, 20:122
trans-[(PPh$_2$)$_2$(CH$_2$)$_2$]$_2$W(N$_2$)$_2$, 20:126
[(PPh$_3$)$_2$N][Fe(CO)$_{11}$H], 20:218
[(PPh$_3$)$_2$N]$_2$[Fe$_3$(CO)$_{11}$], 20:222
(PPh$_3$)$_2$WCl$_4$, 20:124
P(S)[CH$_2$P(C$_3$H$_7$)(C$_6$H$_5$)](C$_6$H$_5$)$_2$, 16:195
[PS$_2$(i-C$_3$H$_7$)$_2$]Mo(CO)$_2$, 18:53
[PS$_2$(i-C$_3$H$_7$)$_2$]$_2$Mo(C$_2$H$_2$)(CO), 18:55
P{Si(CH$_3$)$_3$}(C$_6$H$_5$)$_2$, 17:187
(P$_2$O$_7$)Mn$_2$, 19:121
P$_3$N$_3$Br$_2$F$_4$, 18:197
P$_3$N$_3$Br$_3$F$_3$, 18:197
P$_3$N$_3$Cl$_2$F$_2$[N(CH$_3$)$_2$]$_2$, 18:195
P$_3$N$_3$Cl$_3$[N(CH$_3$)$_2$]$_3$, 18:194
P$_3$N$_3$Cl$_4$[N(CH$_3$)$_2$]$_2$, 18:194
P$_3$N$_3$F$_3$[N(CH$_3$)$_2$]$_3$, 18:195
P$_3$N$_3$F$_4$[N(CH$_3$)$_2$]$_2$, 18:197
(P$_n$O$_{(3n+1)}$H$_2$)(NH$_4$)$_n$, 19:278
Pd(C$_2$H$_4$)[P(C$_6$H$_5$)$_3$]$_2$, 16:127
Pd(C$_2$H$_4$)[P(C$_6$H$_{11}$)$_3$]$_2$, 16:129
Pd(C$_2$H$_4$)[P(OC$_6$H$_4$CH$_3$)$_3$]$_2$, 16:129
Pd(η^3-C$_3$H$_5$)(η^5-C$_5$H$_5$), 19:221
[PdH(BH$_4$){P(C$_6$H$_{11}$)$_3$}$_2$], 17:90
PdHCl[P(C$_6$H$_{11}$)$_3$]$_2$, 17:87

Pd[P(t-Bu)$_3$]$_2$, 19:103
Pd[P(C$_6$H$_5$)(t-Bu)$_2$]$_2$, 19:102
Pd[P(cyclo-C$_6$H$_{11}$)$_3$]$_2$, 19:103
[Pd[P(OCH$_3$)$_3$]$_5$][BPh$_4$]$_2$, 20:77
[Pd$_2$Cl$_2$(t-C$_4$H$_9$NC)$_4$], 17:134
Pd$_2$(η^3-C$_3$H$_5$)$_2$Cl$_2$, 19:220
Pr[C$_3$H$_4$(CH$_3$)(C$_4$H$_9$)]Cl$_2$, 16:114
Pt(acac)$_2$, 20:66
PtBr$_2$(CH$_3$)$_2$, 20:185
PtBr$_2$(CH$_3$)$_2$(C$_5$H$_5$N)$_2$, 20:186
[PtBr$_2$(CN)$_4$][C(NH$_2$)$_3$]·XH$_2$O, 19:11
[PtBr$_2$(CN)$_4$]K$_2$·2H$_2$O, 19:4
[PtBr$_4$]K$_2$, 19:2
[PtBr$_6$]H$_2$, 19:2
[PtBr$_6$]K$_2$, 19:2
Pt(CF$_3$COCHCOCH$_3$)$_2$, 20:67
Pt[(CF$_3$CO)$_2$CH]$_2$, 20:67
cis-[Pt[C(NHC$_6$H$_5$)(C$_2$H$_5$O)]Cl$_2$[P-(C$_2$H$_5$)$_3$], 19:175
trans-[Pt[C(NHC$_6$H$_5$)(NHC$_2$H$_5$)]Cl[P-(C$_2$H$_5$)$_3$]]ClO$_4$, 19:176
cis-[Pt[C(NHC$_6$H$_5$)$_2$]Cl$_2$[P(C$_2$H$_5$)$_3$]], 19:176
[Pt(CN)$_4$]Ba·3H$_2$O, 19:112
[Pt(CN)$_4$]Ba·4H$_2$O, correction, 20:243
[Pt(CN)$_4$]Br$_{0.25}$[C(NH$_2$)$_3$]$_2$·H$_2$O, 19:10, 12
[Pt(CN)$_4$]Br$_{0.3}$K$_2$·3H$_2$O, 19:1, 4, 15
[Pt(CN)$_4$][C(NH$_2$)$_3$]$_2$, 19:11
[Pt(CN)$_4$](C$_6$H$_4$S$_4$)$_2$ (tetracyanoplatinate (II) salt with 2,2'-bi-1,3-dithiolylidene (1:2)), 19:31
[Pt(CN)$_4$]Cl$_{0.3}$K$_2$·3H$_2$O, 19:15
[Pt(CN)$_4$]Cs$_{1.75}$·2H$_2$O, 19:6, 7
[Pt(CN)$_4$]Cs$_2$F$_{0.19}$, 20:29
[Pt(CN)$_4$]Cs$_2$(HF$_2$)$_{0.23}$, 20:26
[Pt(CN)$_4$]Cs$_2$(HF$_2$)$_{0.38}$, 20:28
[Pt(CN)$_4$]Cs$_2$·H$_2$O, 19:6
[Pt(CN)$_4$]K$_{1.75}$·1.5H$_2$O, 19:8, 14
[Pt(CN)$_4$]K$_2$·H$_2$O, 19:3
[Pt(CN)$_4$]Rb$_{1.6}$·2H$_2$O, 19:9
[Pt(CN)$_4$]Rb$_2$(HF$_2$)$_{0.29}$1.67H$_2$O, 20:24
[Pt(CN)$_4$]Rb$_2$(HF$_2$)$_{0.38}$, 20:25
[Pt(CN)$_4$]Rb$_2$[H(SO$_4$)$_2$]$_{0.46}$H$_2$O, 20:20
[Pt[(CS)$_2$(CN)$_2$]$_2$](C$_6$H$_4$S$_4$) (bis[2,3-dimercapto-2-butenedinitrilato(2-)]platinate(1-) salt with 2,2'-bi-1,3-dithiolylidene (1:1)), 19:31
[Pt[(CS)$_2$(CN)$_2$]$_2$](C$_6$H$_4$S$_4$)$_2$ (bis[2,3-dimercapto-2-butenedinitrilato(2-)]

platinate(2-) salt with 2,2'-bi-1,3-dithiolylidene (1:2)), 19:31
Pt[C$_2$(C$_6$H$_5$)$_2$] [P(C$_6$H$_5$)$_3$]$_2$, 18:122
Pt(C$_2$H$_4$)[P(C$_6$H$_5$)$_3$]$_2$, 18:121
Pt(C$_2$H$_4$)$_2$ [P(cyclo-C$_6$H$_{11}$)$_3$], 19:216
Pt(C$_2$H$_4$)$_3$, 19:215
[Pt(C$_2$H$_5$)Cl{P(C$_2$H$_5$)$_3$}$_2$], 17:132
[Pt(C$_2$O$_4$)$_2$]K$_{1.64}$·2H$_2$O, 19:16, 17
[Pt(C$_2$O$_4$)$_2$]K$_2$·2H$_2$O, 19:16
Pt[C$_3$H$_4$(C$_6$H$_5$)$_2$](C$_5$H$_5$N)$_2$Cl$_2$, 16:115
Pt[C$_3$H$_4$(C$_6$H$_5$)$_2$]Cl$_2$, 16:114
Pt[C$_3$H$_5$(CH$_2$C$_6$H$_5$)](C$_5$H$_5$N)Cl$_2$, 16:115
Pt[C$_3$H$_5$(CH$_2$C$_6$H$_5$)]Cl$_2$, 16:114
Pt[C$_3$H$_5$(C$_6$H$_4$CH$_3$)](C$_5$H$_5$N)Cl$_2$, 16:115
Pt[C$_3$H$_5$(C$_6$H$_4$CH$_3$)]Cl$_2$, 16:114
Pt[C$_3$H$_5$(C$_6$H$_4$NO$_2$)](C$_5$H$_5$N)Cl$_2$, 16:115
Pt[C$_3$H$_5$(C$_6$H$_4$NO$_2$)]Cl$_2$, 16:114
Pt[C$_3$H$_5$(C$_6$H$_5$)](C$_5$H$_5$N)Cl$_2$, 16:115
Pt[C$_3$H$_5$(C$_6$H$_5$)]Cl$_2$, 16:114
Pt[C$_3$H$_5$(C$_6$H$_{13}$)](C$_5$H$_5$N)Cl$_2$, 16:115
Pt[C$_3$H$_5$(C$_6$H$_{13}$)]Cl$_2$, 16:114
Pt(C$_3$H$_6$)(C$_5$H$_5$N)Cl$_2$, 16:115
Pt(C$_3$H$_6$)Cl$_2$, 16:114
Pt(C$_8$H$_{12}$)$_2$ (bis(1,5-cyclooctadiene)-platinum), 19:213, 214
PtCl[P(O)(OCH$_3$)$_2$] [P(OH)(OCH$_3$)$_2$] [P-(C$_6$H$_5$)$_3$], 19:98
PtCl$_2$, β-, 20:48
trans-[PtCl$_2$(C$_2$H$_4$)(C$_5$H$_5$N)], 20:181
cis-[PtCl$_2$(C$_6$H$_5$NC)[P(C$_2$H$_5$)$_3$]], 19:174
trans-[PtCl$_2$[P(i-C$_3$H$_7$)$_3$]$_2$], 19:108
trans-[PtCl$_2$[P(n-C$_4$H$_9$)$_3$]$_2$], 19:116
trans-[PtCl$_2$[P(C$_6$H$_5$)$_3$]$_2$], 19:115
trans-[PtCl$_2$[P(cyclo-C$_6$H$_{11}$)$_3$]$_2$], 19:105
Pt(OH)$_2$(CH$_3$)$_2$·1.5H$_2$O, 20:185, 186
Pt[P(C$_2$H$_5$)$_3$]$_3$, 19:108
Pt[P(C$_2$H$_5$)$_3$]$_4$, 19:110
Pt[P(i-C$_3$H$_7$)$_3$]$_3$, 19:108
Pt[P(C$_6$H$_5$)(t-Bu)$_2$]$_2$, 19:104
[Pt[P(OCH$_3$)$_3$]$_5$][BPh$_4$]$_2$, 20:78
Pt(OCOCH$_3$)$_2$ [P(C$_6$H$_5$)$_3$]$_2$, 17:130
Pt[P(C$_6$H$_5$)$_3$]$_2$(CO$_3$), 18:120
Pt[P(C$_6$H$_5$)$_3$]$_4$, 18:120
Pt[P(cyclo-C$_6$H$_{11}$)$_3$]$_2$, 19:105
Pt[P(O)(OCH$_3$)$_2$]$_2$[[As(CH$_3$)$_2$]$_2$-o-C$_6$H$_4$], 19:100
[Pt[S(CH$_2$)$_2$NH$_3$](terpy)](NO$_3$)$_2$, 20:104
[Pt[S(CH$_2$)$_2$OH](terpy)]NO$_3$, 20:103
PtS$_2$, 19:49
[Pt(terpy)Cl]Cl·2H$_2$O, 20:101

PtTe$_2$, 19:49
Pt$_2$Cl$_4$(C$_2$H$_4$)$_2$, 20:181,182
Pt$_2$Cl$_4$(C$_6$H$_5$CH=CH$_2$)$_2$, 20:181, 182
Pt$_2$Cl$_4$(C$_{12}$H$_{24}$)$_2$, 20:181, 183

RbAl$_{11}$O$_{17}$, 19:55
Rb$_{1.6}$[Pt(CN)$_4$]·2H$_2$O, 19:9
Rb$_2$[Pt(CN)$_4$](HF$_2$)$_{0.29}$·1.67H$_2$O, 20:24
Rb$_2$[Pt(CN)$_4$](HF$_2$)$_{0.38}$, 20:25
Rb$_3$[Pt(CN)$_4$][H(SO$_4$)$_2$]$_{0.46}$·H$_2$O, 20:20
Re[CH$_3$C(O)][CH$_3$C(NH$_2$)](CO)$_4$, 20:204
Re[CH$_3$C(O)](CO)$_5$, 20:201
[Re(CO)$_2$Cl$_2$(NO)]$_2$, 16:37
[Re(CO)$_4$Cl]$_2$, 16:35
ReCl$_3$[P(CH$_3$)$_2$C$_6$H$_5$]$_3$, 17:111
ReCl$_5$, 20:41
ReF$_5$, 19:137-139
cis-ReH(CH$_3$C(O)]$_2$(CO)$_4$, 20:200, 202
ReH$_5$[P(CH$_3$)$_2$C$_6$H$_5$]$_3$, 17:64
ReOCl$_3$[P(C$_6$H$_5$)$_3$]$_2$, 17:110
Re$_2$(CO)$_5$Cl$_3$(NO), 16:36
Re$_2$(C$_2$H$_3$O$_2$)$_4$Cl$_2$, 20:46
Re$_3$Br$_9$, 20:47
Re$_3$Cl$_9$, 20:44, 47
Re$_3$H$_3$(CO)$_{12}$, 17:66
Re$_3$I$_9$, 20:47
Re$_4$H$_4$(CO)$_{12}$, 18:60
cis-[RhBr$_2$(en)$_2$]Br, 20:60
Rh(CO)$_2$(η^5-C$_5$Ph$_4$Cl), 20:192
Rh(η^5-C$_5$H$_4$Cl)(η^4-C$_8$H$_{12}$), 20:192
Rh(η^5-C$_5$H$_4$Cl)(cod), 20:192
Rh(η^5-C$_5$Ph$_4$Br)(η^4-C$_8$H$_{12}$), 20:192
Rh(η^5-C$_5$Ph$_4$Br)(cod), 20:192
Rh(η^5-C$_5$Ph$_4$Cl)(η^4-C$_6$H$_{12}$), 20:191
Rh(η^5-C$_5$Ph$_4$Cl)(cod), 20:191
Rh(η^4-C$_8$H$_{12}$)(η^5-C$_5$Cl$_5$), 20:194
trans-[RhCl(CS)[P(C$_6$H$_5$)$_3$]$_2$], 19:204
cis-[RhCl$_2$(en)$_2$]Cl, 20:60
Rh(cod)(η^5-C$_5$Cl$_5$), 20:194
[Rh(en)$_2$(C$_2$O$_4$)]ClO$_4$, 20:58
cis-[Rh(en)$_2$(NO$_2$)$_2$]NO$_3$, 20:59
RhH[(+)-diop]$_2$, 17:81
[Rh(NH$_3$)$_5$(CO$_3$)]ClO$_4$·H$_2$O, 17:152
Rh(NO)(OCOCH$_3$)$_2$[P(C$_6$H$_5$)$_3$]$_2$, 17:129
Rh(NO)[P(C$_6$H$_5$)$_3$]$_3$, 16:33
Rh(OCOCH$_3$)[P(C$_6$H$_5$)$_3$]$_3$, 17:129
[Rh[P(OCH$_3$)$_3$]$_5$][BPh$_4$], 20:78
Rh$_2$Cl$_2$(η^4-C$_6$H$_{10}$)$_2$ (di-μ-chloro-bis(η^4-1,5-hexadiene)dirhodium), 19:219
Rh$_2$Cl$_2$(η^2-C$_6$H$_{12}$)$_4$ (di-μ-chloro-tetra-

Formula Index

kis(η^2-2,3-dimethyl-2-butene)dirhodium), 19:219
$Rh_2Cl_2(\eta^4$-cyclo-$C_8H_{12})_2$, 19:218
$Rh_4(CO)_9(\mu$-$CO)_3$, 20:209
$Rh_4(CO)_{12}$, 17:115
$[Rh_6(\mu_6$-$C)(\mu$-$CO)_9(CO)_6]K_2$, 20:212
$Rh_6(CO)_{16}$, 16:49
$[Rh_{12}(\mu$-$CO)_2(\mu_3$-$CO)_8(CO)_{20}]Na_2$, 20:215
$Ru(CO)_3(C_8H_{12})$, 16:105
$[Ru(CO)_3Cl_2]_2$, 16:5
$[Ru(C_2H_3O_2)(NH_3)_4(NO)](ClO_4)_2$, 16:14
$RuCl(CO)(OCOCH_3)[P(C_6H_5)_3]_2$, 17:126
$[RuCl(NH_3)_4(NO)]Cl_2$, 16:13
$[RuCl(NO)_2\{P(C_6H_5)_3\}_2](BF_4)$, 16:21
$[Ru(en)_3]Cl_2$, 19:118
$[Ru(en)_3]Cl_3$, 19:119
$[Ru(en)_3][ZnCl_4]$, 19:118
$RuH(CH_3CO_2)[P(C_6H_5)_3]_3$, 17:79
$RuH(NO)[P(C_6H_5)_3]_3$, 17:73
$RuH(OCOCH_3)(CO)[P(C_6H_5)_3]_2$, 17:126
$[RuH\{P(C_6H_5)_3\}_2\{(\eta^6$-$C_6H_5)P(C_6H_5)_2\}]$–$[BF_4]$, 17:77
$RuH_2[P(C_6H_5)_3]_4$, 17:75
$[Ru(NCO)(NH_3)_4(NO)](ClO_4)_2$, 16:15
$[Ru(NH_3)_4(N_2O)]Br_2$, 16:75
$[Ru(NH_3)_4(N_2O)]Cl_2$, 16:75
$[Ru(NH_3)_5(N_2O)]I_2$, 16:75
$Ru(OCOCF_3)_2(CO)[P(C_6H_5)_3]_2$, 17:127
$Ru(OCOCH_3)_2(CO)_2[P(C_6H_5)_3]_2$, 17:126
$Ru(OCOCF_3)_2(NO)[P(C_6H_5)_3]_2$, 17:127
$Ru[P(OCH_3)_3]_5$, 20:80
$Ru_2(CO)_6Cl_4$, 16:51
$Ru_3(CO)_{10}(NO)_2$, 16:39
$Ru_3(CO)_{12}$, 16:45, 47

$[[(S)C(CH_3)CH_2C(CH_3)]_2(C_2N_2H_2)]Co$, 16:227
$[(S)C(CH_3)CH_2C(CH_3)]_2(C_2N_2H_4)$, 16:226
$[\mu$-$SC(CH_3)_3]_2Ir_2(CO)_4$, 20:237
$[[S(CH_2)_2NH_2]Pt(terpy)](NO_3)_2$, 20:104
$[[S(CH_2)_2OH]Pt(terpy)]NO_3$, 20:103
$[S(CH_3)C_6H_4]Li$, 16:170
$[S(CH_3)C_6H_4](NH_2)$, 16:169
$[S(CH_3)C_6H_4]P(C_6H_5)_2$, 16:171
$[S(CH_3)C_6H_4]_2P(C_6H_5)$, 16:172
$[S(CH_3)C_6H_4]_3P$, 16:173
$(SCH_3)GeH_3$, 18:185
$[[(SCH_3)_2C_2H_4]_2Mo_2Cl_4]$, 19:131
$[S(CH_3)_2]CoBr[(C=NO)_2H(CH_3)_2]_2$, 20:128

$[\{(SC_6H_5)C(CH_3)\}W(CO)_5]$, 17:99
$(SC_6H_5)GeH_3$, 18:165
$(\mu$-$SC_6H_5)_2Ir_2(CO)_4$, 20:238
$S(GeH_3)_2$, 18:164
$(SH)C(C_4H_3S)$=$CHC(O)CF_3$, 16:206
$S(NH)[N(SH)]$, 18:124
$[[S(O)(CH_3)_2]Cr]Br_3$, 19:126
SOF_4, 20:34
$[S(O_3)NH_2]Ag$, 18:201
SO_3PNa_3, 17:193
$[(SO_4)_2H]_{0.46}Rb_3[Pt(CN)_4]\cdot H_2O$, 20:20
$(S)P[CH_2P(C_3H_7)(C_6H_5)](C_6H_5)_2$, 16:195
$S[Si(CH_3)_3]_2$, 19:276
$S[SiH(CH_3)_2]_2$, 19:276
$S[SiH_2(CH_3)]_2$, 19:276
$S(SiH_3)_2$, 19:275
$[S_2CN(CH_3)_2]_2Co(NO)$, 16:7
$[S_2CN(C_2H_5)_2]_2Mo(NO)_2$, 16:235
$[S_2CN(C_2H_5)_2]_2Fe(NO)$, 16:5
cis-$(S_2O_6)[Co(NH_3)_4(H_2O)(OH)]$, 18:81
cis-$(S_2O_6)[Cr(en)_2(H_2O)(OH)]$, 18:84
$(S_2O_6)_2[Cr_2(en)_4$-μ-$(OH)_2]$, 18:90
cis-$(S_2O_6)[Cr(NH_3)_4(H_2O)(OH)]$, 18:80
$(S_2O_6)_2[(en)_4Co_2$-μ-$(OH)_2]$, 18:92
$[S_2P(i$-$C_3H_7)_2]Mo(CO)_2$, 18:53
$[S_2P(i$-$C_3H_7)_2]_2Mo(C_2H_2)(CO)$, 18:55
S_3N, 18:124
$(S_4N)[n$-$C_4H_9)_4N]$, 18:203, 205
S_4N_4, 17:197
$[S_5N_5][AlCl_4]$, 17:190
$[S_5N_5][FeCl_4]$, 17:190
$[S_5N_5][SbCl_6]$, 17:189
S_7NH, 18:203, 204
$Sb[C_6H_4As(CH_3)_2]_3$, 16:187
$[Sb(C_6H_5)_3]_3Cu(NO_3)$, 19:94
$[SbF_6]Hg_{2.91}$, 19:26
$[Sb_2F_{11}]_2Hg_3$, 19:23
$Sc[[(CH_3)_3Si]_2N]_3$, 18:115
$[\{SeC(NH_2)_2\}Co](ClO_4)_2$, 16:84
$[\{SeC(NH_2)_2\}Co](SO_4)$, 16:85
$[\{SeC(NH_2)_2\}Hg]Br_2$, 16:86
$[\{SeC(NH_2)_2\}Hg]Cl_2$, 16:85
$[\{[(SeC(NH_2)_2Hg]Cl_2\}_2$, 16:86
$Se(GeH_3)_2$, 20:175
$SeOF_5H$, 20:38
SeO_2F_2, 20:36
SeO_4H_2, 20:37
$Se[Si(CH_3)_3]_2$, 20:173
$Si(CH_3)F_3$, 16:139
$Si(CH_3)_2F_2$, 16:141

[Si(CH$_3$)$_3$]$_2$NH, 18:12
[Si(CH$_3$)$_3$]$_2$S, 19:276
[Si(CH$_3$)$_3$]$_2$Se, 20:173
[Si(CH$_3$)$_3$]$_2$Te, 20:173
SiCr(CO)$_3$(η^5-C$_5$H$_5$)H$_3$, 17:104
[SiH(CH$_3$)$_2$]$_2$S, 19:276
SiHI(CH$_3$)$_2$, 19:271
[SiH$_2$(CH$_3$)]$_2$S, 19:276
(SiH$_3$)(CH$_3$)C$_5$H$_4$, 17:174
(SiH$_3$)C$_5$H$_5$, 17:172
SiH$_3$I, 19:268, 270
(SiH$_3$)$_2$S, 19:275
SiI(CH$_3$)$_3$, 19:272
SiMo(CO)$_3$(η^5-C$_5$H$_5$)H$_3$, 17:104
Si[N(CH$_3$)$_2$](CH$_3$)$_3$, 18:180
SiO$_2$, gels, 20:2
SiP(CH$_3$)$_3$(C$_6$H$_5$)$_2$, 17:187
SiW(CO)$_3$(η^5-C$_5$H$_5$)H$_3$, 17:104
cyclo-Si$_5$(CH$_3$)$_{10}$, 19:265
cyclo-Si$_6$(CH$_3$)$_{12}$, 19:265
Sm[Fe(CN)$_6$]·4H$_2$O, 20:13
Sn(CH$_3$)$_3$(C$_5$H$_5$), 17:178
[Sn(CH$_3$)$_3$]$_2$O, 17:181
Sn(C$_2$H$_5$)(OCH$_2$CH$_2$)$_3$N, 16:230
SnC$_8$H$_{17}$O$_3$N, 16:230
Sn(TaS$_2$), 19:47
spcUO$_2$, 20:97

[Ta(CO)$_6$][As(C$_6$H$_5$)$_4$], 16:71
[Ta(CO)$_6$]K{(CH$_3$OCH$_2$CH$_2$)$_2$O}$_3$], 16:71
TaCl$_5$, 20:42
TaS$_2$ (2H(a)phase), 19:35
(TaS$_2$)(NH$_3$), 19:42
(TaS$_2$)Na$_X$, 19:42, 44
(TaS$_2$)Sn, 19:47
(TaS$_2$)$_2$B, 19:42
(TaS$_2$)$_2$(C$_5$H$_5$N), 19:40
TeF$_4$, 20:33
Te(GeH$_3$)$_2$, 20:175
Te[Si(CH$_3$)$_3$]$_2$, 20:173
[(terpy)PtCl]Cl·2H$_2$O, 20:101
[(terpy)Pt[S(CH$_2$)$_2$NH$_3$]](NO$_3$)$_2$, 20:104
[(terpy)Pt[S(CH$_2$)$_2$OH]]NO$_3$, 20:103
Te$_2$O$_7$, 17:155
Th(C$_5$H$_5$)$_3$Cl, 16:149
cis-[TiBr(C$_5$H$_7$O$_2$)$_2$], 19:142
Ti(CH$_3$)Br$_3$, 16:124
Ti(CH$_3$)Cl$_3$, 16:122
Ti[[(CH$_3$)$_3$Si]$_2$N]$_3$, 18:116

Ti(η^5-C$_5$H$_5$)$_2$(BH$_4$), 17:91
[Ti(C$_5$H$_5$)Cl$_2$]$_X$, 16:238
Ti(C$_5$H$_5$)Cl$_2$[P(CH$_3$)$_2$(C$_6$H$_5$)]$_2$, 16:239
cis-[TiCl$_2$(C$_5$H$_7$O$_2$)$_2$], 19:146
TiCl$_3$[P(CH$_3$)H$_2$]$_2$, 16:98
TiCl$_3$[P(CH$_3$)$_2$H]$_2$, 16:100
TiCl$_3$[P(CH$_3$)$_3$]$_2$, 16:100
TiCl$_3$[P(C$_2$H$_5$)$_3$]$_2$, 16:101
cis-[TiF$_2$(C$_5$H$_7$O$_2$)$_2$], 19:145
Ti[Mn(CO)$_5$]$_3$, 16:61
TlAl$_{11}$O$_{17}$, 19:53

U(C$_5$H$_5$)$_3$Cl, 16:148
U(η^8-cyclo-C$_8$H$_8$)$_2$, 19:149, 150
UCl$_6$, 16:143
UF$_5$, 19:137-139
UO$_2$ spc, 20:97

V[[(CH$_3$)$_3$Si]$_2$N]$_3$, 18:117
V(C$_{26}$H$_{14}$N$_8$)O(VOHp), 18:48
VCl$_4$, 20:42
VCuO(daaen)(H$_2$O), 20:95
(VF$_4$O)Ni·7H$_2$O, 16:87
VO(C$_{44}$H$_{28}$N$_4$) (vanadium(IV), oxo–
 [5,10,15,20-tetraphenyl-21H,23H-
 porphinato(2-)]-), 20:144
VOHp, 18:48
VO(tpp), 20:144
[V$_{10}$O$_{28}$](NH$_4$)$_6$·6H$_2$O, 19:140, 143
[V$_{10}$O$_{28}$]Na$_6$·17H$_2$O, 19:140, 142

trans-[WBr(CO)$_4$(CC$_6$H$_5$)], 19:172, 173
WBr$_5$, 20:42
W(CO)$_2$(η^5-C$_5$H$_5$)(NO), 18:127
W(CO)$_3$(η^5-C$_5$H$_5$)SiH$_3$, 17:104
trans-[W(CO)$_4$I(CS)][N(n-C$_4$H$_9$)$_4$], 19:186
[W(CO)$_5${C(CH$_3$)(OCH$_3$)}], 17:97
[W(CO)$_5${C(CH$_3$)(SC$_6$H$_5$)}], 17:99
W(CO)$_5$[C(C$_6$H$_5$)$_2$], 19:180
W(CO)$_5$[C[N(CH$_3$)$_2$]C$_6$H$_5$], 19:169
W(CO)$_5$[C(OCH$_3$)(C$_6$H$_5$)], 19:165
W(CO)$_5$[C[OSi(CH$_3$)$_3$]C$_6$H$_4$-p-(CH$_3$)],
 19:167
W(CO)$_5$(CS), 19:183, 187
W(η^5-C$_5$H$_5$)(CH$_3$)(NO)$_2$, 19:210
WCl(η^5-C$_5$H$_5$)(NO)$_2$, 18:129
WCl$_4$[(PPh$_2$)$_2$(CH$_2$)$_2$], 20:125
WCl$_4$(PPh$_3$)$_2$, 20:124
trans-W(N$_2$)$_2$[(PPh$_2$)$_2$(CH$_2$)$_2$]$_2$,
 20:126

[Zn(Bzo$_4$[16]octaeneN$_4$)][ZnCl$_4$], 18:33
Zn(CH$_3$)$_2$, 19:253
[Zn(C$_{28}$H$_{20}$N$_4$)][ZnCl$_4$], 18:33
[ZnCl$_4$][Ru(en)$_3$], 19:118
ZnCr$_2$O$_4$, 20:52
ZnH$_2$, 17:6
[ZnH$_3$]Li, 17:10
[ZnH$_3$]Na, 17:15
[ZnH$_4$]Li$_2$, 17:12
ZnK$_2$Cl$_4$, 20:51
[Zn$_2$(CH$_3$)$_2$H$_3$]Na, 17:13
Zr(η^5-C$_5$H$_5$)$_2$H$_2$, 19:224, 225
ZrCl(η^5-C$_5$H$_5$)$_2$H, 19:226
Zr$_2$OCl$_2$(η^5-C$_5$H$_5$)$_4$, 19:224